THE TENT CATERPILLARS

THE CORNELL SERIES IN ARTHROPOD BIOLOGY

EDITED BY *George C. Eickwort*

Army Ants: The Biology of Social Predation
by William H. Gotwald, Jr.

The Tent Caterpillars
by Terrence D. Fitzgerald

THE TENT CATERPILLARS

Terrence D. Fitzgerald

Department of Biology
State University of New York at Cortland

Comstock Publishing Associates A DIVISION OF
Cornell University Press | ITHACA AND LONDON

First published 1995 by Cornell University Press

Printed in the United States of America.
Color plates printed in Hong Kong.

♾ The paper in this book meets the minimum requirements of the American National Standard for Information Sciences—Permanence of Paper for Printed Library Materials, ANSI Z39.48–1984.

Library of Congress Cataloging-in-Publication Data

Fitzgerald, Terrence D.
The tent caterpillars / Terrence D. Fitzgerald.
p. cm. — (Cornell series in arthropod biology)
Includes bibliographical references and index.
ISBN 0-8014-2456-9 (alk. paper)
1. Tent caterpillars. I. Title. II. Series.
QL561.L3F58 1995
595.78'1—dc20 94-48048

To Christine

Contents

The color plates follow page 16.

Publisher's Foreword

The field of entomology is undergoing a renaissance as modern behavioral and ecological approaches are applied to the study of insects and their relatives. Recognizing the significance of this development and the need for books that consider arthropods from a modern evolutionary viewpoint, the late George C. Eickwort initiated the Cornell Series in Arthropod Biology.

The volumes in the series focus on the behavior and ecology of a particular taxon, varying in rank from class to genus. Written by scientists who are making the greatest advances in expanding our knowledge of arthropod biology, the books are comprehensive in their scope, not only detailing the subject animals' behavioral ecology but also summarizing their evolutionary history and classification, their development, important aspects of their morphology and physiology, and their interactions with the environment. Each volume can thus serve as a primer for students and scholars wishing to study that group of animals.

But beyond applying modern critical thought to the biology of arthropods, the series, we hope, will engender the great enthusiasm for entomology that George Eickwort spread to all who knew him.

Preface

Tent caterpillars occur throughout the temperate regions of the world and are among the most familiar of insects. The caterpillars colonize both herbaceous plants and trees, often achieving huge numbers during intermittent outbreaks. They are capable of defoliating tens of thousands of hectares of forest, and much of the scientific research on these insects has centered on techniques for containing them. Their significance as pest species certainly cannot be overstated, yet tent caterpillars are much more than mere folivores. They differ most markedly from other economically significant caterpillars, such as the gypsy moth, in that they are social. Their colonial behavior has intrigued scientists throughout much of this century, and we now know far more about tent caterpillars than we do about any other social caterpillar. Indeed, the eastern tent caterpillar is the best known of all the social caterpillars and, so far, the most social of all caterpillars known.

My intent in writing this book was to consider all aspects of the biology of tent caterpillars and to create a reference that would be as comprehensive as the literature would allow. Thus, although I mostly discuss the biology and behavioral ecology of the caterpillars, I have included chapters on their economic impact and the century-long warfare we have waged against them. There have been no reviews of tent caterpillars since the publication of Frederick Stehr and Edwin Cook's monograph on the Nearctic species in 1968. To my knowledge there has been no previous attempt to provide a comprehensive treatment of the biology and ecology of any single caterpillar or group of caterpillar species. After having read nearly 500 papers during the four years that I spent researching and writing this book, I can say with some confidence that there are few other caterpillars whose overall biology has been so extensively studied. That we know so much about tent caterpillars is due not only to their economic importance and social behavior but also to the fact that field populations are easy to find and, during periods of abundance, the caterpillars can be readily collected in large numbers. To our benefit, tent caterpillars have served as convenient subjects for generations of biologists.

In preparing to write this book, I reviewed both the applied and basic literature of the nineteenth and twentieth centuries. The book contains numerous references to the Eurasian literature, but major emphasis is placed on the Nearctic species. Indeed, most of our knowledge of tent caterpillars is derived from studies of a few species, notably the eastern, western, and forest tent caterpillars in North America and the lackey moth in Eurasia. Tent caterpillars share many basic features of their biology with other species of caterpillars, and readers will find much of the material on the basic biology relevant to other caterpillars as well.

Many people have contributed directly or indirectly to this book. I am especially grateful to the late George Eickwort and to Robb Reavill, first, for suggesting that I write this book and, second, for their editorial assistance and encouragement during the preparation of the manuscript. Margo Quinto skillfully copyedited the manuscript, George Whipple offered expert advice on the preparation of the illustrations and designed the book with care, and Helene Maddux shepherded the manuscript through the production process with unfailing good humor. Timothy M. Casey, Ann Hajek, Barbara Joos, Lauren Schroeder, Frederick Stehr, and Robert Vander Meer provided helpful reviews of individual chapters. Nancy Stamp read the entire manuscript and offered many useful comments and suggestions. Lawrence Abrahamson, Alan Adams, Douglas Allen, J. M. Cano, M. D. Chisholm, James Costa, Janice Edgerly, John Franclemont, K. C. Joshi, Daniel Robison, and Dessie Underwood graciously provided manuscripts. Eileen Williams and Ellen Paterson were extremely helpful in filling my many interlibrary loan requests, and Dawn Van Hall provided photographic support. I would also like to give special thanks to John Simeone for first drawing my attention to the study of insects. I am particularly indebted to the National Science Foundation for supporting much of my research on tent caterpillars.

TERRENCE D. FITZGERALD

Cortland, New York

THE TENT CATERPILLARS

1

Seasonal History

It is early spring, the temperature at dawn near freezing. The first rays of the morning sun strike a small shelter constructed from silk in the crotch of a cherry tree, burning off a light layer of snow that has fallen overnight. Within the tent, a tightly clustered family of tent caterpillars stirs to life as the larvae absorb the warming rays of the sun. Cold and cloudy weather have kept them confined to the tent, and they haven't eaten for several days. The product of a single egg mass, the colony consisted of some three hundred siblings when the eggs hatched two weeks before, but their numbers have already dwindled. Some were lost while feeding away from the tent, chilled to the point of immobility by a sudden change in the weather, then driven from the tree by wind and sleet. Others were killed by a warbler that tore open the tent and then snatched them from the structure.

When the basking caterpillars are warm enough, they make their way, one by one, through a series of small openings in the tent that lead to the surface. There they mill about sluggishly in the morning air. From time to time a few caterpillars hesitantly venture a short distance onto a branch that juts from the top of the tent, but they soon turn back. This behavior is repeated over and over by other caterpillars until hunger eventually defeats timidity and a small contingent, clustered tightly together, pushes forward onto the branch. They advance slowly in the cold morning air, barely warm enough to move. When they reach a patch of leaves at the end of the branch, they arrange themselves side by side and begin to feed along the margin of a young, partially eaten leaf. Others arrive in successive waves, and soon the feeding site is overcrowded, forcing the latecomers to spill onto adjacent leaves. In twenty minutes the first of the caterpillars to reach the leaves have fed to repletion and, driven by an instinct opposite that which brought them to food, but every bit as strong, they return immediately to the tent.

An hour later, all of the caterpillars have fed and reassembled at the tent. Although the air is still cold, the larvae are basking in the full morning sun. Packed tightly together just under the surface of the tent the aggregate has warmed well above the ambient temperature. They appear lifeless, but their guts are driven to full activity by the heat of the sun, and they are busy processing their fibrous meals. Twice more that day the caterpillars repeat this ritual of collecting and processing, and by the time they are mature, each will have consumed some fifteen thousand times its initial weight in leaves, nearly fifty percent of which will have passed through the gut undigested.

Three weeks later the caterpillars have grown to more than a thousand times their original mass. Their tent, expanded during daily episodes of communal spinning, is now large and conspicuous. Tachinid flies and predatory stink bugs have joined the caterpillars on the tent, and they wait for an opportunity to attack. Many caterpillars have already succumbed, and the tent is littered with the shrunken cadavers of larvae killed during earlier attacks by braconid wasps. The caterpillars are clearly vulnerable, but they are not defenseless. Having detected the buzz of a tachinid's wings, a larva responds by rapidly swinging its body from side to side, creating a moving target and deflecting the attack. Others join in the display, and before long the entire colony is alerted to danger. Some of the larvae seek shelter within the tent, but the agitated displays of the caterpillars are sufficient to keep the fly at bay for the moment.

After six weeks of feeding, the caterpillars have cropped nearly all of the leaves of their natal tree. The orderly pattern of feeding and rest that has marked the life of the colony until now has given way to constant and near-frenetic activity. Driven by hunger, the caterpillars strike off independently on prolonged and largely fruitless forays in search of food. They repeatedly attempt to leave the tree but make little progress through the thick vegetation that covers the ground. Densely matted silk surrounds the base of the tree, and short trails radiate in all directions.

Three days after its last meal, a caterpillar searching on the ground discovers the fallen stem of a goldenrod and, making rapid progress along its length, finds a small cherry tree located a little over a meter from the tent. Here it feeds to repletion, then methodically retraces its path back to the tent, all the while marking the substrate with a pheromone secreted from the tip of its abdomen. Hungry caterpillars searching aimlessly at the base of the tree react excitedly when they detect the pheromone and immediately stream onto the trail. Caterpillars resting on the tent are alerted to the discovery of food when the sated caterpillar returns. Within twenty minutes the whole colony has converged at the new source of food.

Had the newly discovered tree been more substantial, the caterpillars might have finished their larval life together commuting between their tent and the distant cherry tree. But, as is often the case, the energy demands of the colony greatly exceed the capacity of the tree to sustain it, and shortly after the last leaves are eaten, the social bond that holds the colony together disintegrates. Within days the caterpillars abandon their tent and disperse in all directions.

Many of the caterpillars find little or no additional food and eventually spin cocoons within which they either die or metamorphose to small, largely inviable pupae. Others continue to feed in isolation and, when fully grown, seek out suitable sites in which to construct their cocoons. Some of these cocoons eventually produce moths, but most yield only parasitic flies and wasps. In all, only a handful of the several hundred caterpillars in the original colony survive to breed; the rest are consumed by predators and disease, or eventually fall victim to inclement weather, parasitoids, and insufficient food. From these few survivors, three new egg masses are produced, more than enough to assure that in only a few years the descendants of this single family will number in the tens of thousands.

This brief natural history of the eastern tent caterpillar portrays some of the more significant events that occur during the lifetime of a typical colony of these remarkable insects. Although we know considerably less of the natural history of many of the other species of tent caterpillars, all are social species, and their natural histories are likely to be highlighted by the four collective behaviors that characterize colonies of the eastern tent caterpillar: shelter building, thermoregulation, cooperative foraging, and predator defense. In addition, as far as we know, all tent caterpillars share the pattern of seasonal development illustrated in Figure 1.1 and described in the following paragraphs.

Reproductively mature moths eclose from their cocoons in late spring or summer. The males are more active fliers than the females, and they are attracted by the sex pheromones the females secrete. Mating takes place within hours of female eclosion, and the gravid females disperse in search of oviposition sites soon thereafter, often completing their life's functions in less than a day. The female produces a single egg mass, which she attaches to a stem of the host plant. Embryogenesis proceeds rapidly, and within a few weeks fully formed caterpillars are found within the chorions of their eggs. The tiny larvae lie sequestered until the following spring, undergoing an obligatory period of diapause, which is broken only by prolonged exposure to cold.

Tent caterpillars are among the earliest of the spring-feeding insects, and hatching of the eggs is tied closely to the phenology of the host. The caterpillars typically chew their way out of their eggs just as the buds on their natal plant are bursting and the first leaves of spring are beginning to appear. The newly eclosed siblings cluster on the surface of the egg mass until most have eclosed, then set off in search of food and a suitable aggregation site.

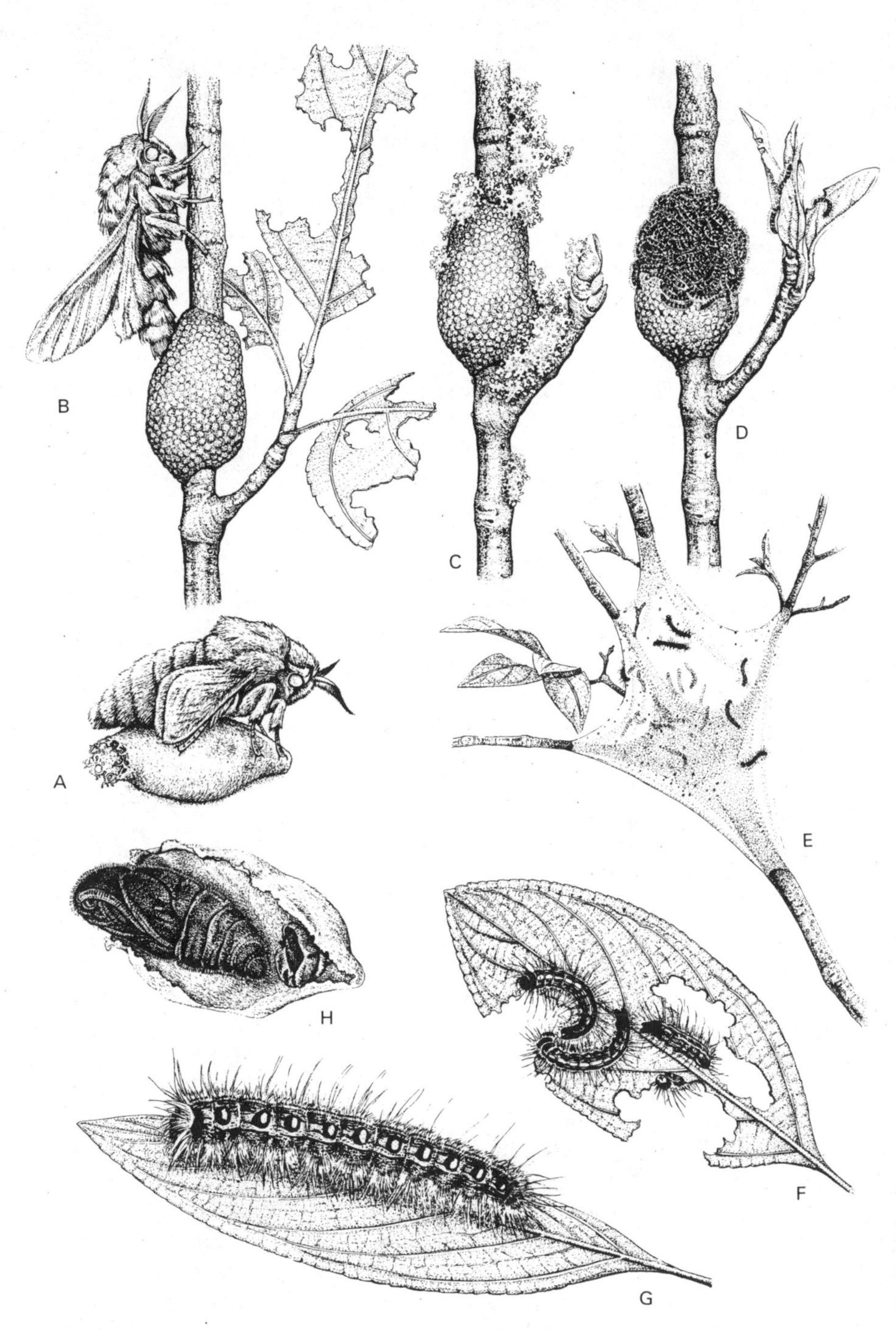

Figure 1.1. Seasonal life history of the eastern tent caterpillar. (A) Adult eclosion, June–July. (B) Oviposition, June–July. (C) Overwintering egg mass, July–April. (D) Larval eclosion, April–May. (E) Tent, May–June. (F) Group foraging, May–June. (G) Mature caterpillar, June. (H) Pupa, June. Other species of tent caterpillar follow this same general plan of seasonal development but differ in the details of their shelter-building and foraging behaviors. (Drawings by Edward Rooks.)

Some species of tent caterpillar begin to construct a tent within a few days of eclosion; for those species, the tent becomes a permanent aggregation site to which the colony returns after each bout of feeding. Other species build a series of small tents that are used for molting then abandoned. The forest tent caterpillar is the only species known that does not build a tent.

Tent caterpillars are active foragers, and they move in loose processions from aggregation sites to distant feeding sites several times a day, traveling single file or several abreast. During these forays, the caterpillars lay down silk strands to gain secure purchase on smooth plant surfaces and a trail pheromone that is secreted from the tip of the abdomen. The caterpillars also search independently for food, and successful foragers can alert colony mates to the discovery of a new food find by laying down a chemical recruitment trail.

Tent caterpillars may continue to aggregate and to forage together until they are full-grown, or they may disperse in the penultimate instar, finishing their growth in isolation from their siblings. The duration of the larval stage varies according to climatic conditions, but most commonly this stage lasts about eight weeks. Mature caterpillars seek pupation sites under overhangs or in other protected areas. After spinning a cocoon, the caterpillar metamorphoses to the pupal stage, and in about two weeks the adult ecloses.

2

Systematics, Geographic Distribution, and the Evolution of Sociality

Systematics and Geographic Distribution

Tent caterpillars belong to the Lasiocampidae, a moderately sized family of about 1000 moth species distributed throughout the temperate and tropical regions of the world. Although most genera occur in the Old World tropics, the South American fauna is species-rich, with a particularly large diversity of moths in the genus *Euglyphus* (Franclemont 1973). Lasiocampids are closely related to the Old and New World silk moths in the families Bombycidae and Saturniidae. These three families of moths, and the less well known moths of the family Apatelodidae, are assigned to the superfamily Bombycoidea. All Old and New World tent caterpillars are currently placed in the genus *Malacosoma*:

Order Lepidoptera
- Suborder Ditrysia
  - Superfamily Bombycoidea
    - Family Lasiocampidae Harris
      - Subfamily Lasiocampinae Harris
        - Tribe Malacosomatini Aurivillius
          - Genus *Malacosoma* Harris

Eleven additional genera of the Lasiocampidae, consisting of some 30 species, occur in North America (Franclemont 1973). Thirty-four species of lasiocampids are found in Europe (Goater 1991).

Tent caterpillars were originally placed in the genus *Bombyx*. Hübner (1820) established the genus *Malacosoma* (from Latin *malakos*, soft, and *soma*, body) for the tent caterpillars in 1820. In 1828, Curtis proposed the name *Clisiocampa* (*klisia*, chamber; *kampe*, caterpillar), apparently unaware that Hübner had already proposed a generic name for the group. Although the genus designation *Malacosoma* clearly had priority, tent caterpillars were widely referred to as *Clisiocampa* through the early 1900s. Because the gender of *Clisiocampa* is feminine, many of the species names lacked gender agreement when eventually transferred to *Malacosoma*, which is neuter. Thus, it is common to find in the early North American literature reference to *M. americana*, *M. californica*, and *M. constricta*, all of which have the feminine ending *a* for the species name instead of the appropriate neutral *um* ending. The use of feminine endings for specific names is still common in the Eurasian literature, where one finds frequent reference to species such as *M. neustrium* and *M. indica*.

The genus *Malacosoma* is distributed throughout the Holarctic region. Most species occur in the Palearctic realm, including Eastern and Western Europe, North Africa, the former Soviet Union, India, China, and Japan. Though the taxonomy of the genus is far from settled, a review of the literature indicates that there are 26 described species (Table 2.1). Of these, six are Nearctic in distribution, and the rest are Palearctic. Some of the widely distributed species show considerable variation in characters and are further divided into subspecies. The ranges of the North American and European species have been determined (Figs. 2.1, 2.2), but the distribution and taxonomic status of the Asian forms are much less certain.

North American Species of *Malacosoma*

In an exhaustive project undertaken while a Ph.D. candidate at the University of Minnesota, Frederick Stehr traveled countless miles throughout the western United States to study and collect tent caterpillars during the field seasons of the years 1960–1962. To facilitate his studies, he constructed a large mobile insectary with space for rearing cages, a workbench, and storage areas, and he set up house in a pickup camper adjacent to his field study sites (Fig. 2.3). His studies of the morphological, ecological, and behavioral characteristics of field populations led him to conclude that there are six valid species of *Malacosoma* in the Americas (Stehr and Cook 1968). Three of those species are further separated into subspecies (Tables 2.1, 2.2).

The North American species *M. disstria*, *M. constrictum*, *M. tigris*, and

Table 2.1. The *Malacosoma* of the world

Species[a]	Distribution[b]	Taxonomic reference[c]
M. alpicolum alpicolum Staudinger 1870 = *intermedia* Milliere 1875	Mountainous regions in Europe	1
M. alpicolum mixtum Rothschild 1925	Mountainous regions in North Africa	1
M. americanum (Fabricius) 1793	Eastern half of USA, southeastern Canada	6
M. autumnarium Yang 1978	China	4
M. betula Hou 1980	China	4
M. californicum (Packard) 1864	Populations of the western tent caterpillar with characters too inconsistent to warrant recognition as subspecies occur throughout much of western North America	6
M. californicum ambisimile (Dyar) 1893 = *ambisimile* (Dyar) 1893	California: narrowly distributed in region just south of San Francisco Bay	6
M. californicum californicum (Packard) 1864	California: immediate vicinity of San Francisco Bay	6
M. californicum fragile (Stretch) 1881 = *fragile* (Stretch) 1881	Southern half of Nevada and extending into Utah, Arizona, and central California	6
M. californicum lutenscens (Neumoegen & Dyar) 1893 = *lutescens* (Neumoegen & Dyar) 1893	Central USA and Canada	6
M. californicum pluviale (Dyar) 1928 = *pluviale* (Dyar) 1893	Pacific Northwest, western Canada and extending eastward in a narrow belt, possibly as far as the Atlantic coast	6
M. californicum recenseo (Dyar) 1928	California: distributed in a narrow zone lying between the Sierra Nevada and the Central Valley of California	6
M. castrensis castrensis (Linnaeus) 1758 = *halophila* Stauder 1915	Europe (with the exception of the southern Balkans), North Africa	1
M. castrensis shardaghi Daniel 1951	Southern Balkans and western Turkey	1
M. constrictum austrinum Stehr 1968	Southwest corner of California, northern Baja	6

Table 2.1—*cont.*

Species[a]	Distribution[b]	Taxonomic reference[c]
M. constrictum constrictum (Henry Edwards) 1874	Western California, Oregon, and southwest Washington	6
M. dentatum Mell 1938 = *neustrium dentata* Mell 1938	China	2
M. disstria Hübner 1820	Most of USA and southern Canada	6
M. fasciatum Class 1932	China	3
M. flavomarginatum Poujade 1886	Tibet	5
M. franconicum franconicum [Denis & Schiffermüller] 1775 = *dorycnii* Milliere 1864 = *panormitana* Turati 1909 = *calabricum* Stauder 1921 = *joannisi* Viette 1965	Europe	1
M. incurvum aztecum (Neumoegen) 1893	Mexico	6
M. incurvum incurvum (Henry Edwards) 1882 = *fragilis incurva* (Henry Edwards) 1882	Central and southern Arizona	6
M. incurvum discoloratum (Neumoegen) 1893	Utah and extending into western Colorado, southern Nevada, and northern Arizona	6
M. indicum (Walker) 1855	India	2
M. insignis de Lajonquiere 1972	China	2, 4
M. kirghisica Staudinger 1879	China, southern Russia	4, 5
M. laurae laurae de Lajonquière 1977	Spain	1
M. liupa Hou 1980	China	4
M. luteum luteum (Oberthür) 1878 = *orientalis* Oberthür 1916 = *brunneoolivaceae* Rothschild, Hartert & Jordan	North Africa	1
M. neustrium neustrium (Linnaeus) 1758 = *flavescens* Grünberg 1912 = *mauginii* Turati 1924	Europe, western Turkey, North Africa	1
M. neustrium testacum Motschulsky 1860	Eastern Europe, Asia	2, 3
M. paralellum Staudinger 1887 = *neustria parallela* Strand 1882	Eurasia	2
M. rectifasicum de Lajonquiere 1972	China	2

Table 2.1—*cont.*

Species[a]	Distribution[b]	Taxonomic reference[c]
M. robertsi de Lajonquière 1972	India	2
M. tibetanum Hou 1980	China	4
M. tigris (Dyar) 1902	Southern Utah, western Colorado, Arizona, New Mexico, western Texas, and Mexico	6
M. vulpes (Hampson) 1900	Pakistan	2

[a]Synonyms given are the names accepted just prior to the taxonomic revision listed as the reference.

[b]See also Figs. 2.1 and 2.2.

[c]1, de Freina & Witt 1987; 2, Lajonquière 1972; 3, Inoue et al. 1982; 4, Hou 1980; 5, Collier 1936; 6, Stehr and Cook 1968.

M. americanum are the most distinctive of the Nearctic species and are readily separable. Stehr's attempts to interbreed these species failed to produce any hybrids. *M. americanum* males appeared to be able to fertilize the eggs of *M. californicum* females, but no populations of species having characters intermediate between *californicum* and *americanum* are known to occur in nature (Stehr and Cook 1968). The western tent builders—*M. californicum, M. incurvum,* and their subspecies—are more difficult to distinguish, and their taxonomy is not yet fully resolved. Breeding studies in which various combinations of the subspecies of *californicum* and *incurvum* were caged together indicated that they readily hybridize, but, on the basis of other considerations, Stehr considered the moths to be distinct species.

The western tent caterpillar, *M. californicum,* consists of a large central population and six highly variable subspecific populations located on the periphery of the range of the central population (Fig. 2.1). Four of these subspecies, *ambisimile, pluviale, lutescens,* and *fragile* were formerly recognized as full species. *pluviale* has the broadest distribution of the subspecies, and Franclemont (1973) stated that, in his opinion, *pluviale* is as distinct from the other subspecies as *californicum* is from *incurvum.* However, the fact that none of the adult forms of the subspecies, including *pluviale,* can be separated except by associating them with collection sites attests to the high degree of variability and introgression among the populations. Identification to the subspecies level requires reference to a melange of characteristics including, particularly, larval color patterns and superficial properties of the egg mass.

According to Stehr and Cook (1968), *M. tigris* has the most southern

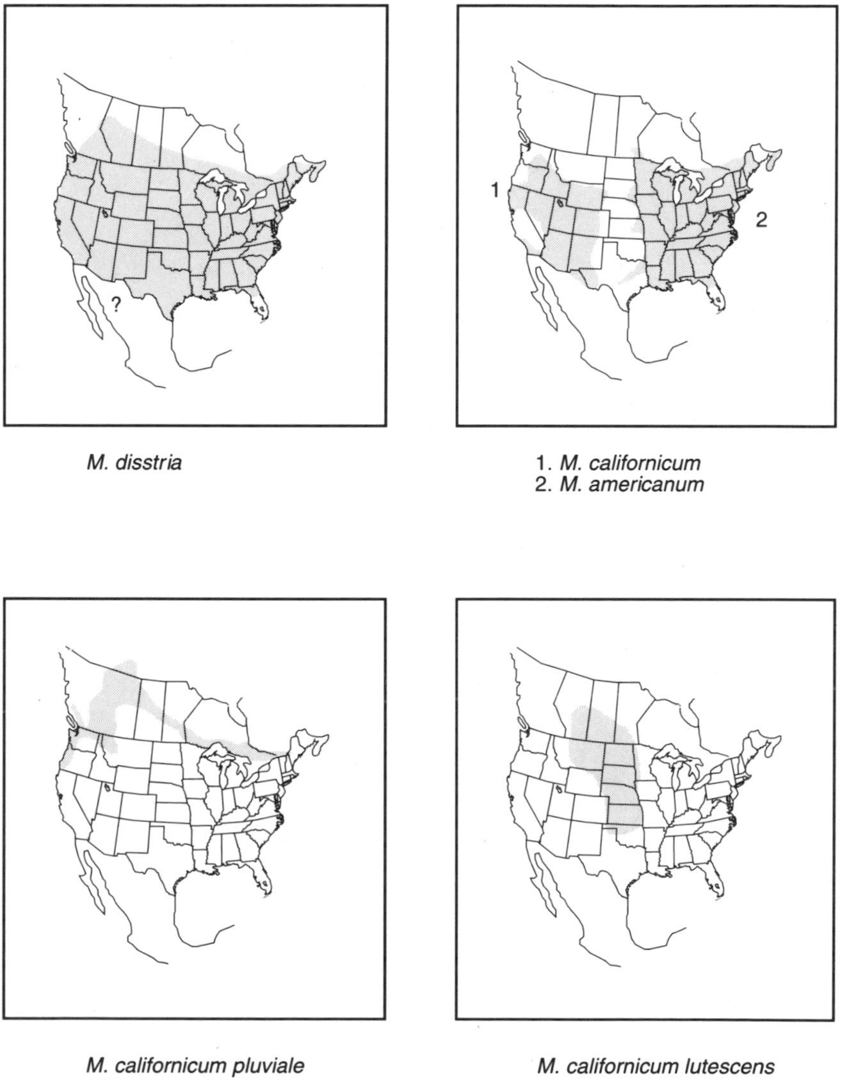

M. disstria

1. *M. californicum*
2. *M. americanum*

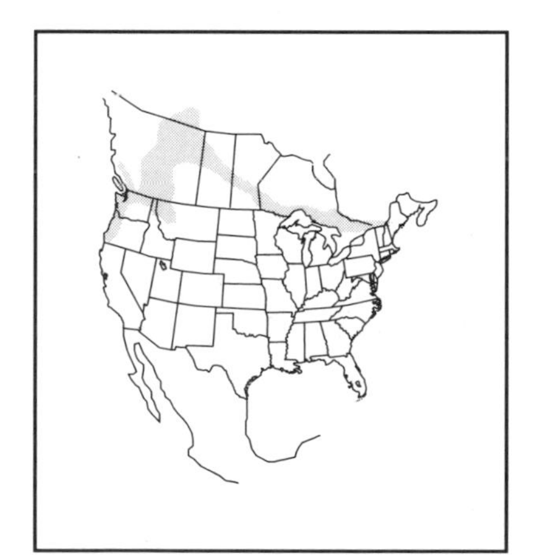

M. californicum pluviale

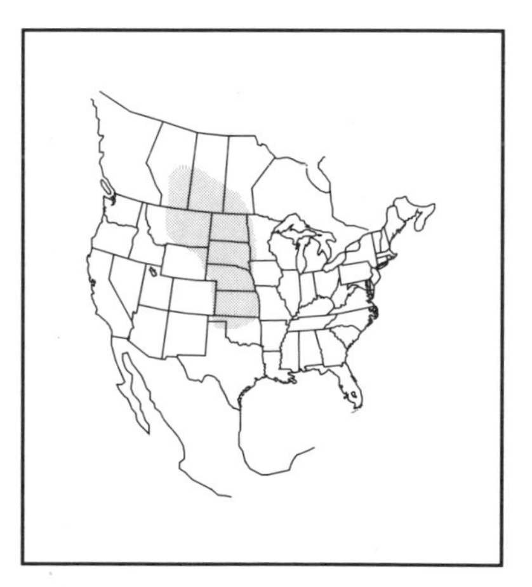

M. californicum lutescens

Figure 2.1. Distribution of Nearctic species of *Malacosoma*. Distribution boundaries are approximate and based for the most part on the locality data reported in Stehr and Cook 1968. Readers interested in locating populations of particular species in the field should consult that publication for the precise locations of their collection sites. *(Figure continues overleaf.)*

M. californicum fragile

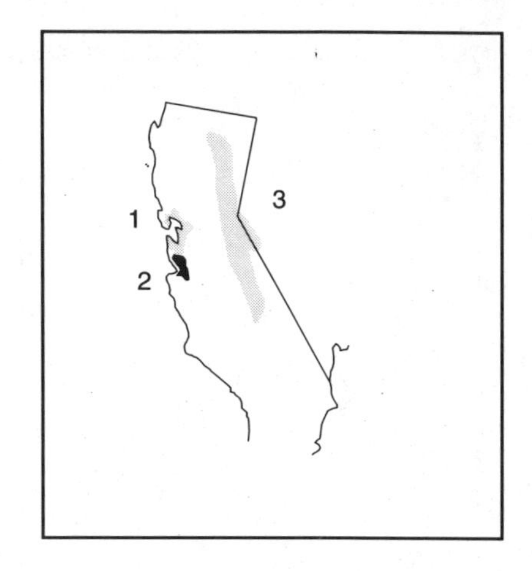

1. *M. californicum californicum*
2. *M. californicum ambisimile*
3. *M. californicum recenseo*

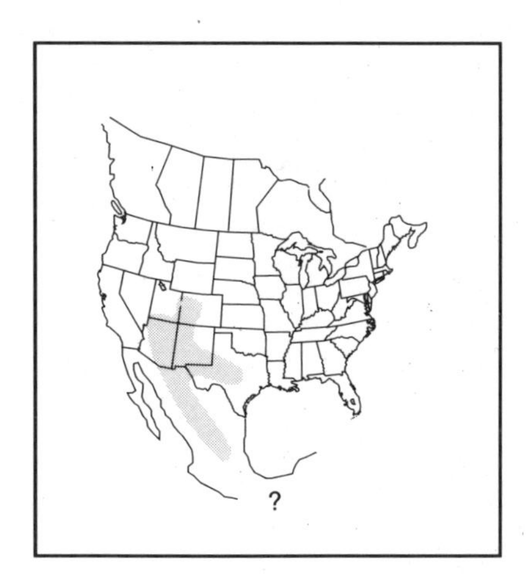

M. tigris

1. *M. constrictum constrictum*
2. *M. constrictum austrinum*
3. *M. incurvum incurvum*
4. *M. incurvum discoloratum*
5. *M. incurvum aztecum*

Figure 2.1—*cont.*

distribution of any North American tent caterpillar. They recorded three specimens collected at San Cristobal de las Casas, Chiapas, Mexico, in 1940 and speculated that this species also occurs in more southerly regions that support its host trees at elevations that are cool enough to enable it to break diapause. The northern distributional limits of *M. califoricum pluviale* extend above the 60th parallel, and an infestation of an unknown species, tentatively identified as *M. californicum*, was recently recorded from Anchorage, Alaska (Holsten and Eglitis 1988).

The identification of the larval forms of the various species of *Malacosoma* depends to a good extent on the color patterns of the caterpillars. These patterns are highly variable both within and among species, and no simple set of diagrams can provide the information needed to separate the species. Reference to the distribution maps reproduced here (Fig. 2.1) will serve to limit the possible identity of species whose collection locality is known. The maps together with a knowledge of host tree species (see Table 5.1) and larval behavioral characteristics will, in many cases, suffice for identification. For more definitive identification, and for western tent caterpillars whose ranges overlap, reference can be made to the keys and to the accompanying full color plates of the caterpillars provided in the taxonomic works of Stehr and Cook (1968), reproduced by Franclemont (1973). Diagnostic characteristics of the adults are also found in those works. Keys to the species based on characteristics of the egg masses are provided by Stehr and Cook (1968).

Eurasian Species of *Malacosoma*

In their work on the Bombycoidea of the western Palearctic, de Freina and Witt (1987) recognized six species of *Malacosoma*, some with subspecies (Table 2.1, Fig. 2.2). The ranges of four of those species extend into eastern Europe, and the subspecies *M. neustrium testacum* ranges across all of Asia. The latter species is the most widely distributed of all *Malacosoma*.

Studies conducted at the turn of the century showed that several of the European species are capable of hybridizing (see Tutt 1906a and references therein). Pairings of *M. neustrium* males and *M. castrensis* females, *M. neustrium* males and *M. franconicum* females, and *M. franconicum* males and *M. castrensis* females produced partially fertile egg masses. Pairings between *M. castrensis* and *M. franconicum* and between *M. alpicolum* and *M. castrensis* produced only infertile eggs. Successful interspecific pairings produced varying degrees of egg fertility and offspring viability. Typically, fewer than 50% of the eggs were fertile, but, occasionally, pairings produced egg fertility rates as high as 90%. Developmental problems were common, and successful individuals of one

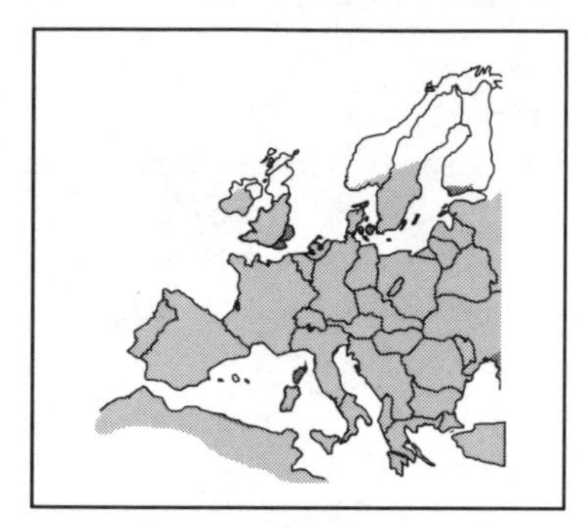

M. neustrium neustrium

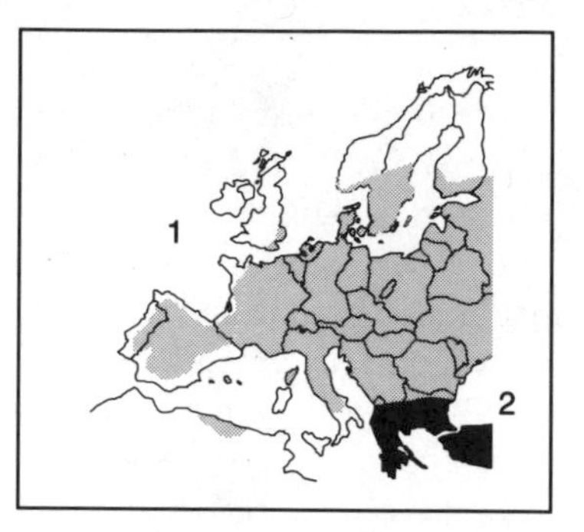

1. *M. castrensis castrensis*
2. *M. castrensis shardaghi*

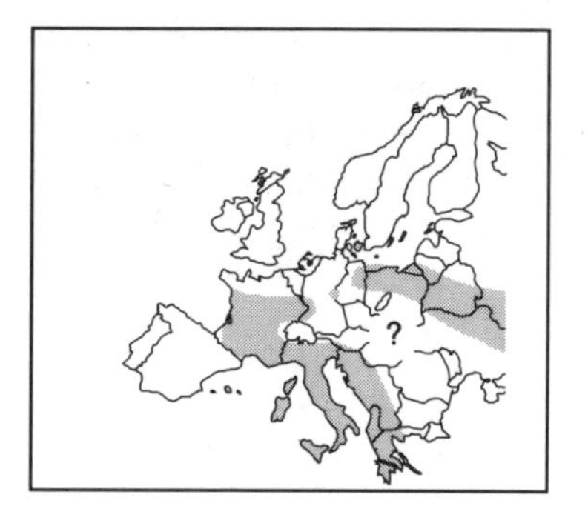

M. franconicum

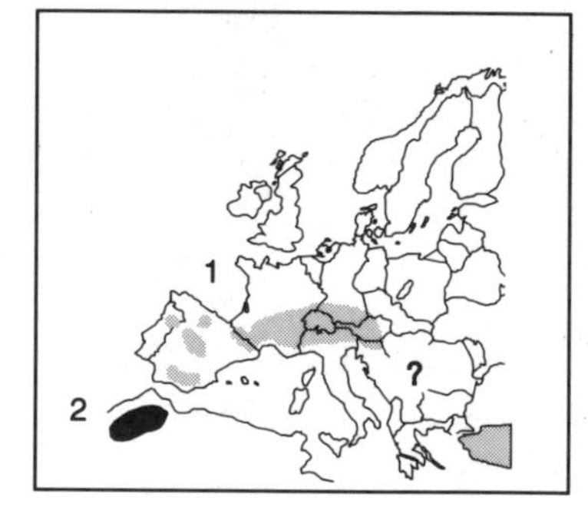

1. *M. alpicolum alpicolum*
2. *M. alpicolum mixtum*

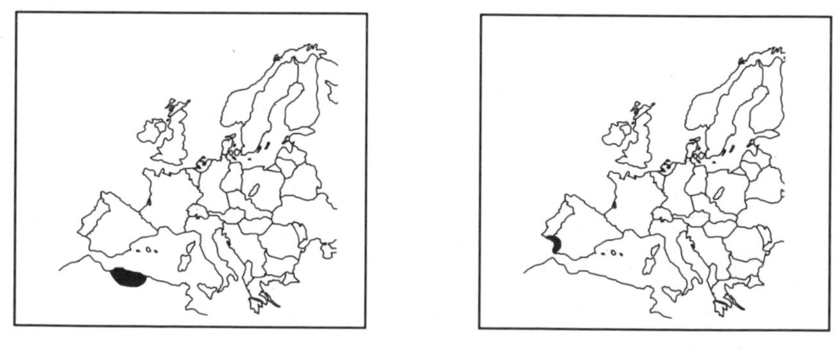

M. luteum *M. laurae*

Figure 2.2. Distribution maps for western Palearctic species of *Malacosoma*. (Distribution based on de Freina and Witt 1987.)

Figure 2.3. Mobile insectary used by Frederick Stehr while conducting field studies of tent caterpillars during the 1960s. (From Stehr and Cook 1968.)

gender often emerged out of synchrony with those of the other. In one such case involving the pairing of *M. neustrium* males and *M. castrensis* females (Bacot 1902), the females emerged three weeks before the males. By forcing males, Bacot was able to make f_1 hybrids of these species. Although the females went through the physical motions associated with egg laying and even secreted cement from the accessory glands, they failed to lay eggs. These studies show that there is incomplete reproductive isolation among some European *Malacosoma* under laboratory conditions, but there have been no reports of hybrid swarms in nature.

Fourteen additional species (Table 2.1) have been described from Asia, most of them from China. The distributional limits of those species are largely uncharted. Most species are known only from adults collected at light traps. Keys to the Asian forms are not available, but the taxonomic works listed in Table 2.1 provide descriptions of the diagnostic features of the adult forms.

Phylogenetic Relationships

There have been no comprehensive studies of the phylogenetic relationships of the genera of the Lasiocampidae or of the various species

Table 2.2. Common names of American species of tent caterpillars approved by the Entomological Society of America

Eastern tent caterpillar	*Malacosma americanum* (Fabricius)
Forest tent caterpillar	*M. disstria* Hübner
Pacific tent caterpillar	*M. constrictum austrinum* Stehr *M. constrictum constrictum* (Henry Edwards)
Sonoran tent caterpillar	*M. tigris* (Dyar)
Southwestern tent caterpillars	*M. incurvum incurvum* (Henry Edwards) *M. incurvum aztecum* (Neumoegen) *M. incurvum discoloratum* (Neumoegen)
Western tent caterpillars	*M. californicum* (Packard) *M. californicum ambisimile* (Dyar) *M. californicum californicum* (Packard) *M. californicum fragile* (Stretch) *M. californicum lutescens* (Neumoegen & Dyar) *M. californicum pluviale* (Dyar) *M. californicum recenseo* Dyar

Source: Stoetzel 1989.

of *Malacosoma*. Studies of the family by Franclemont (1973) indicate that tent caterpillars have no near relatives in the Nearctic fauna.

Stehr and Cook (1968) proposed that there are two distinct groups of American tent caterpillars. One group consists of the species *disstria, tigris,* and *constrictum,* and the other of *americanum, incurvum,* and *californicum.* On the basis of cladistic analysis, using the European species *franconicum, alpicolum,* and *castrensis* as the outgroup, Darling and Johnson (1982) concluded that there are three monophyletic groups in North America. A cladogram, derived from morphological and biological characteristics of North American species and *M. neustrium* (Table 2.3), is shown in Figure 2.4 and is in approximate agreement with the conclusions of Stehr and Cook (1968). The exact relationship of these groups remains uncertain, and only one possible relationship, based on the morphology of the seventh sternite of the male (character 10), is illustrated here.

Sociality of Caterpillars

Although most species of caterpillar live solitary lives, larval gregariousness is not uncommon among the families of the Lepidoptera (Fitzgerald and Peterson 1988, Fitzgerald 1993a). About 5% of the larvae of North American butterflies form assemblages of 10 or more caterpillars

Plate 1.

A. Second-instar larvae of the forest tent caterpillar, *Malacosoma disstria*, molting en masse on a silk mat spun on the upper surface of a partially consumed aspen leaf.

B. Larvae of the Sonoran tent caterpillar, *M. tigris*, resting together on a thinly spun molting tent.

C. The silk tent of the eastern tent caterpillar, *M. americanum*, within which the colony aggregates and molts.

D. Tents of the southwestern tent caterpillar, *M. incurvum*, on a tree defoliated by multiple colonies.

Plate 2.

A. The eusocial wasp *Polistes fuscatus* (Vespidae) tearing apart a larva of the eastern tent caterpillar it has killed. The caterpillar had just returned to the tent after feeding and was attacked as it entered the structure. The exposed gut of the caterpillar is fully packed with leaf fragments. The fly is an opportunistic feeder.

B. Nonpigmental coloration of the cuticle of the eastern tent caterpillar. The white middorsal stripe and the subdorsal blue areas are produced by the selective filtering of light by transparent microtubercles.

C. Larvae of the eastern tent caterpillar hanging by their prolegs from the shaded side of their tent to facilitate convective heat loss and prevent overheating.

D. A foraging column of eastern tent caterpillars moving over the surface of their host tree.

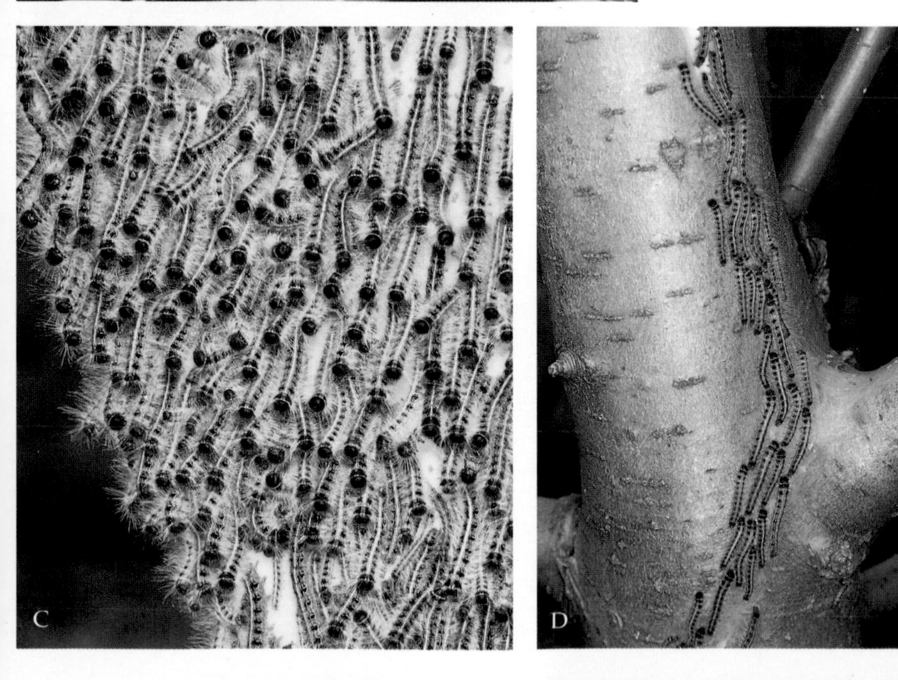

Table 2.3. Characters used to construct the cladogram for *Malacosoma* shown in Figure 2.4

Character	Ancestral trait	Derived trait
1	Large, permanent tents	Small, molting tents
2	Multiple hosts	Only on *Quercus*
3	Helical egg mass	Clasping egg mass
4	Upper posterior corner of ovipositor valves not produced, rounded, or angulate in lateral view	Upper posterior corner of ovipositor valves produced as flattened lobes, strongly angled in lateral view
5	Upper posterior corner of ovipositor valve rounded or only slightly produced	Upper posterior corner of ovipositor valve strongly produced
6	Male epiphysis of foreleg small or absent	Male epiphysis large and sickle-shaped
7	Larval setal group L2 with 3 setae	Larval setal group L2 with 2 setae
8	Larval setal group D1 with 5 setae	Larval setal group D1 with 4 setae
9	Large, permanent tents	Tent absent
10	Rear margin of male 7th sternum strongly and coarsely serrate	Rear margin of male 7th sternum sinuous, smooth, or slightly denticulate

Source: Reprinted, courtesy of the Entomological Society of Washington, from Darling and Johnson 1982.

Note: The European species *M. franconicum*, *M. castrensis*, and *M. alpicolum* were used as the outgroup.

(Stamp 1980). Less is known of the behavior of the caterpillars of moths, but life history data collected during surveys of Canadian forest species showed that nearly 8% of the larvae of the 392 species studied aggregate in groups of eight or more caterpillars for at least part of the larval stage (Herbert 1983). Analysis of data for moths found on the British Isles indicates that about 4% of 783 species lay their eggs in batches (Herbert 1983), but moth larvae may disperse immediately after hatching, so this value probably overestimates the number of species that remain gregarious as larvae. The aggregative tendency is particularly well developed in the Lasiocampidae. In addition to the tent caterpillars, the American *Gloveria howardi* and *Eutachyptera psidii* (Fig. 2.5) and species of the Old World genus *Eriogaster* form large larval aggregations and live in communal shelters.

Zoologists consider same-species groups of vertebrates with such aggregative tendencies to be social animals, but the term *social* is used

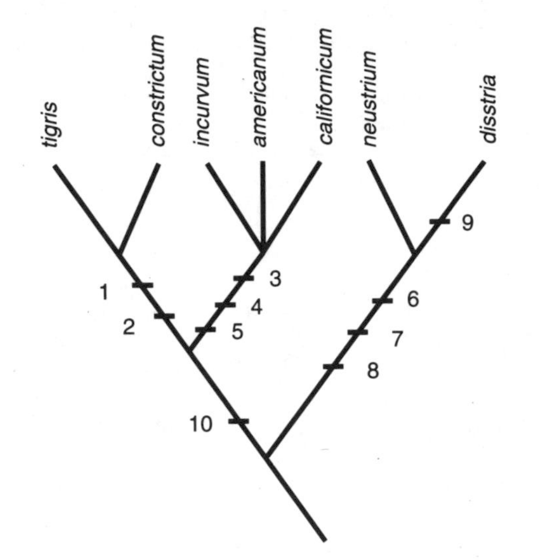

Figure 2.4. A cladogram for North American species of *Malacosoma*. Species on branches above numbered bar are presumed to have the derived trait for the character represented by that number (see Table 2.3). Relationships among the three monophyletic groups are tentative and only one relationship, based on character 10, is illustrated here. (Redrawn, by permission of the Entomological Society of Washington, from Darling and Johnson 1982.)

much more restrictively by entomologists. Entomologists tend to equate the terms *eusocial* (true-social) and *social*, and have traditionally considered as social only those species that possess the three defining qualities of eusociality: the simultaneous presence of parents and offspring, cooperative brood care, and reproductive division of labor. Those characteristics are possessed by the termites, ants, and some bees and wasps, but tent caterpillars possess none of them. In the evolutionary hierarchies proposed by students of eusocial insects, colonies of tent caterpillars have been classified as mere aggregates or, at best, as presocial insects (Eickwort 1981).

Although the repertoire of gregarious caterpillars of moths and butterflies is not nearly as extensive as that of the eusocial insects, numerous studies have shown that they do engage in cooperative interactions. These include leaf-shelter building, which involves communal efforts in tying and folding leaves; communal spinning to create mats, foliage enveloping webs and tents; communal basking to facilitate thermoregulation; collective efforts to breach tough and spiny plant surfaces; antipredator group displays; group exploration and trail laying; and recruitment to food and resting sites (Fitzgerald 1993a).

Figure 2.5. A silk shelter of *Eutachyptera psidii* suspended from a branch of an oak tree. Some shelters have been reported to reach a length of a meter or more.

Early accounts of sociality among butterfly larvae were given by Samuel H. Scudder (1889) and J. W. Tutt (1906b, 1907), but the British entomologist Frank Balfour-Browne was one of the first to draw attention to the cooperative nature of caterpillar colonies. In a paper presented at the Third International Congress on Entomology in Zurich in 1925 he took issue with the myrmecologist William Wheeler's restrictive treatment of the social insects in the latter's now-classic book *Social Life among the Insects* (1923). Balfour-Browne suggested that social insects be defined more broadly as species "in which the members of the family do something for the common good" (Balfour-Browne 1925:334). More recently E. O. Wilson (1975:7) proposed a similar definition, arguing that the essence of sociality is "reciprocal communication of a cooperative nature" and that "the terms *society* and *social* need to be defined broadly in order to prevent the exclusion of many interesting phenomena." In the spirit of these more expansive definitions we can consider tent caterpillars and other colonial caterpillars to be social insects, and I refer to them as such throughout this book.

It is clear, however, that tent caterpillars and other social lepidopterans lack both the complexity and richness of social interactions that characterize eusocial insects. A measure of the relative degree of sociality expressed by colonies of tent caterpillars and colonies of the domestic honey bee, one of the most social of all animals, is presented in Table 2.4. The societies are evaluated by reference to the 10 qualities of sociality defined by E. O. Wilson (1975). The honey bee clearly outranks tent caterpillars in nearly all the categories. On the other hand, tent caterpillar societies are markedly more sophisticated than the nonadaptive feeding aggregations of many other insect larvae, such as those formed by blow flies (Table 2.4). Indeed, if weight is given to the importance of information flow, the integration of behavior, and the proportion of time spent in cooperative interactions, tent caterpillar societies fare well when compared with the societies of many vertebrates. An aggregate of birds or mammals that cooperated to build a communal shelter, foraged in unison, and shared information regarding the location of food patches, were one to exist, would indeed be considered a highly social unit.

Balfour-Browne (1925) attempted to trace the evolutionary origins of the sociality of caterpillars. He suggested that insight into the evolutionary process could be had by observing the ways in which extant species of caterpillars use their silk to facilitate aggregation. He envisioned an evolutionary hierarchy in which caterpillars that spun simple two-dimensional resting mats on the surface of their food gave rise to species that constructed three-dimensional feeding webs that enveloped their food. These, in turn, gave rise to species that constructed home webs

Table 2.4. The expression of 10 qualities of sociality in tent caterpillars, the honey bee, and blow fly larvae

Quality of sociality	Expression in tent caterpillars	Expression in honey bee	Expression in blow fly larvae on carcass
Colony size	Colonies typically contain a few hundred individuals, but composite colonies of thousands are not uncommon.	Mature colonies consist of tens of thousands of individuals.	Typically hundreds of individuals.
Demography	Nonadaptive.	Adaptive.	Nonadaptive.
Cohesiveness	Colonies initially are highly cohesive, both when resting and foraging, but cohesiveness tends to weaken or disappear altogether as colonies mature.	Highly cohesive throughout life span of the colony.	Highly cohesive.
Pattern and amount of connectedness	Unpatterned, signals directed at random among siblings; highly connected.	Patterned; hierarchical, highly connected.	No connectedness.
Permeability	Permeable. Colonies usually consist of single families, but there is no indication of kin recognition. At high population density, colonies consisitng of multiple families are common.	Very low or zero permeability; bees from other colonies usually are not tolerated.	High; aggregates may consist largely of unrelated individuals.
Compartmentalization	Not compartmentalized.	Not compartmentalized.	Not compartmentalized.
Differentiation of roles	Although early reports suggested a weak division of labor within colonies of some species, more recent studies do not.	Caste system consisting of a queen, drones, and workers.	None.

Table 2.4–*cont.*

Quality of sociality	Expression in tent caterpillars	Expression in honey bee	Expression in blow fly larvae on carcass
Integration of behavior	Bouts of tent building and trail laying are synchronized.	Highly integrated at all levels of colony life.	No integration.
Information flow	Active communication with extra-silk pheromones to hold foraging columns together and to lead tentmates to food finds and resting sites. Passive tactile or acoustical signals may lead to aggregation and spinning on the surface of the tent and may warn of predators.	Highly developed systems of active communication functioning in multiple contexts based upon tactile, acoustical, and particularly chemical signals.	Probably none or at most passive tactile signals.
Time devoted to social activities (cooperative interactions)	When not resting in contact, caterpillars in young colonies engage in group basking, tent building, group exploration, collective antipredator defense and en masse migrations in search of food and resting sites. As colonies mature, the caterpillars of some species disperse and complete their development in isolation. Adults are not social.	When not resting, worker bees are continually engaged in social behavior. Larvae are nonactive.	No cooperative interactions are known to occur.

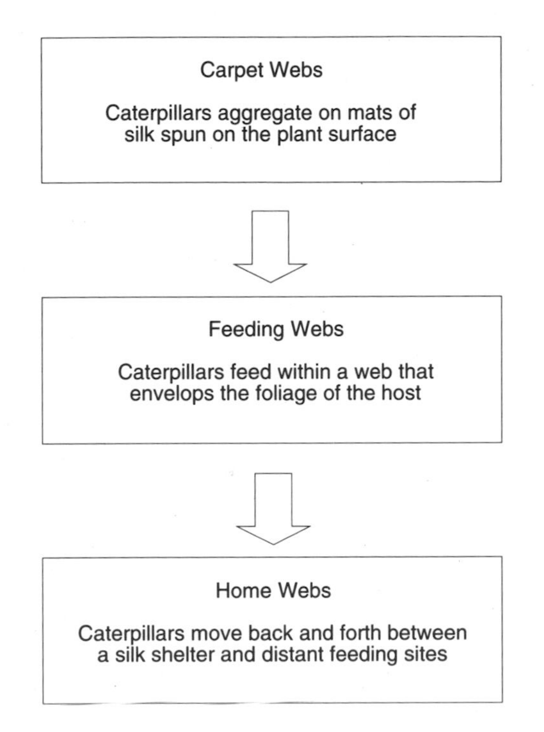

Figure 2.6. An evolutionary hierarchy of sociality of caterpillars as proposed by Balfour-Browne (1925).

distinct from their feeding sites (Fig. 2.6). Although there is no reason to suspect that caterpillars that occupy feeding webs are more social than those that rest on carpet webs, the construction of a home web does indeed appear to represent a major advance in the evolution of sociality among caterpillars. This is true, not because of the relative complexity of the structure, but because the use of a permanent or semipermanent home web requires the occupants to practice central-place foraging, a foraging pattern that appears to necessitate a relatively high degree of communication among siblings.

All central-place foragers, however, are not home-web builders. The larvae of the saturniid moth *Hylesia lineata*, for example, initiate forays to distant feeding sites from a simple, two-dimensional resting mat (Janzen 1984). Thus, although Balfour-Browne presents a plausible pathway for the evolution of increasingly complex silk structures, such structures represent only one character state that must be considered when attempting to trace the evolutionary origins of the sociality of caterpillars. Unfortunately, 70 years after the publication of his paper we still know very little of the behavior and ecology of social caterpillars, and it is possible to speculate in only the most tentative way on the factors that might have influenced the evolution of tent caterpillar sociality.

Recent inquiries into the origins of lepidopteran sociality have been concerned largely with elucidating the origins of the batch-laying habit, because social aggregates of caterpillars originate from eggs that are laid in proximity. Stamp (1980), for example, suggested several factors that favor cluster oviposition: selection pressure from predators that capture flying females; inclement weather that limits egg-laying opportunities; scarcity of males that require females to search extensively for mates; and scarcity of adult food plants. Eggs laid in batches may also be less vulnerable to desiccation and to predation than solitary eggs, particularly if they are distasteful and aposematically colored. Cluster oviposition is particularly likely to arise when larval food plants are patchily distributed and adult encounters with them are infrequent (Stamp 1980, Young 1983).

The advantages of aggregation to caterpillars are numerous. These include group facilitation in overcoming plant defenses; the facilitation of predator defense, foraging, thermoregulation, and shelter building; the enhancement of aposematism; and the conservation of host tree foliage (Fitzgerald and Peterson 1988, Fitzgerald 1993a and references therein). The specific advantages to aggregation identified for the tent caterpillars are treated in other chapters of this book. These include the facilitation of tent building, thermoregulation, and food finding as discussed in Chapters 6 and 7. The significance of aggregation and synchronized foraging as both passive and active defensive strategies to minimize depredation by natural enemies is considered in Chapter 8.

Courtney (1984) argued that that cluster oviposition arises first and sets the stage for the evolution of larval adaptations that could lead initially to the maintenance of contact and eventually to sociality. He argued that batch-laying butterflies have higher realized fecundity than butterflies that lay eggs singly and that selection pressure for realized fecundity, in itself, is the chief factor underlying the evolution of cluster oviposition. Moreover, he viewed the apparent benefits accruing to aggregated larvae as consequences rather than causes of the evolution of egg clustering.

The larvae of many caterpillars, however, disperse after hatching, and it is clear that, even though they are the product of a single egg mass, posthatch behavioral mechanisms must operate to keep siblings together. Shared response to silk, or to extra-silk substrate markers, appears to be the proximate basis for aggregation by caterpillars. Furthermore, although proximity of eggs may indeed be a prerequisite for the development of incipient caterpillar societies, the success of aggregated siblings can be expected to affect the evolution of adult oviposition patterns. If we assume that the genes regulating the ovipositional habit lie dormant in the caterpillar and that aggregated larvae enjoy increased fitness over solitary caterpillars, then selective pressures should lead to populations of adults genetically predisposed to lay their eggs close enough together to allow siblings to find one another. Selective pressures directed solely at the adult or egg (Stamp 1980, Young 1983) would be expected to moderate the impact of larval survivorship on the evolution of ovipositional patterns.

In a recent attempt to reconstruct the first stages in the evolution of the sociality of caterpillars, Fitzgerald and Costa (1986) and Fitzgerald and Peterson (1988), like Balfour-Browne (1925), recognized that silk served as the likely catalyst for the process. Sociality of caterpillars perhaps is derived from the behavior of a solitary ancestral species that marked its feeding arena with silk or used silk to mark trails between a resting and feeding site. Such a silk-marked substrate, encountered by a second caterpillar searching for food, could serve to contain the visitor and serve as an initial stimulus to aggregate.

Some species of solitary lasiocampids, such as the lappet moth *Tolype velleda,* may occur in aggregates of several individuals (Franclemont 1973). It is unknown, however, whether *Tolype,* or any extant species of solitary lasiocampid, uses silk or extra-silk pheromones to mark trails or resting sites, though it is highly probable. Stehr (1987) found that all species of lasiocampid caterpillars he examined had an "anal point" located on the venter between the anal prolegs. This structure has been implicated in the trail-marking behavior of tent caterpillars (see Chapter 6) and may serve a similar function in other species, including solitary forms.

Marking behavior by caterpillars is not uncommon and the larvae of several solitary species found in families other than the Lasiocampidae are known to establish silk resting mats and to mark trails between resting and frequently visited feeding sites. Moreover, the "trapping" effect of the chemical trails of tent caterpillars is well documented. Close colonies of the eastern tent caterpillar often coalesce when their trail systems join (see Fig. 5.12), and forest tent caterpillars have been reported to rest and forage with colonies of the eastern tent caterpillar after wandering onto their trail system (Fitzgerald and Edgerly 1979a). Tutt (1900) reported an instance of tent sharing between a colony of *M. neustrium* and the tent-building moth *Euproctis.* The tent caterpillars were observed to expropriate and expand a *Euproctis* tent and to feed along-

Table 2.5. Some gregarious caterpillars that form large aggregates and are known to be, or are suspected of being, central-place foragers for all or part of the larval stage.

Family	Species	Tent maker?	Central-place foraging?	Reference site	Reference
Papilionidae	*Papilio anchisiades*	N	Y	Mexico	Ross 1964
Pieridae	*Mylothris chloris*	N	Y	Nigeria	E. B. Britton 1954
	Eucheria socialis	Y	Y	Mexico	Westwood 1836, Kevan & Bye 19
	Neophasia terlooti	Y	?	North America	Behr 1869
	Aporia crataegi	Y	?	England	Tutt 1907
Nymphalidae	*Brassolis isthmia*	N[a]	Y	Panama, Costa Rica	Dunn 1917; Young 1986
	Nymphalis antiopa	N	?	North America	Scudder 1889
Lasiocampidae	*Gloveria howardi*	Y	?	Arizona	Franclemont 1973
	Eutachyptera psidii	Y	?	Mexico	Franclemont 1973
	Eriogaster lanestris	Y	Y	England	Balfour-Browne 19
	Eriogaster amygdali	Y	Y	Lebanon	Talhouk 1975
Saturniidae	*Hylesia lineata*	N	Y	Costa Rica	Janzen 1984
	Hylesia acuta	Y	Y	Mexico	Wolfe 1988
Thaumetopoeidae	*Thaumetopoea* spp.	Y	Y	Israel	Halperin 1990
	Thaumetopoea pityocampa	Y	Y	Europe	Fabre 1991
	Anaphe spp.	Y	?	Africa	Pinhey 1975
Lymantriidae	*Euproctis chrysorrhoea*	Y	?	Great Britain	Carter and Hargreaves 198
Eupterotidae	*Hyposoides radama*	Y	?	Madagascar	Sharp 1970

Note: Colonies that practice central-place foraging are likely to use advanced systems of tra based communication that facilitate cooperative foraging, but the possibility has not yet be investigated for any of the species listed here.

[a]Construct communal shelter from leaflets of coconut.

side the resident caterpillars. Recently, I observed a colony of *M. tigris* caterpillars resting in contact with the larvae of an unidentifed species of gregarious nymphalid caterpillar on *Quercus*.

Promoted by cluster oviposition and mutual response to simple cues associated with silk or substrate markers, incipient colonies of caterpillars could form the basis for the evolution of more complex forms of social interaction. As the colonies increased in size, selective forces such as those consequent on the local exhaustion of the food resource, predation, disease, interspecific competition, host phenology, and the degree of polyphagy could have led to the evolution of specific social patterns of aggregation and foraging as discussed in Chapter 6. Colonies shaped by a subset of these selective forces that favored fixed-base foraging could be expected to have evolved more complex societies based on communicative systems that facilitate cooperative foraging and other adaptive interactions.

So far as is presently known, tent caterpillars sit at the pinnacle of caterpillar social evolution. It is highly probable, however, that as our knowledge of the behavior of other gregarious caterpillars increases we will find species that are equally social or have even more advanced social behavior. The candidate species most likely to be defined by advanced social interactions are those that form large aggregations and forage from a fixed base. Too little is presently known of the foraging behavior of social caterpillars to say with certainty which species meet those requirements, but some likely candidates are listed in Table 2.5. Future studies of these and other caterpillar species with similar patterns of foraging behavior will add enormously to our knowledge of caterpillar sociality.

3

The Caterpillar and the Pupa

The anatomy and physiology of caterpillars are largely dedicated to foraging and food processing. Indeed, the voracious appetite of caterpillars has led to their characterization as little more than walking digestive tracts. But caterpillars are much more than that, for they must also respond to a broad array of environmental variables that directly affect their day-to-day survival. Among the most prominent of those variables are the components of climate, particularly temperature and humidity; the chemical, physical, and distributional properties of their food supply; and the ever-present threat of predators and parasitoids. The overt features of the anatomical and physiological systems that enable tent caterpillars to cope with these and other components of their environment are shared with other caterpillars, but many aspects of their anatomy and physiology are fine-tuned to the specific set of circumstances that affect them.

Anatomy and Physiology of the Caterpillar

Although the tagmata (major body sections) of the caterpillar are not as evident or as functionally divergent as those of the adult, the body of the caterpillar consists of a distinct head, thorax, and abdomen (Fig. 3.1). The head of the caterpillar is endowed with visual, tactile, and chemical receptors that mediate feeding and trail following. The mouthparts are largely adapted to harvesting leaves and spinning silk. Short, segmented legs occur on each of the three segments of the thorax. The

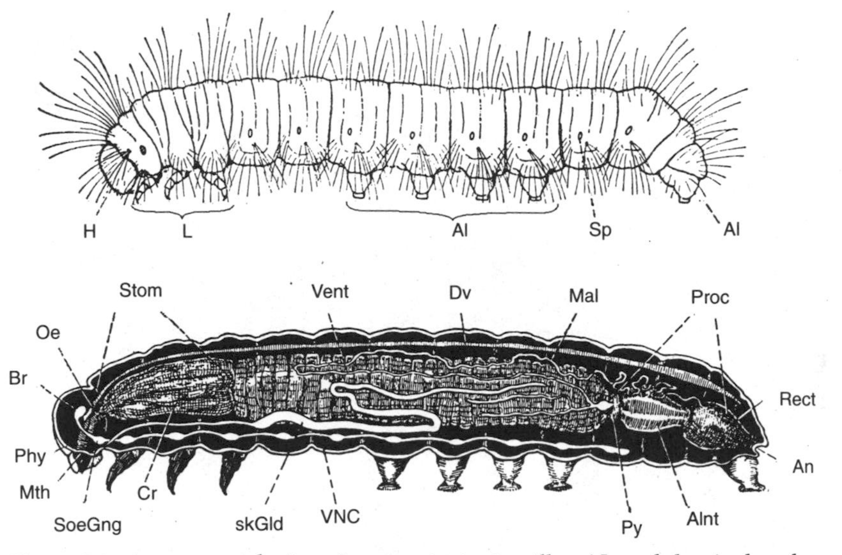

Figure 3.1. Anatomy of the larval eastern tent caterpillar. AL = abdominal prolegs, aInt = anterior intestine, An = anus, Br = brain, Cr = crop, Dv = dorsal blood vessel, H = head, L = thoracic legs, Mal = malpighian tubules, Mth = mouth, Oe = esophagus, Phy = pharynx, Proc = proctodeum or hindgut, Py = pylorus, Rect = rectum, SkGld = silk gland, SoeGng = subesophageal ganglion, Sp = spiracle, Stom = stomodeum or foregut, Vent = ventriculus or midgut, VNC = ventral nerve cord. (From Snodgrass 1961.)

abdomen has 10 segments and bears five pairs of fleshy, unsegmented prolegs.

The body wall of the caterpillar is soft, distensible, and able to accommodate the sizable increase in mass that occurs during a stadium. Functional pairs of spiracles, the external openings of the tracheal system, are found on the prothorax and on the first eight abdominal segments. A structure, apparently unique to lasiocampid caterpillars, termed an anal point by Stehr (1987), occurs on the ventral surface of the last abdominal segment between the anal prolegs (see Fig. 6.7). The structure appears to be involved in the trail-marking behavior of tent caterpillars (Chapter 6), but its function has not been determined in other genera of the Lasiocampidae. The cuticle of older caterpillars is brightly colored and bears elongate, dorsal and lateral setae (Fig. 3.2). Unlike the larvae of the lasiocampids *Gloveria* and *Eutachyptera*, tent caterpillar larvae lack the urticating hairs that cause dermal irritation when they come in contact with human skin.

The internal organs of the caterpillar include those associated with the nervous system, the body wall muscles, the digestive tract with its

Figure 3.2. Sixth-instar larvae of the eastern tent caterpillar. Elongate dorsal and lateral setae and patterned cuticles, consisting of white, blue, and orange colors, are characteristic of tent caterpillar larvae.

associated Malpighian tubules, the tracheal system, the circulatory system, and the fat body. The nervous system of the caterpillar consists of an anterior brain, a nerve cord that runs along the ventral midline of the thorax and abdomen with segmental ganglia, and sensory and motor neurons that extend from the central nervous system to the receptors and muscles. Well-developed systems of muscles line the body wall and are innervated by motor neurons programmed to produce the complex, stereotypical movement patterns associated with locomotion, feeding, silk spinning, and evasive behaviors. The digestive tract fills most of the hemocoel (blood cavity) and consists of discrete regions dedicated to the storage, digestion, and egestion of food. Like all other organ systems of the body, the gut is bathed in hemolymph, which is circulated slowly under the influence of overt body movements and the tubelike dorsal heart. The hemocoel is suffused with tracheae that not only transport the gases associated with respiration but, like ligaments, also support and suspend the gut at the center of the body. Malphigian tubules arising from the gut wall filter the hemolymph for waste products, which are then either stored or excreted into the lumen of the hindgut.

Ingestion, Digestion, and Excretion

The mouthparts and digestive tract of the caterpillar are largely dedicated to processing leaves, the sole food source of the larva. The definitive mouthparts of the tent caterpillar consist of the labrum and its undersurface, the epipharynx; the mandibles; hypopharynx; maxillae; and labium. The last three structures are modified in caterpillars to form a spinneret and, except for the sensory structures of the maxillae, are less involved in the process of ingestion than are the mandibles and labrum.

The labrum, or upper lip, is deeply notched at its anterior edge. The labral notch is a common feature of phytophagous insects and functions as a guide, channeling the edge of a leaf between the cutting surfaces of the mandibles, which lie just below the lip. Powerful adductor muscles that occupy much of the head capsule drive the mandibles. As the caterpillar chews, fragments of leaf are drawn into the oral cavity, or cibarium, and enter the digestive tract through the mouth, the anterior opening of the pharynx. The epipharynx, which forms the forward part of the roof of the cibarium, is innervated by eight sensory neurons that are thought to be involved in gustation, but nothing is known of their sensitivity (Dethier 1980a). Other sensory neurons involved in feeding

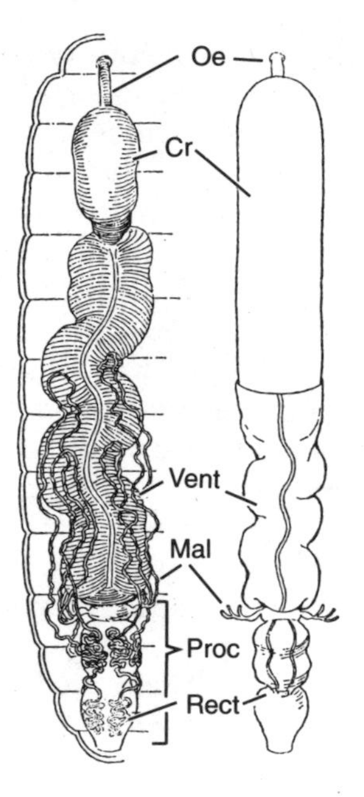

Figure 3.3. Condition of the gut of the eastern tent caterpillar before (left) and immediately after (right) feeding. During a bout of feeding, the caterpillar fully packs its crop (Cr) with leaf particles, extending it to its maximum dimensions. While the caterpillar rests or basks in the sun at the tent site, the meal is slowly passed to the midgut (Vent) and digested, and the crop returns to its former dimensions. Mal = Malpighian tubules, Oe = esophagus, Proc = proctodeum or hindgut, Rect = rectum. (From Snodgrass 1961.)

occur on the antennae and maxillae (see Chemoreception, in this chapter).

The alimentary tract of the eastern tent caterpillar consists of a foregut, midgut, and hindgut (Figs. 3.1, 3.3). The foregut, or stomodeum, consists of a pharynx, esophagus, and expandable crop. Leaf fragments taken in by a caterpillar pass from the pharynx to the esophagus and then into the crop, where they are temporarily stored. The meal taken by tent caterpillars is typically large enough to cause the crop to expand forward and backward, displacing the midgut (Fig. 3.3). After feeding, caterpillars rest at the tent, often basking in the sun to raise their body temperature to facilitate digestion (see Chapter 7). During this process, the muscles of the gut wall set up peristaltic contractions that transport macerated leaf fragments from the crop to the midgut, where the useable components are digested and absorbed. As food is processed, the midgut elongates to accommodate the meal, and the emptied crop contracts to its former dimensions. When the meal has been fully digested, the midgut contains a brownish liquid filled with numerous air bubbles.

The eastern tent caterpillar digests its food slowly or not at all at body temperatures below 15°C (Casey et al. 1988). When the temperature is cool and the sky is cloudy, the caterpillars may be unable to elevate their body temperatures very far above this threshold temperature, and they process their food much more slowly than they do on sunny days. Nonetheless, caterpillars with only partially emptied crops still forage, packing as much new material into their crops as they can. Caterpillars maintained experimentally under a marginal temperature regime in the laboratory forage as often as caterpillars that have access to a radiant heat source, but they grow much more slowly (Casey et al. 1988).

Most of the food that tent caterpillars process consists of cellulose and other nondigestible components for which the caterpillars possess no appropriate enzymes. In the anterior intestine, these residues are compressed into pellets by muscles of the intestinal wall. The fecal pellets are then passed to the rectum and to the outside through the anus. Tent-building species process most of their food while aggregated on or in their tents, and the structures eventually become thoroughly littered with the fecal pellets. As they near maturity, tent caterpillars often disperse throughout the treetops, and, during outbreaks, they collectively pass huge quantities of fecal pellets that create an audible plinking as they rain down from the canopy.

Nitrogenous wastes and other end products of metabolism are removed from the hemolymph by the excretory system, which consists of six Malpighian tubules that rest close to the posterior part of the midgut and the anterior intestine (Figs. 3.1, 3.3). Three tubules arise from a common duct on either side of the anterior intestine at the point where it joins the midgut. The distal ends of the tubules are embedded in the walls of the rectum to form a cryptonephridium, an arrangement that appears to facilitate the resorption of water from digestive wastes that pass through the rectum. The tubules are completely emptied during each molt (Takahashi et al. 1969).

In all but the ultimate instar, the Malphigian tubules appear white in the freshly dissected caterpillar. During the last larval stadium, the lumen of the tubules becomes occluded with white or yellow crystalline bodies. The material is excreted into the intestine so that by the time the caterpillar has ceased feeding and is preparing to spin its cocoon a considerable quantity of the material has accumulated. The pastelike material is passed from the anus and worked into the walls of the completed cocoon, where it serves to stiffen the silk (Snodgrass 1930). If a cocoon is torn apart, the material billows up as a fine powder. The possibility that the crystals may offer some protection from pupal predators has not been investigated, but the crystals may be irritating to humans (Stehr and Cook 1968). An elemental analysis of the crystalline bodies found in the cocoon silk of *Malacosoma neustrium testacum* indicates that the material is approximately 90% calcium oxylate monohyrdate by weight (Ohnishi et al. 1968). Studies by Takahashi et al. (1969) indicate that most of the oxalic acid in the crystals is derived from the food of the caterpillar, though it is unknown how the insoluble material moves through the gut wall.

The metabolic cost of feeding has not yet been measured for tent caterpillars. Studies of other species of caterpillars indicate that the metabolic rate of feeding caterpillars may be as much as 100% as great as that of the resting insect (Aidley 1976, McEvoy 1984). Nonetheless, only a small amount of energy is burned during the digestive process. The energy expended by the cinnabar moth during feeding amounts to only 3% of the total energy value of the ingested food (McEvoy 1984).

Spinneret and Silk Glands

Tent caterpillars lay down silk to secure their footing as they move over the branches of their host tree. Frequently used branches, and particularly those near a colony's resting site, accumulate considerable quantities of silk and eventually bear broad and conspicuous trails. Many species also construct silk tents, some of which are among the largest shelters built by any noneusocial insect (see Chapter 7). Both trail systems and tents require large amounts of silk, but they are the products of collective efforts that involve hundreds of individuals, each of which contributes a relatively small quantity of silk over an extended period of time. In contrast, each caterpillar must eventually build its own cocoon, a process that requires the caterpillar to secrete a considerable quantity of silk in a short span of time. Ruggiero and Merchant (1986) estimated that approximately 80% of the silk produced by *M. americanum* is used for cocoon formation. The caterpillars conserve the raw materials from which the cocoon is formed by contributing little or no silk to the tent and trails during their last stadium. Approximately 3% of the assimilated energy of the caterpillars of *M. americanum* (Ruggiero and Merchant 1986) and *M. neustrium testacum* (Shiga 1976a) is channeled to silk production.

Silk is produced and stored in labial glands and secreted to the outside through a spinneret located at the bottom of the head. The labial glands of the tent caterpillar are elongate structures that extend far back into the abdomen (Figs. 3.1, 3.4A). Each gland has a prominent reservoir located near its center. The anterior ends of the glands consist of narrowed ducts that unite where they enter the silk press (Fig. 3.4B), a structure that regulates the release of the silk and determines the shape of the emerging strands. The ducts of two small accessory glands, the glands of Filippi, anastomose with the anterior silk ducts. The function of these glands is unknown in any species of caterpillar.

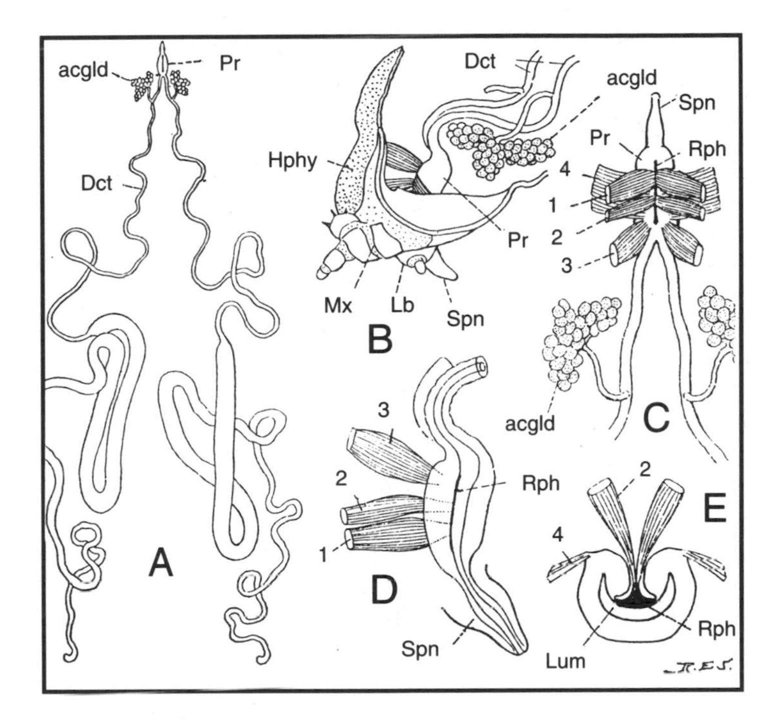

Figure 3.4. Silk-spinning apparatus of the eastern tent caterpillar. acgld = gland of Filippi, Dct = ducts of silk gland, Hphy = hypopharynx, Lb = labium and associated palpus, Lum = lumen of silk press, Mx = maxilla, Pr = silk press, Rph = raphe, Spn = spinneret. (From Snodgrass 1961.)

The maxilla and labium, which are separate structures in most insects, are united in caterpillars to form a complex structure that bears the spinneret (Fig. 3.4B). Inconspicuous labial palps occur on either side of the spinneret. Their function in tent caterpillars has not been investigated. The spinneret is small and spinelike and is directed downward and backward from the tip of the labium so that it contacts the substrate as the caterpillar moves. Internally, the duct of the spinneret is continuous with the opening of the silk press (Fig. 3.4C, D), a structure that is thought to be derived from the salivarium of other insects by the fusion of the tips of the hypopharynx and the labium. The dorsal wall of the press has a sclerotized bar, or raphe. Three sets of muscles that arise from the hypopharynx are attached to the raphe. An additional set of muscles, attached to the lateral walls of the press, originate on the labium. The dorsal and lateral walls of the press can be lifted by these muscles to increase the diameter of the lumen (Fig. 3.4D, E), creating a vacuum that draws the liquid silk forward from the glands.

When the tip of the spinneret is touched to the bark of a tree or a leaf surface, the sticky outer coating of the silk causes the strand to adhere to the substrate. If the caterpillar then swings its head or otherwise moves away from the point of adhesion, a continuous strand of silk is drawn from the spinneret and trails behind the larva. The strand of the tent caterpillar can be drawn from the spinneret by grasping and pulling it with a forceps. Because the strand can be drawn despite the resistance of the caterpillar, the insect must be unable to stop the flow of the silk through any internal mechanism. The caterpillars apparently terminate a bout of spinning by forcibly breaking or biting the strand.

The exact chemical makeup of the silk of tent caterpillars has not been ascertained, but, in the main, its composition is likely to resemble that of related species. Dried cocoon silk of the silkworm, *Bombyx mori*, consists of approximately 73% fibroin and 22% sericin; the remainder is largely wax and inorganic components (Iizuka 1966). The proteins fibroin and sericin occur in an aqueous solution in the silk glands at a concentration of 15–30%. The fibroin is synthesized in the rough endoplasmic reticulum of the cells that line the lumen of the posterior portion of the silk gland and moves to the lumen as fibroin globules. As the secretion moves down the gland it becomes progressively dehydrated, and the globules adhere to each other, eventually forming microfibrils. Sericins secreted from the middle section of the silk gland surround the fibroin core as it moves down the gland. The microfibrils align as they pass through the silk press, and in the process the strands harden. The silk emerges from the spinneret as a pair of fibroin strands coated with sticky sericins (Fig. 3.5). In the silkworm, each strand consists of 900–2000 aligned microfibrils (Sehnal and Akai 1990).

Central Nervous System

The central nervous system of the larval tent caterpillar, like that of other insects, consists of a bilobed brain and a ventral nerve cord with its associated ganglia (Fig. 3.6). The nerve cord, originating behind the brain, is divided into two lateral connectives that pass around the esophagus, then extend along the midventral surface of the body to the seventh abdominal segment. A ganglion occurs just beneath the esophagus, and additional ganglia occur in each of the three segments of the thorax.

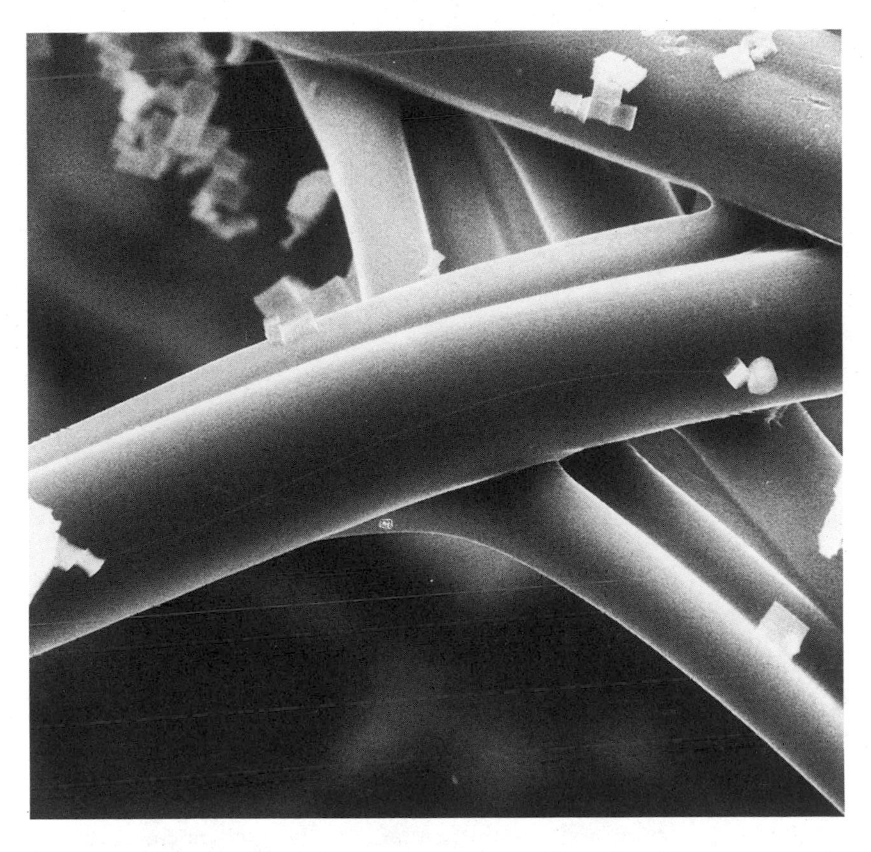

Figure 3.5. Cocoon silk of the eastern tent caterpillar. The silk issues from the spinneret as paired strands. Calcium oxylate crystals distributed among the strands are passed from the anus of the spinning caterpillar and serve to stiffen the silk fibers. (Courtesy of Janice Edgerly-Rooks.)

Eight ganglia occur in the abdomen. The eighth abdominal ganglion, which lies close to the seventh ganglion, is considered to be a compound ganglion, and prominent nerves extend from it to the posterior regions of the abdomen.

Motor and sensory neurons issuing from the brain innervate the simple eyes, or stemmata, the antennae, and the labrum. Frontal connectives arising from the brain lead to a frontal ganglion that gives rise to the recurrent nerve. The recurrent nerve extends posteriorly over the digestive tract, giving rise to the stomatogastric nervous system. This system innervates the anterior regions of the alimentary tract. The corpora car-

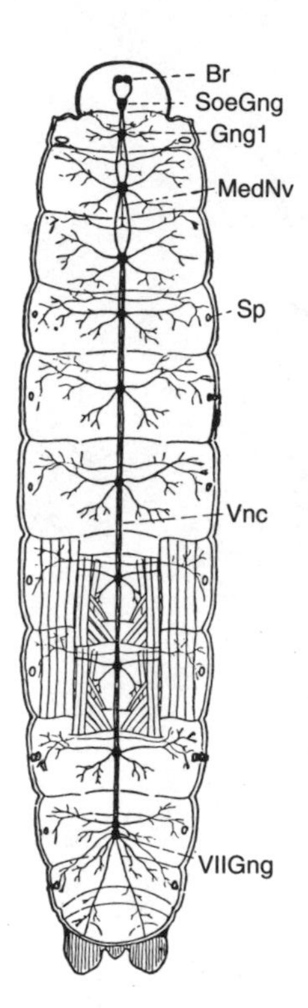

Figure 3.6. Nervous system of the eastern tent caterpillar. Br = brain, Gng_1 = first thoracic ganglion, MedNv = median nerve, SoeGng = subesophageal ganglion, Sp = spiracle, VIIIGng = eighth abdominal ganglion, Vnc = ventral nerve cord. (Reprinted from R. E. Snodgrass, *Principles of Insect Morphology*. Copyright © 1993 by Cornell University. Used by permission of the publisher, Cornell University Press.)

diaca and the corpora allata, endocrine glands that regulate growth and development, communicate with the brain via the corpora cardiaca connectives.

Nerves arising from the subesophageal ganglion contain both sensory and motor neurons that innervate the muscles of the cervix, mandibles, hypopharynx, maxillae, and labium. Lateral branches of the segmental ganglia principally innervate the muscles and sensory structures of the segment in which they are found, though some of the neurons found in a segment may arise from the ganglion of the preceding segment.

Tent caterpillars also have median nerves arising from each of the 11 ganglia of the thorax and abdomen (Fig. 3.7). Lateral branches of these nerves appear to innervate the ventilatory system of the segment immediately following. Thus, the fine fibers that arise from the lateral branches of the median nerve of the metathorax terminate in the vicinity of the trachaea and closing muscles of the spiracles of the first abdominal segment.

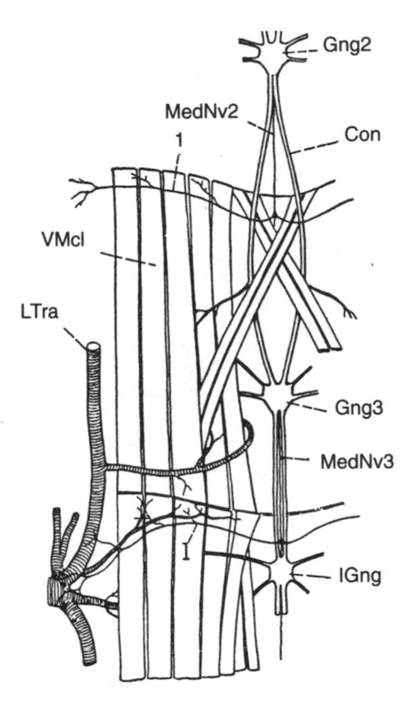

Figure 3.7. Section of the ventral nerve cord of the eastern tent caterpillar showing median nerves. Con = ventral nerve cord, Gng2 and Gng3 = second and third thoracic ganglia, IGng = first abdominal ganglion, 1 = lateral branch of median nerve, Ltra = lateral tracheal trunk, MedNv2 and MedNv3 = second and third median nerves, VMcl = ventral longitudinal muscle. (Reprinted from R. E. Snodgrass, *Principles of Insect Morphology*. Copyright © 1993 by Cornell University. Used by permission of the publisher, Cornell University Press.)

Photoreception

The eyes of the caterpillar consist of six stemmata that occur on each side of the head just behind the antennae (Fig. 3.8). Although these structures superficially resemble the ocelli found on adult and nymphal insects, internally they more nearly resemble the ommatidium of the compound eye. Like the ommatidium, the light-gathering dioptric apparatus of each stemma of *Malacosoma* consists of a dome-shaped, superficial cornea and an underlying crystalline lens (Fig. 3.9). The cornea of the first and sixth stemmata of *Malacosoma* appears tripartite, whereas the lenses of the other stemmata are unitary (Singleton-Smith and Philogene 1981). The sensory apparatus consists of retinular cells, which bear the light-sensitive rhabdom on their inner surfaces, and the ensheathing corneagen cells. There are seven retinular cells in each stemma. The axons of the retinular cells join with those of the other associated stemmata, and the 42 fibers travel to the tritocerebrum of the brain ensheathed as the stemmal nerve (Singleton-Smith and Philogene 1981). The dioptric apparatus is cast along with the cuticle of the head during each molt, but the sensory apparatus is retained and enlarged.

Studies of other species of caterpillars show that the lenses of the

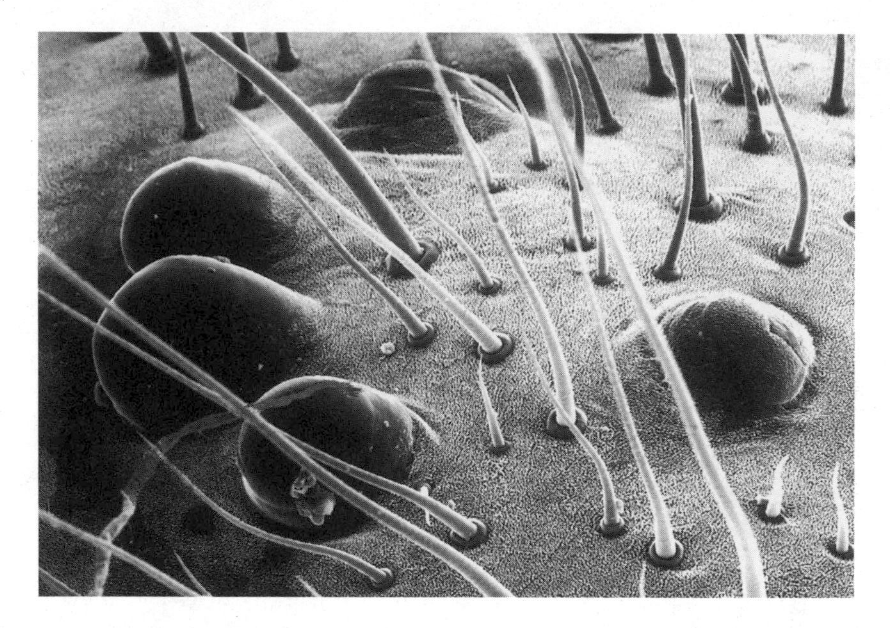

Figure 3.8. Five of the six stemmata forming the eye of the larval eastern tent caterpillar. (Courtesy of Janice Edgerly-Rooks.)

stemmata form real inverted images that are projected onto the receptor cells. Owing to the sparse innervation of each stemma, however, the capacity of the dioptric apparatus to form images would appear to be of little significance. Each stemma is likely to capture only a single point of light and by itself provide the caterpillar with no information about the form of an object.

On the basis of his detailed studies of the woolly bear caterpillar, Dethier (1942, 1943) suggested that the 12 spots of light collected by the two sets of stemmata might allow the caterpillar to form a mosaic image of an object much like that supposed to be formed by the compound eye, albeit much coarser. Indeed, caterpillars scan nearby objects by swinging their heads, a tactic that appears to allow them to ascertain the crude shape or orientation of the object. Although there have been no investigations of the visual acuity of tent caterpillars, studies of other species have demonstrated that larvae can see dark vertical shapes against light-colored backgrounds, much like branches would appear against the sky. Hundertmark (1937) found that the first-instar larvae of the nun moth, *Lymantria monacha,* released in the center of a field bordered by upright wooden blocks typically moved directly to the widest blocks. And studies of the gypsy moth, *L. dispar* (Roden et al. 1992)

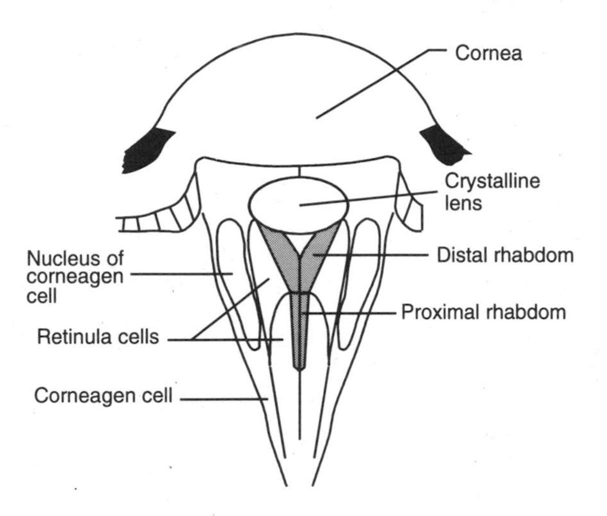

Figure 3.9. Stemma of a caterpillar. (Redrawn, by permission of Wistar Institute Press, from Dethier 1943.)

indicated that the caterpillars assess not only the diameter of objects but also their height and can see horizontal branches and vertical stems at distances of 2–3 m.

A handful of studies have shown that caterpillars have color vision. Using electrophysiological recording techniques, Ishikawa (1969) found that the stemmata of the larvae of the silkworm, *B. mori*, are particularly sensitive to ultraviolet light and to greenish colors, suggesting that the caterpillar has two receptor systems with different spectral sensitivities. Similar conclusions regarding spectral sensitivity were drawn in two other studies of color vision. Weiss et al. (1944) investigated the orientation of 13 species of lepidopterous larvae to colored filters, and Saxena and Goyal (1978) studied larvae of the swallowtail butterfly *Papilio demoleus*. There have been no attempts to study color vision in *Malacosoma*, but the caterpillars are likely to have spectral sensitivities similar to those of the species studied. Tent caterpillars are also likely to be visually sensitive to their tent and well-established trunk trails, both of which may appear particularly prominent when bathed in sunlight.

Some insects certainly are able to detect the plane of polarization of sunlight and use it as an orientational cue on cloudy days. Although the direct rays of the sun are not polarized, sunlight becomes polarized as it passes through the earth's atmosphere. Because the plane in which most of the light waves vibrate is related to the position of the sun, the sun can be used as a compass even when its position is obscured by cloud cover, provided the animal can find an unobstructed patch of blue sky.

The ability to orient by reference to polarized light has been particularly well documented for the honey bee. Early experiments by the German ethologist Karl von Frisch showed that the orientation of the dance of worker bees could be deflected a given number of degrees by rotating a polarizing filter between a dancing worker and a patch of blue sky (Winston 1987). Wellington et al. (1951) reported similar results when they placed a polarizing filter between the sky and mature tent caterpillars moving over the ground in search of pupation sites. When the axis of the Polaroid screen was rotated through 90°, the investigators reported that the path of the caterpillars changed by a similar amount. Dethier (1989), who studied the response of arctiid caterpillars to polarized light, postulated that the sensitivity of caterpillars to directional light may enable them to walk in relatively straight paths, a strategy that on average may be the best for discovering patchily distributed resources.

Although tent caterpillars are strongly photopositive (see Chapter 7), it is unknown if colonies depend at all on directional information from the sun to facilitate movement between their resting sites and distant feeding sites as do ants and honey bees. Unlike those other insects, tree-dwelling tent caterpillars orient in three dimensions, and their choices of pathways are largely defined by the branching pattern of the host tree. Consequently, the larvae must make many changes in direction relative to the position of the sun during a foraging bout, and their sensitivity to chemical trails must often, if not always, take priority over their sensitivity to light when they are away from their tent.

Chemoreception

The receptors of the head that mediate feeding are a relatively conservative feature of caterpillar morphology, and even distantly related species have antennae and maxillae that bear the same number of sensilla and innervating neurons. Electrophysiological recordings from the antennae and maxillae of several species of caterpillar show that these structures may also share a conservative physiology in that they are relatively nonspecific in their responsiveness. Furthermore, different sensory neurons lack a commitment to specific substances such as sugar, salts, and acids. Instead, a broad array of chemical substances, including extracts from nonfood plants, elicit action potentials from many of the neurons that innervate these receptors, though they often show a hierarchy of response so that some compounds stimulate the neurons to a greater extent than others.

Antennae

Three-segmented antennae occur at the sides of the mandibles. The antennae are the main olfactory receptors of the caterpillar, and each bears three sensilla basiconica. Together, the sensilla of each antenna are innervated by 16 neurons (Schoonhoven and Dethier 1966). Because the presence of elongated tactile setae prevents the sensilla from contacting the substrate, caterpillars sample host plants at a distance. This sampling of plant volatiles appears to be the first level of discrimination in a process that enables a caterpillar to separate host plants from nonacceptable plants. Taste receptors associated with the sensilla of the maxillae and epipharynx are subsequently brought into play.

Electrophysiological recordings from the antennal receptors show that the unstimulated neurons fire at a background frequency of 1–10 Hz (Dethier 1980a). These rates increase to frequencies as high as 65 Hz when a leaf is held a few millimeters from the antennal receptor. Studies of the eastern tent caterpillar by Dethier (1980a) show that, at least among the neurons innervating the medial sensillum of the antenna, there is no clear-cut peripheral filtering of the sort that occurs in some of the pheromone detectors of moths. The neurons of the sensillum are relatively nonspecific, firing in response not only to host plants but to a broad range of nonhost species such as elm, oak, red maple, parsley, geranium, and milkweed as well. Different neurons, however, respond differently to a given plant; some show no response, and others fire at a high or moderate rate. Furthermore, the overall pattern of the response of the neurons is unique for different plant species, suggesting that the ability of the caterpillars to discriminate among plants may depend on the number of cells that fire or on the ratios of the action potential frequencies of the responding neurons. Host plant discrimination may also depend on other factors, such as the relative response latency of the receptors or adaptation rates and variation in the interspike intervals of firing neurons, not yet assessed in the tent caterpillar. The broad response spectra of these neurons indicate that information regarding the suitability of a food substance is encoded in the collective pattern of neuronal response, which is then integrated and decoded in the central nervous system.

Investigations of lepidopterous larvae other than tent caterpillars have indicated that there may be a universal antennal sensitivity in caterpillars to cold and relative humidity (Dethier and Schoonhoven 1968). But no definitive thermoreceptors or hygroreceptors have yet been identified from the antennae of tent caterpillars. Indeed, Dethier (1980a) found that the sensory neurons of the medial sensillum of the antenna of the eastern tent caterpillar, which in other caterpillars have been shown to be humidity detectors, were nonresponsive to either cold or water. The importance of behavioral thermoregulation for tent caterpillars, however, argues for the presence of such thermoreceptors. Likewise, receptors sensitive to changes in humidity would facilitate the larva's ability to locate recesses of the tent where humidity was conducive to molting. Dethier and Schoonhoven (1968) and Dethier (1980b) suggested that these receptors might also enable caterpillars to determine the quality of leaves. Evaporation from the chewed edges of leaves creates a local drop in temperature of a magnitude well within the sensitivity of caterpillar cold receptors. In contrast, wilted leaves lose little water and produce little or no local temperature change. Normal transpirational losses from intact leaves with concomitant local temperature shifts could likewise provide caterpillars with information concerning the turgidity (palatability) of the leaf.

Mouthparts

The undersurface of the labrum, the epipharynx, which forms the front part of the roof of the oral chamber, is innervated with eight sensory neurons that are involved in gustation (Dethier 1980a), but nothing is known of their sensitivity. In addition, each maxilla of the tent caterpillar has a palpus and a galea endowed with receptors. The palpus is directed downward and contacts the substrate as the caterpillar moves about. Each palpus bears a cluster of eight small sensilla basiconica at its tip (Fig. 3.10). Studies of species other than tent caterpillars indicate that five of these sensilla probably function as contact chemoreceptors, and the others may be olfactory receptors (Schoonhoven 1972). One or more of the sensilla of tent caterpillars are sensitive to the trail pheromone 5β-cholestane-3,24-dione (Roessingh et al. 1988). Studies of *M. neustrium* and *M. americanum* have shown that ablation of the antennae and galeae have no effect on the ability of the caterpillars to follow trails prepared from the synthetic trail pheromone. Removal of the maxillary palps, however, results in complete loss of the ability to follow chemical trails (Roessingh et al. 1988). As discussed in Chapter 6, unilateral ablation of the maxillae results in pronounced circling toward the intact side.

The role that the palpal sensilla of tent caterpillars play in feeding is unknown. Although they undoubtedly play some role in gustation, stud-

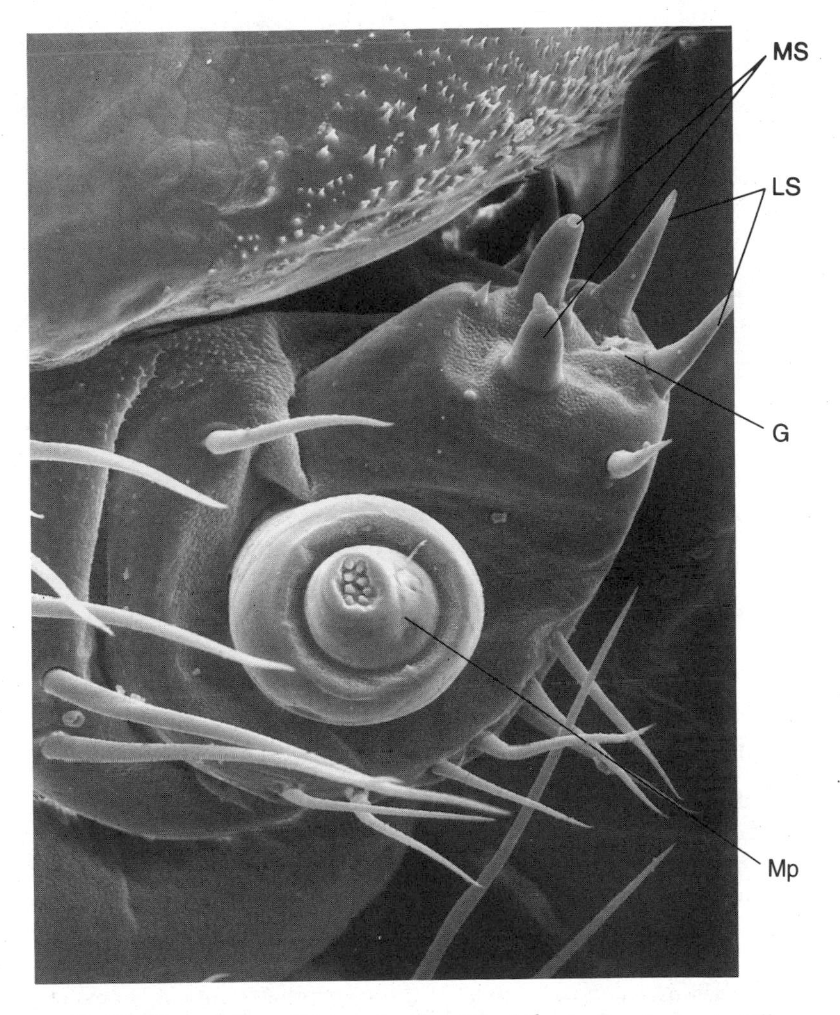

Figure 3.10. Maxillary palpus (Mp) and galea (G) of the eastern tent caterpillar. Sensilla at the tips of these structures monitor chemical stimuli during feeding and trail following. LS, MS = lateral and medial sensilla styloconica. (Courtesy of Janice Edgerly-Rooks.)

ies of other species of caterpillars indicate that the receptors of the galeae and epipharynx play the dominant role. Each galea of the maxillae bears a pair of sensilla styloconica that are located so as to contact the food that the caterpillar takes into its mandibles. Each sensillum is innervated by four neurons. In almost all species of caterpillars studied, one of these neurons is particularly sensitive to sucrose. The other neurons respond to various substances including other sugars, salts, amino acids, and chemical stimulants and chemicals that serve as deterrents and are unique to particular plants.

Dethier and Kuch (1971), and Dethier (1972) assessed the sensitivity of the medial and lateral sensilla styloconica of the galea of the eastern tent caterpillar to a wide array of chemicals commonly found in the food plants of caterpillars. Like the neurons of the antennal chemoreceptors, the neurons of the sensilla of the galea show broad responsiveness and overlapping sensitivities. The medial sensillum responds to sodium chloride, proline, valine, alanine, glutamic acid, aspartic acid, arginine, cystine, β-sitosterol, and ascorbic acid. The sensillum is also responsive to sucrose, inositol, fructose, glucose, and galactose, listed in descending order of sensitivity. The lateral sensilla responds to sodium chloride, phosphate, sucrose, fructose, glucose, inositol, sorbitol, proline, valine, alanine, and arginine. Neither sensillum shows responsiveness to amygdalin, a cyanogenic glycoside that occurs in cherry, the principal host tree of the caterpillar. Response spectra of the sensilla when bathed in the sap of host and nonhost plants are very similar and provide little insight into the sensory basis for host plant discrimination.

Coloration of the Body Wall

The larvae of the various species of tent caterpillars have distinctive color patterns that form the primary basis for their separation. The basic colors shared by all species of *Malacosoma* are black, white, blue, and yellow or orange. These colors are either pigmental or structural in nature.

The earlier larval instars are largely black. Scanning electron microscopy has revealed that much of the caterpillar's cuticle is covered with minute, pyriform microtubercles, approximately 1.5 mm in diameter and 4–6 μm in height (Fig. 3.11). The microtubercles contain dark pigment and are underlain by a pigmented layer. The pigment absorbs almost all the incident light striking the cuticle so that the larva appears black or dark brown (Fig. 3.12).

As the caterpillar matures, various colors appear and become arranged in prominent species-specific patterns. Of particular interest are the blues and nonpigmental "surface whites," which Byers and Hinks (1973) and Byers (1975) found are structural colors produced by the selective filtering of incident light by transparent microtubercles. The major areas of structural coloration of *M. americanum* are the white

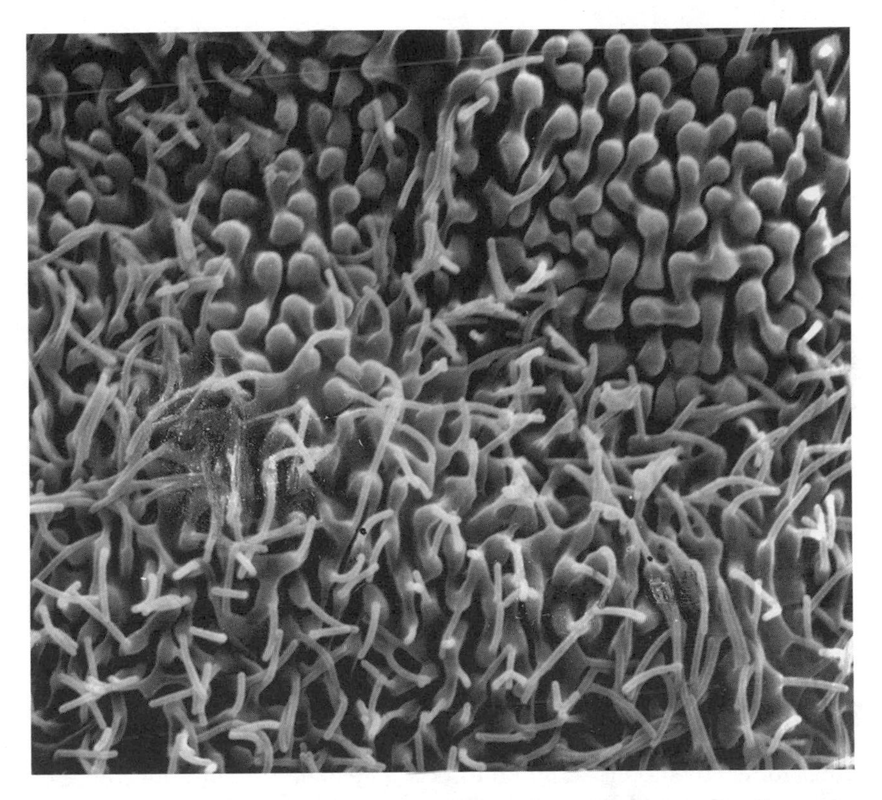

Figure 3.11. Fine structure of the integument of the eastern tent caterpillar at the juncture of the black cuticle (above) and the white middorsal stripe (below). Reflection of incident light that strikes the filamentous mat causes the cuticle in the lower area to appear white. (Courtesy of Janice Edgerly-Rooks.)

middorsal stripe and the subdorsal blue areas (Plate 2B). The white center of the dorsal "keyholes" and the blue subdorsal stripe of *M. disstria* are structural colors.

The surface of the cuticle in these areas of blue and white bears transparent filaments, approximately 0.3 μm in diameter and 10 μm in length, arranged to form a mat 3–6 μm deep (Fig. 3.12). The light-scattering properties of the cuticle in these areas of white and blue is due to differences in the refractive indices of air and the filamentous mat, a process known as a Tyndall effect. If the mat is wetted with a material having an index of refraction similar to that of the filaments (ca. 1.57), the middorsal stripe appears transparent rather than white.

The diameter of the filaments is sufficiently small that they reflect blue light more than other components of incident light. If the filamentous

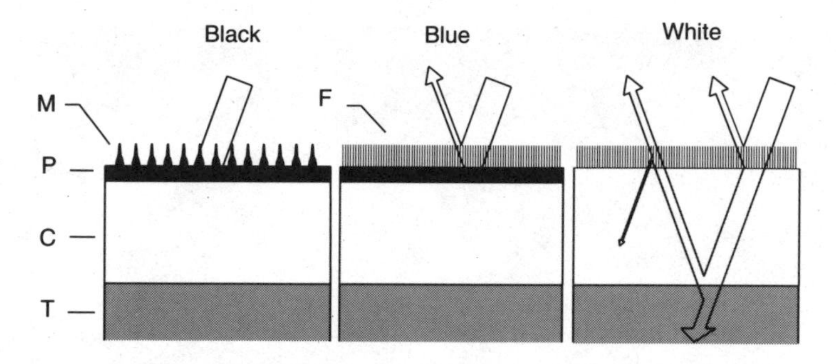

Figure 3.12. Reflection and absorption of light by the body wall of the eastern tent caterpillar. C = cuticle, F = filamentous mat, M = microtubercles, P = pigment layer, T = tissue underlying cuticle. (Redrawn from J. R. Byers, "Tyndall Blue and Surface White of Tent Caterpillars, *Malacosoma* spp.," *Journal of Insect Physiology* 21 [1975]:401–415, © 1975, with permission of Pergamon Press Ltd., Headington Hill Hall, Oxford OX3 BW, UK.)

mat is underlain by a black pigment layer, short wavelengths are reflected, and transmitted light is absorbed so that the cuticle appears blue. If the pigment layer is absent, some of the transmitted light, in addition to the blue, is reflected back so that the cuticle appears white rather than blue (Fig. 3.12).

The blue produced by the Tyndall effect is distinct from the iridescent blues more commonly found on insects. The metallic blue iridescence of the wings of the Morpho butterfly, for example, is attributable to the interference of light waves that are reflected from closely spaced surfaces on the scales. Tyndall colors appear flat rather than shiny, and they do not change in appearance when viewed from different angles as do iridescent colors. The larvae of tent caterpillars are the only immature insects so far known to exhibit Tyndall colors, and they are the only insects that utilize true cuticular processes to produce the effect (Byers 1975).

Locomotion

Tent caterpillars have three sets of thoracic legs and five sets of abdominal prolegs (Fig. 3.1). The thoracic legs are much shorter than those of the adult, but they have the full segmentation of a typical insect leg. Nonsegmented abdominal prolegs of approximately the same length as the thoracic legs occur on abdominal segments 3–6, and a pair of anal prolegs occurs at the tip of the abdomen. Muscles attached to the lateral walls or to the distal surface of the proleg serve to retract the appendage

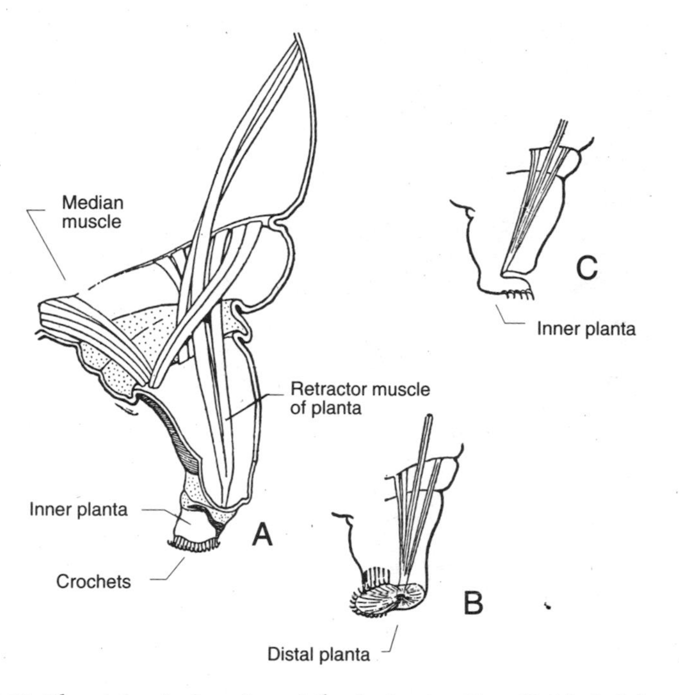

Figure 3.13. The abdominal proleg of the tent caterpillar. (Relabeled from R. E. Snodgrass, *Principles of Insect Morphology*. Copyright © 1993 by Cornell University. Used by permission of the publisher, Cornell University Press.)

into the body (Fig. 3.13A). There are no extensor muscles; the fleshy appendages are extended away from the body by the pressure of the hemolymph.

The tip of the proleg consists of a fleshy lobe, termed the *planta*, that has both an inner and a distal surface. Hooklike crochets embedded along the mesal edge of the plantae allow the caterpillar to obtain a secure hold on the substrate (Fig. 3.14). The crochets and plantae take on various adaptive configurations depending on the nature of the substrate. These configurations can be simulated by letting the human hands serve as a pair of prolegs. The tips of the fingers represent the crochets, the palms of the hands the inner plantar surfaces, and the backs of the hands, the distal plantae. The grasp of the prolegs on a small twig is simulated by holding a cylinder between the palms of both hands while bending the fingers so that the tips press tightly against the surface. In the caterpillar, the median muscles serve to press the crochets against the twig, and contraction of the retractor muscles disengages them (Fig. 3.13A).

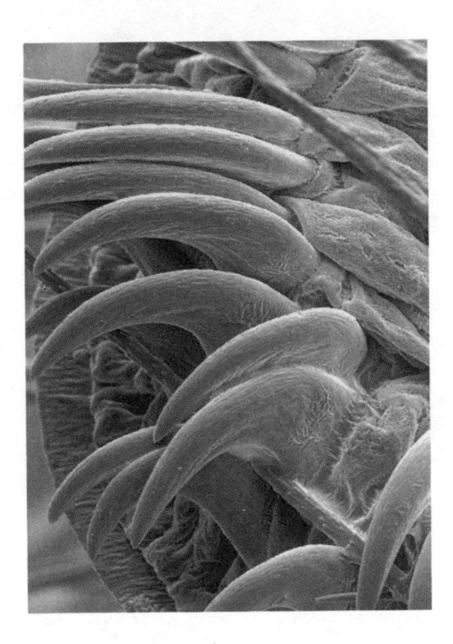

Figure 3.14. Crochets of the proleg of the eastern tent caterpillar. (Courtesy of Janice Edgerly-Rooks.)

Movement on flat surfaces that are smooth and hard can be simulated by pressing the backs of the hands flat against a tabletop so that the slightly curved fingers of each hand are in line and directed inward toward each other. In this, the normal position of the appendages, the tips of the crochets are held above the substrate, and the caterpillar walks on the distal plantae (Fig. 3.13B). In this position, slight contraction of the retractor muscles acts to pull the center of the distal plantar surface upward and creates a vacuum that increases purchase on smooth substrates.

The orientation of the tips of the prolegs when the caterpillar walks on a rough, flat surface can be simulated by pressing the palms of the hands against a tabletop so that the flat surfaces of the inner wrists face each other and the fingers are directed outward. In this position the crochets (fingertips) can be pressed tightly against the surface to gain purchase. The caterpillar brings the tips of the prolegs to this position by contracting the retractor muscles. This contraction causes the inner surface of the plantae to be pulled downward and the crochets to rotate 180° toward the outside of the body (Fig. 3.13C).

Unlike the legs of the adult, the short legs of caterpillars do not provide the motive force for walking. Instead, the caterpillar is driven forward by the synchronous contraction of longitudinal bands of intersegmental body wall muscles. The contraction of the muscles against the

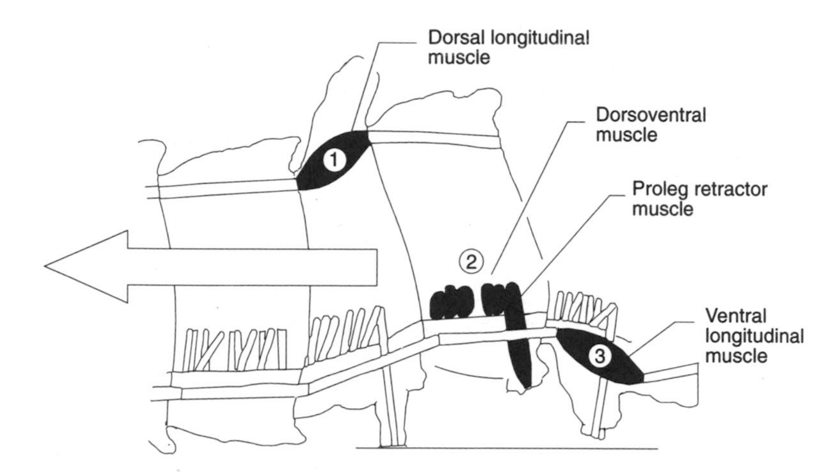

Figure 3.15. Sequence of muscular events underlying the generation of a body wave. These cyclic waves provide the motive force for locomotion in caterpillars. (Redrawn, by permission of Academic Press, from Hughes 1965.)

incompressible fluid contents of the hemocoel sets up successive cycles of body waves that drive the insect forward and carry the legs along in the process.

The body waves that propel the caterpillar move from posterior to anterior (Fig. 3.15). A wave is propagated when (1) the dorsal longitudinal muscles of a segment contract, causing the segment immediately behind it to be drawn forward and in the process lifted from the substrate. At the same time, (2) the dorsoventral muscles and the retractors of the prolegs of the affected segment contract, lifting the segment and drawing the prolegs closer to the body. The ventral longitudinal muscles of the segment then contract (3), shortening the ventral surface of the segment and drawing the prolegs forward. When reextended, the prolegs are firmly planted in a more forward position than they previously held. The wave continues in this manner until it has passed over the entire body, then the cycle is reinitiated in the last abdominal segment. In the tent caterpillar, only a single body wave is present at any one time, and the prolegs of any one segment move forward only once during a given cycle. In contrast, the head is driven forward by hydraulic pressure and is in near-continual rather than cyclic movement (Casey 1991).

The speed at which foraging caterpillars move depends on body length and body temperature (T_b). These factors, in turn, influence stride frequency (the number of body wave cycles per second) and stride length (the distance the caterpillar advances per body wave cycle), which

Table 3.1. Locomotor parameters of third- to sixth-instar *Malacosoma americanum* larvae measured in New Jersey over a 25-day period in April and May

Parameter	Mean ± SE	Range
Body temperature (°C)	17.7 ± 0.11	7.0–30.5
Body length (cm)	2.5 ± 0.03	0.8–5.5
Speed (cm/s)	1.03 ± 0.02	0.2–4.5
Stride frequency (Hz)	1.57 ± 0.02	0.4–3.2
Stride length (cm)	0.6 ± 0.01	0.2–1.4

Source: Data are based on 870 observations. Reprinted from B. Joos, *Physiological Zoology* 65 (1993):362–371, © 1993 by the University of Chicago, by permission of the University of Chicago Press.

directly determine the speed of locomotion. Average values for these parameters for eastern tent caterpillars under field conditions are given in Table 3.1.

Joos (1992) showed that the stride frequency (*SF*) of free-foraging eastern tent caterpillars is temperature-dependent, with a Q_{10} of 1.9 (for every 10°C rise in temperature, stride frequency increases by a factor of 1.9). In contrast, stride length is little affected by T_b ($Q_{10} = 1.2$) but greatly so by body length. The caterpillar's speed thus depends on both T_b and body length (*BL*) and is estimated by the equation:

$$\log_{speed} = -0.947 + 0.0704 \ (\log BL) + 0.034\,(T_b)$$

The amount of time it takes caterpillars to move to feeding sites can vary markedly at different times of the day. A 2.5-cm-long fifth-instar caterpillar moving over the 5-m round-trip distance between its tent and feeding site during the early morning hours when its body temperature is only 7°C requires 22 minutes to complete the trip. The same insect warmed to 30°C during an afternoon foray would require only 4 minutes to cover the same distance. Travel time may, therefore, constitute a significant fraction of the total duration of a foraging bout when the caterpillars are cold. Plots of this equation for other body lengths and for temperatures likely to be experienced by third-instar through mature sixth-instar eastern tent caterpillars are shown in Figure 3.16.

The incremental cost of transport for a caterpillar is estimated to be $0.524\ J{\cdot}g^{-1}{\cdot}m^{-1}$, regardless of body temperature (Casey 1983). Energy budgets for free-foraging fifth- and sixth-instar eastern tent caterpillars moving 5 m between their tent and food indicate that locomotion accounts for 1–8% of the total metabolism of the caterpillar (Table 3.2). The

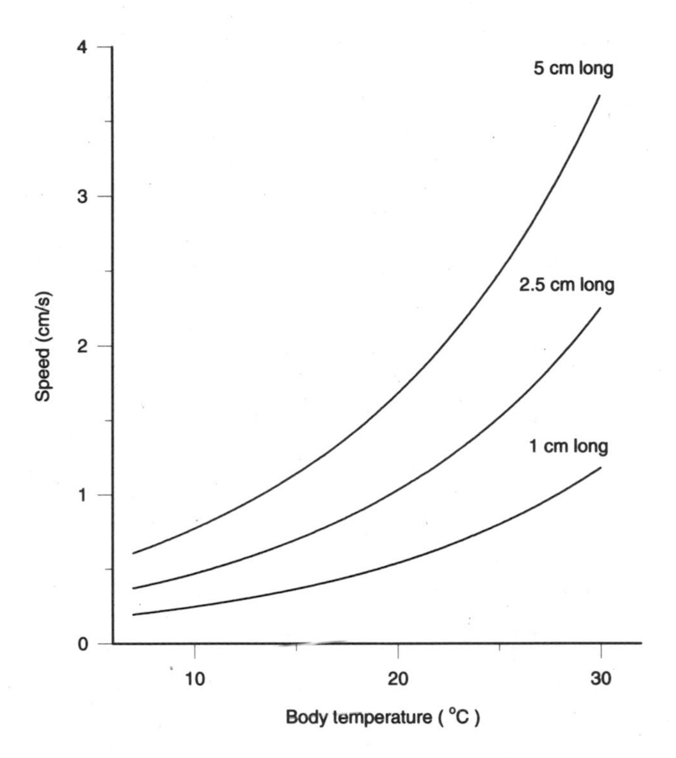

Figure 3.16. Relationship between body temperature and the speed of locomotion for eastern tent caterpillars of three different sizes. (Data from Joos 1992.)

fractional cost varies with both T_b and body mass. Less of the total metabolism of warm caterpillars is attributable to locomotion than is that of cool caterpillars. Warm caterpillars have a higher resting metabolism than cool caterpillars, and the cost of locomotion, which is temperature-independent, accounts for a smaller percentage of their total metabolism. In addition, field colonies of sixth-instar eastern tent caterpillars move to feeding sites only once a day, whereas earlier instars typically forage three times each day (Chapter 6). Thus, the absolute energy expenditure for locomotion is less during the last stadium (Table 3.2).

Caterpillar locomotion is relatively inefficient. The minimum cost of transport for a tent caterpillar is estimated from treadmill studies to be 26.1 ± 9.1 ml $O_2 \cdot g^{-1} \cdot km^{-1}$ (T. M. Casey, pers. comm., 1992). This is four to five times as great as the cost of transport for a vertebrate or for an insect with a hard exoskeleton. The actual cost per stride is no greater for a caterpillar than for those other animals, but, because of the bio-

Table 3.2. Estimated energy budget for *Malacosoma americanum* caterpillars during the fifth and sixth stadia

Body temperature (°C)	Total metabolism (J)		Cost of locomotion (J)		Locomotion as percent of total	
	5th stadium	6th stadium	5th stadium	6th stadium	5th stadium	6th stadium
24.0	369	893	28.5	26.0	7.7	2.9
28.3	712	1885	28.5	26.0	4.0	1.4
30.1	843	2244	28.5	26.0	3.4	1.2
34.4	1041	2706	28.5	26.0	2.7	1.0

Source: Based on travel 5 m to and from tent at mean daytime body temperatures. Data from B. Joos (in litt., 1993).

mechanics of the hydrostatic skeleton, the distance moved per stride is only 25–30% as great as that of comparable animals (Casey 1991). It is this factor that accounts for most of the inefficiency of caterpillar locomotion. In addition, animals with hard skeletons can store energy in elastic components of their body to generate momentum when moving. Walking caterpillars fail to build inertia because of the dampening effect of the hydrostatic skeleton. Each leg must be started from a completely stationary position during each body wave cycle.

Number of Instars and Duration of Larval Stage

The larval stage of the various species of *Malacosoma* consists of five or six instars. Hodson (1941) reported that under field conditions *M. disstria* had five larval instars but that six sometimes occurred when caterpillars were reared in the laboratory. *M. disstria* larvae reared inside on aspen or tupelo had only five larval instars (Muggli and Miller 1980, Smith et al. 1986), but the same species reared on an artificial diet by Lyon et al. (1972) had six larval instars. *M. californicum* and its subspecies may have either five (Clark 1956a, Stelzer 1968, Mitchell 1990) or six larval instars (Baker 1969, Page and Lyon 1973, Robertson and Gillette 1973). Of the Eurasian species, *M. indicum* has been reported to have six larval instars (Joshi and Agarwal 1987) and *M. parallelum,* five (Li 1989). *M. neustrium testacum* has been reported to have either five or six instars (Shiga 1976a).

Colonies of tent caterpillars can grow very rapidly, particularly when reared in the laboratory, and they typically molt synchronously with only a brief lull in activity. Unless colonies are closely monitored, it is easy to overlook an instar, and there are reports in the literature in which it is apparent that that is what occurred. One way to be certain that no instars are overlooked was first suggested by Dyar (1890). He determined that lepidopterous larvae grow geometrically. The head capsule widths of 27 species of caterpillars he studied increased at each molt by an average factor of 1.47. Subsequent studies by others have tended to validate his observations, and the principle has become known as Dyar's rule.

If the natural logarithms of the head capsule widths of a species that follows Dyar's rule are plotted for each instar, the data fall in a straight line. The antilog of the slope of the line, as determined by regression analysis, is the value that best describes the rate at which successive instars increase in size. Published data on the head capsule widths of various species indicate that tent caterpillars conform well to Dyar's rule (Fig. 3.17). For tent caterpillars with six larval instars, the rate of increase (*RI*) in size of the head capsule between molts ranges from 1.42 to 1.55 (mean = 1.48). The ratio of increase reported in Figure 3.17 for *M. americanum* (1.55) is close to that reported by Dyar (1890) for the same species (1.51). Only two sets of reliable data were found for species with five larval instars, and the *RI* for those two averaged 1.68 (Fig. 3.17). The applicability of the rule is not limited to head capsule dimensions. As shown in Figure 3.18, the increase in length of the eastern tent caterpillar from one stadium to the next is also geometric (*RI* = 1.64).

The larval stage of Old and New World species of *Malacosoma* ranges from as few as 6 weeks to as long as 11 weeks (mean = 8 weeks) under field conditions. The duration of the larval stadia of tent caterpillar that are maintained under ideal conditions in the laboratory is much shorter. When colonies of eastern tent caterpillars are held indoors at 22°C and allowed to forage ad libitum on an unlimited supply of young cherry leaves, the six larval stadia are completed in as few as 25 days (Fig. 3.19).

The date of larval eclosion is strongly related to weather. Tent caterpillars eclose early in the spring, just as the leaves of their host trees first appear. Thus, populations at the southern extremes of the caterpillars' range may eclose in early February, and those in more northerly regions may not eclose until May. And populations at high elevations eclose much later than those at lower elevations (Fig. 3.20). Populations of caterpillars that eclose early in the season in response to unusually warm weather are likely to encounter adverse weather conditions later on. In that situation, growth is slowed and the larval stage is prolonged.

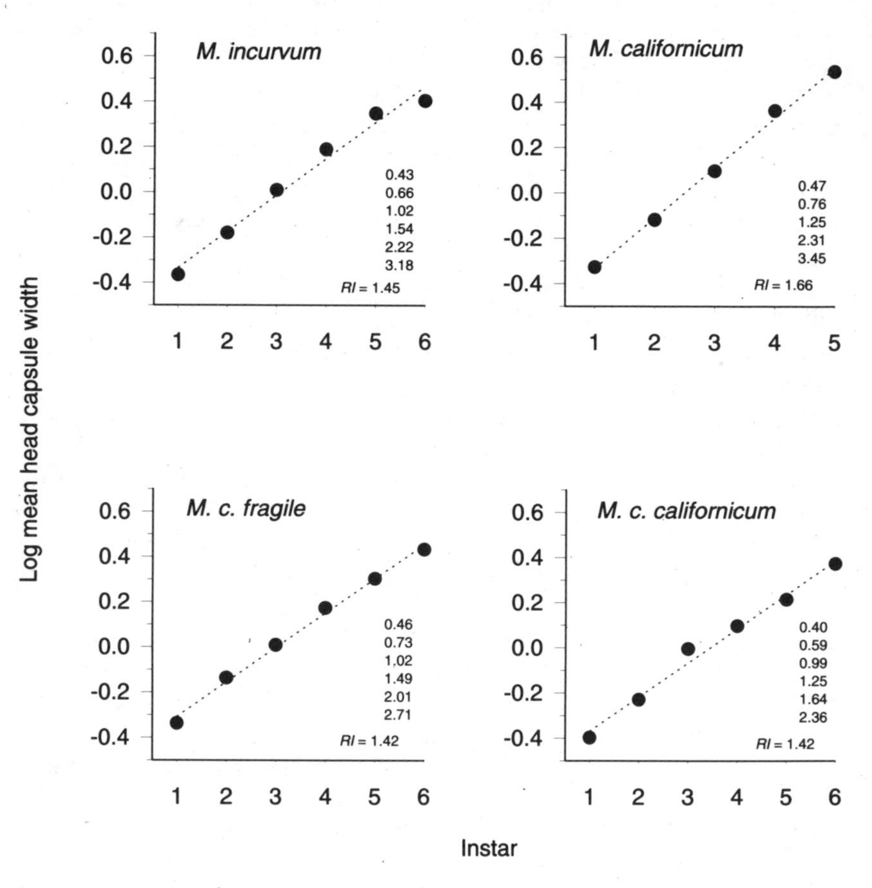

Figure 3.17. Logarithmic plots of mean head capsule widths over instar for seven species of *Malacosoma*. The mean width (mm) of the head capsule for each instar and the rate of increase in size (*RI*) between instars are shown for each species. (Sources of data: *M. incurvum*, Baker 1970; *M. californicum*, Mitchell 1990; *M. californicum fragile*, Baker 1969; *M. californicum californicum*, Robertson and Gillette 1973; *M. disstria*, five instars, Muggli and Miller 1980; *M. disstria*, six instars, Lyons et al. 1972; *M. americanum*, T. D. Fitzgerald and D. G. Battaglia, unpubl. data; *M. indicum*, Joshi and Agarwal, 1987.)

Conversely, populations of caterpillars whose eclosion is delayed because of adverse weather are more likely than early emergers to encounter sustained favorable weather and to complete their larval development rapidly. Thus, it is not unusual to find that in any given geographic area adults begin to fly about the same time year after year, even though hatching dates may vary greatly.

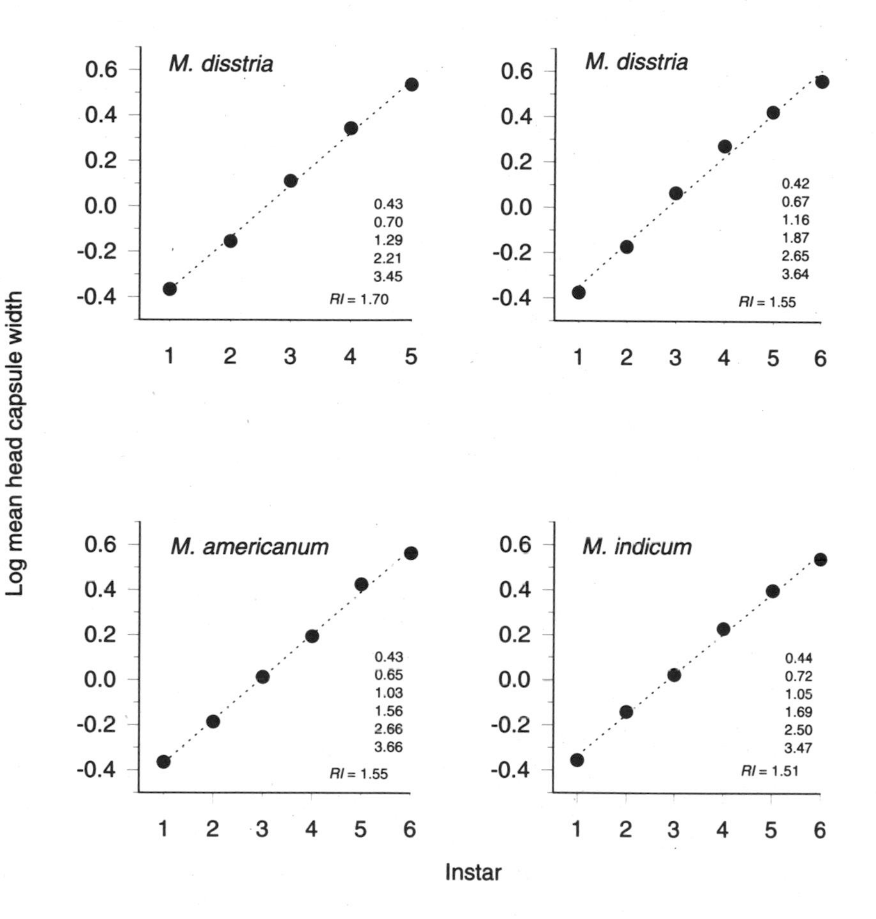

Cocoon Formation and Pupation

Mature tent caterpillar larvae seek secure locations under ledges or in crevices to construct their cocoons. The cocoons of most species are single-walled and elliptic. The cocoon of *M. disstria,* however, is double-walled, consisting of an outer sheet of silk enclosing an elliptical inner cocoon. *M. disstria* is unusual also in that it often forms its cocoons in folded leaves still remaining on trees (Fig. 3.21). The process of cocoon formation in this species was described by Sippell (1957). The mature caterpillar selects a leaf, then spins silk over its upper surface. Contraction of the silk strands causes the leaf to curl, partially surrounding the caterpillar. The larva then spins a wall of silk to seal

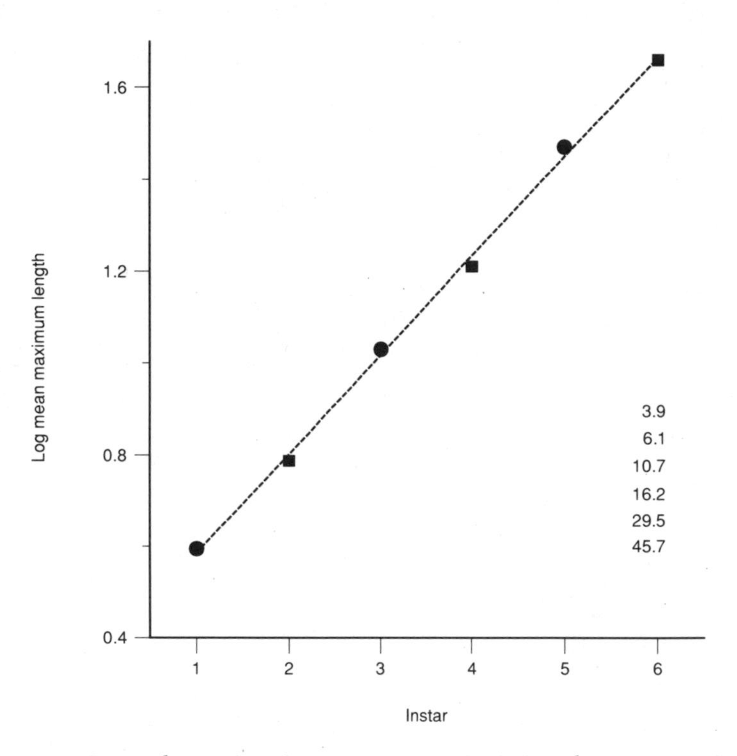

Figure 3.18. Logarithmic plot of mean maximum body length over instar for *Malacosoma americanum* caterpillars (N = 20 per instar). Mean maximum size (mm) achieved during each of the six larval stadia is also shown. (T. D. Fitzgerald and D. G. Battaglia, unpubl. data.)

any remaining gaps, and in the process completely encloses itself within. It then spins the inner dense cocoon, which is similar to that formed by other species.

Snodgrass (1930, 1961) described in detail the process of metamorphosis that occurs inside the cocoon of *M. americanum,* leading to the formation of the pupa. Soon after the cocoon is completed, the caterpillar shrinks markedly and loses most of the setae that covered its body (Fig. 3.22). The external body parts of the pupa begin to form within the cuticle of the shrunken caterpillar. If the cuticle of the larva is removed after several days to reveal the pharate pupa, the insect appears as depicted in Figure 3.22B. At this stage of development, the cuticle of the pupa has been laid down, the abdomen has no legs, the wing pads have everted, and the antennae and thoracic legs are elongated. The compound eyes of the pupa have begun to form in the region of the head

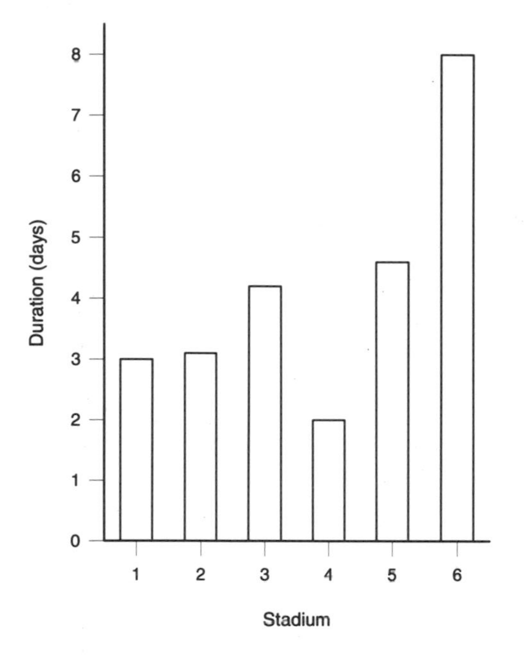

Figure 3.19. Duration of the larval stadia of a colony of eastern tent caterpillars reared under optimal conditions in the laboratory. It is not known whether the relatively short fourth stadium is a chracteristic of the species or a peculiarity of this one colony. (T. D. Fitzgerald and D. G. Battaglia, unpubl. data.)

formerly marked by the stemmata. The mandibles are shrunken to mere rudiments, and the spinneret is lost. The process of external change continues until the visible structures of the pupa have reached their full size, at which time a gluelike material secreted from the body wall coats the appendages and creates a smooth external finish. Soon after, the larval cuticle splits open, and the pupa, only one-third the length of the mature larva, wiggles out.

The hemolymph of the newly formed pupa is greatly thickened by the products formed by the dissolution of unneeded parts of larval organs and by the rupturing of the fat cells. It appears as a creamy, yellow liquid. The major organs of the caterpillar—the alimentary tract, the nervous system, and the tracheal tubes—are not dissolved but are modified to the adaptive needs of the pupa and adult. The reproductive organs grow from their reduced size in the caterpillar. In contrast, the fat body and some of the body wall muscles are dissolved. The disin-

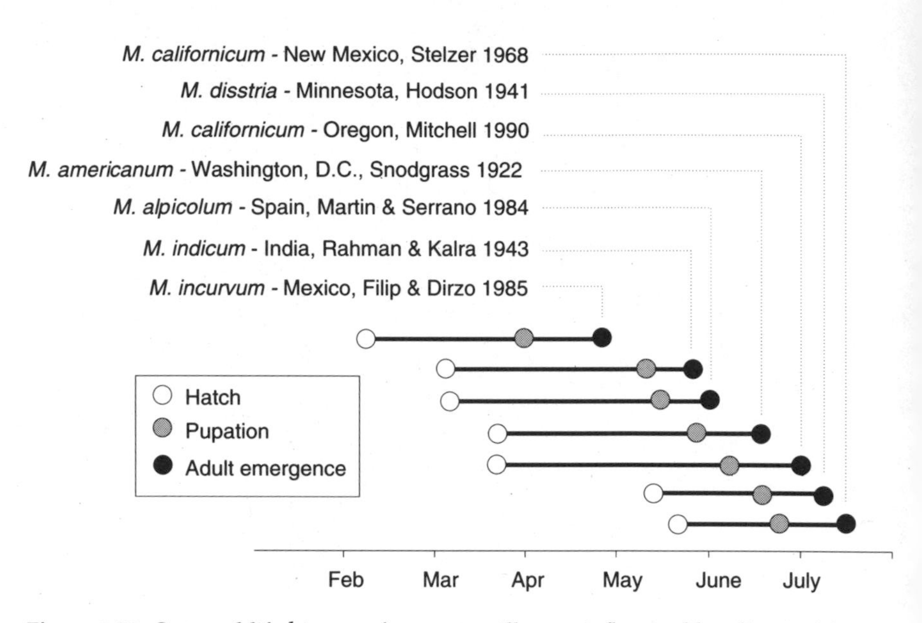

Figure 3.20. Seasonal life history of tent caterpillars as influenced by climate. *Malacosoma* species that occur in northern regions (Minnesota) or at high elevations in southern regions (New Mexico) emerge much later than species from warmer areas, but they complete their development faster.

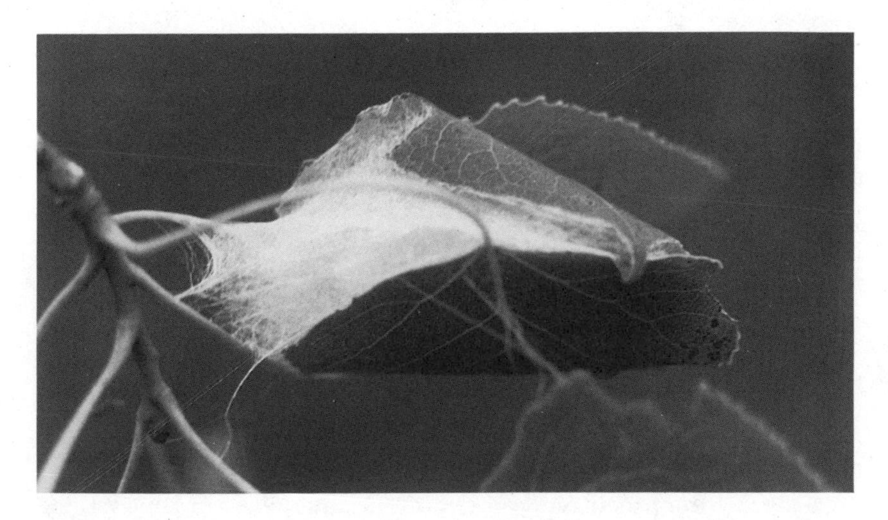

Figure 3.21. Cocoon of the forest tent caterpillar spun within folded leaves of aspen.

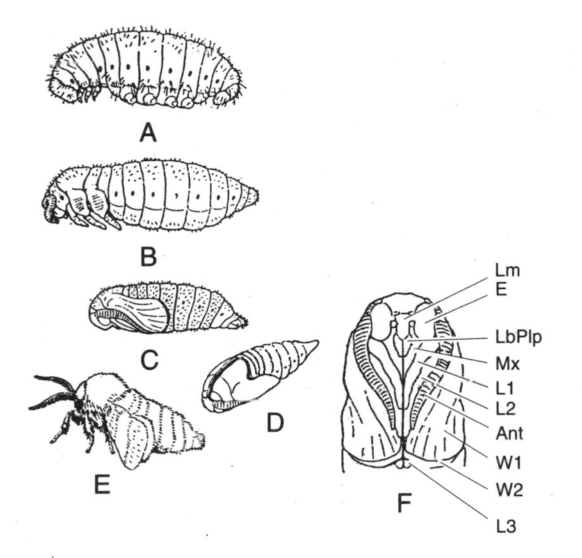

Figure 3.22. Stages in the metamorphosis of the eastern tent caterpillar. (A) Shrunken form of larva as it appears in the cocoon. (B) Pupa just before casting larval cuticle. (C) Fully formed pupa. (D) Split pupal cuticle after emergence. (E) Newly emerged adult. (F) Ventral aspect of pupa. Ant = antenna, E = eye, L1–3 = thoracic legs, LbPlp = labial palpus, Lm = labrum, Mx = maxillary palpus, W1,2 = wings. (Relabeled from Snodgrass 1961.)

tegration of the fat body and the rupturing of the fat cells fuel the process of transformation. The oblique and transverse muscles that firmed the caterpillar's body against the hemolymph to create the hydrostatic skeleton are unneeded in the adult and are largely dissolved. Their products are reused to form the unique muscle systems of the adult. The longitudinal muscles persist into the pupal stage and allow the three unfused segments of the abdomen to create the movements that free the pupa of the larval cuticle.

When the adult is fully formed, the pupal cuticle splits open along the dorsal midline of the thorax, and the moth works its way out (Fig. 3.22D). The eclosing moth, still within the cocoon, secretes a liquid that weakens the bonds between the silk strands, and it pushes its way out of the structure as it completes the process of ecdysis. The unidentified chemical responsible for the digestion of the silk, apparently secreted from the atrophied silk glands (Snodgrass 1922), stains the cocoon brown at the exit site.

The duration of the pupal stage is much less variable than that of the larval stage. Pupation occurs late in the season when the temperature is

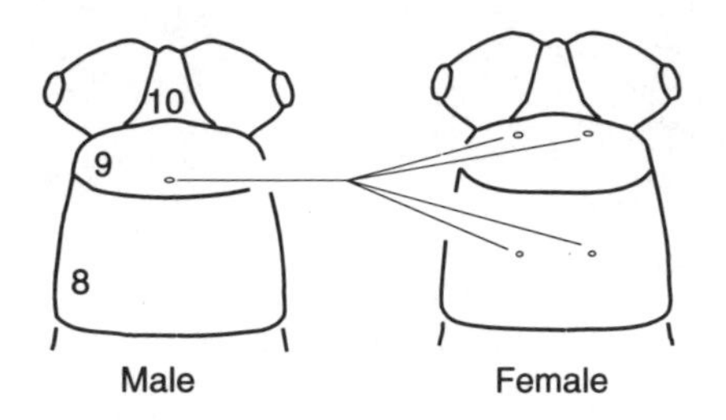

Figure 3.23. Position of external pits on the ventral surface of the abdomen of male and female tent caterpillar larvae. (Redrawn from Stehr and Cook 1968.)

likely to be less variable from year to year and the stadium typically lasts from two to three weeks.

Sexing Larvae and Pupae

Both the larvae and pupae of *Malacosoma* can be sexed by reference to external features. Female larvae grow larger than males. Shiga (1976a), for example, found that fully grown sixth-instar males of *M. neustrium testacum* had an average mass of about 425 mg (range = 350–480 mg), and females averaged about 650 mg (525–750 mg). The gender of *Malacosoma* larvae can also be determined by the number and placement of shallow pits on the ventral surface of the abdomen. Males have a single pit on the ninth segment, and females have two on the eighth segment and two on the ninth (Fig. 3.23) (Stehr and Cook 1968). The pits are difficult to see in the earlier instars. Male pupae also are significantly smaller than female pupae (Fig. 3.24). Male pupae have a single, sealed genital aperture on the ninth segment. Females have a sealed, slitlike aperture on the eighth segment and an oval aperture on the ninth (Fig. 3.25).

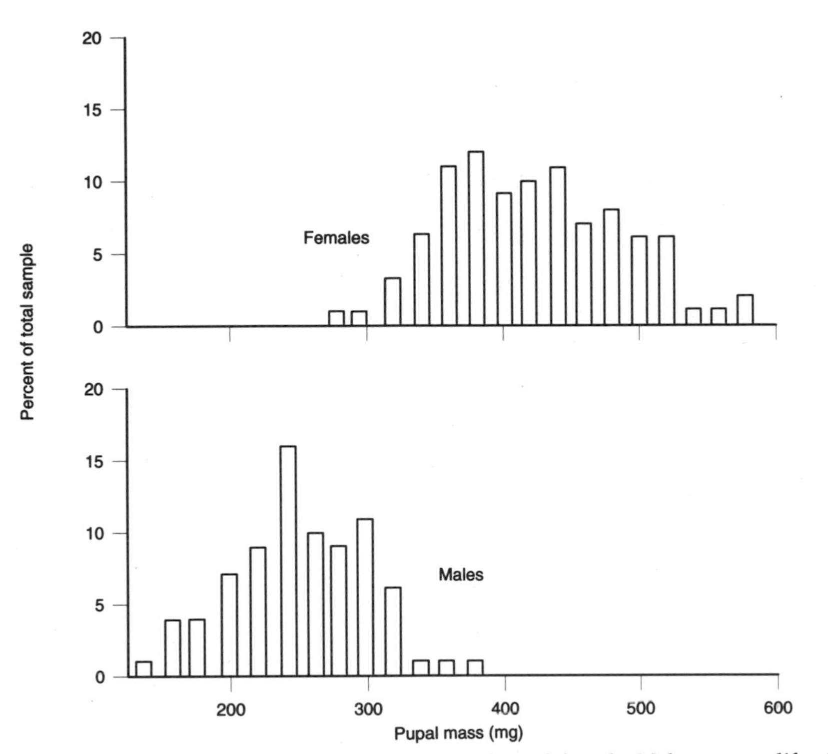

Figure 3.24. Distribution of pupal mass of male and female *Malacosoma californicum*. (Data from Stelzer 1968.)

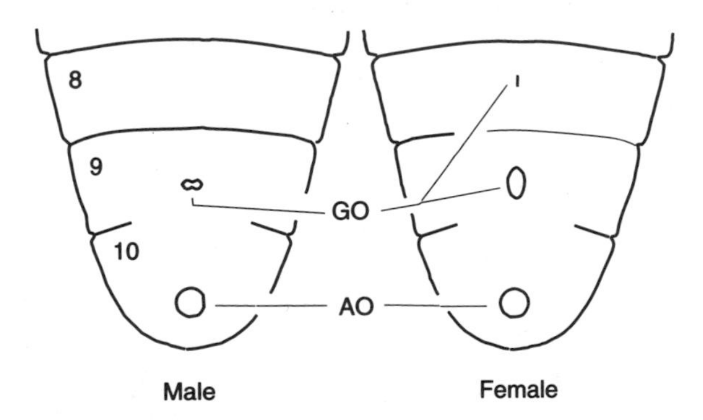

Figure 3.25. Anal (AO) and genital (GO) openings of pupae of *Malacosoma disstria*. (Redrawn, by permission of the Entomological Society of America, from Muggli 1974.)

4

The Moth, the Egg Mass, and the Pharate Larva

Tent caterpillar larvae sequester all the energy required to fuel the activities of the nonfeeding adult. The abdomen of the newly emerged male moth is packed with the fat needed to power flights in search of females. The abdomen of the female is filled with mature oocytes formed from the fat reserves of the pupa. The female is reproductively mature at the moment of emergence, and she advertises her maturity by secreting a sex pheromone that draws males to her. Mating typically occurs within hours of emergence, and oviposition occurs soon thereafter. The gregarious habit of the tent caterpillar larva allows the female to lay all her eggs in a single batch, saving both time and energy that would otherwise be required to search for multiple hosts. Thus, a typical female may emerge, call males, copulate, oviposit, and, being completely spent, die in less than a day. This brevity of life and concomitant rush to fulfill the reproductive functions is not uncommon among moths. Of all the life stages, it is the fecund female that has the greatest reproductive value. Selective processes have favored females that transfer their oocytes to the relative security of the egg mass as quickly as possible. But the tempo of the tent caterpillar's life slows markedly thereafter. The pharate larvae that form several weeks after the eggs are laid are the most durable of the insect's life stages. They lie quiescent within the chorions of their eggs for up to 10 months, physiologically adapted to endure the extremes of climate that occur during the long interval between summer and spring.

The Moth

The larvae of the tent caterpillar and the tents they create are familiar to people wherever they occur, but few would recognize the tent caterpillar moth. Indeed, like most moths, the adults are most active at night and, except for their affinity for electric lights, rarely make their presence known. The adult is a medium-sized moth with a wing span of 2.5–4.5 cm and body length of 1.0–1.8 cm (Stehr and Cook 1968, de Freina and Witt 1987). The moth is densely covered with hairlike scales. In *Malacosoma americanum*, these scales have a pile depth of approximately 2 mm and create a furlike body covering (Casey 1981; Fig. 4.1). The dorsal surfaces of the forewings of all species of *Malacosoma* each bear two oblique bands of colored scales that contrast to a greater or lesser extent with the rest of the wing. Male moths typically have smaller bodies and wings than females, and the males of some species may be colored differently than the females (see Stehr and Cook 1968 for photographs of North American species). Both sexes have bipectinate antennae, but the teeth of the antennae are noticeably longer in the male (Fig. 4.2). All species examined have 31 sets of chromosomes (Stehr and Cook 1968, Ennis 1976).

When the scales are removed from the head of the moth, the insect can be seen to have greatly reduced mouthparts (Fig. 4.3). As is the case with almost all moths, the mandibles, which are the primary implements of the caterpillar, are completely lacking in the adult. Many species of moths bear a prominent proboscis with which they imbibe liquids, but adult *Malacosoma* lack even that appendage, and the maxillae from which the proboscis would otherwise be formed are reduced to small lobelike structures. Aside from the prominent antennae, three-segment labial palpi, covered with long scales, are the only conspicuous appendages of the head. The adult has prominent compound eyes but lacks ocelli.

The tent caterpillar moth does not feed during its brief life but retains the basic features of the alimentary tract that it had as a larva. The foregut and midgut are much smaller than those of the larva (Fig. 4.4). In the newly emerged adult, the dorsal vesicle of the esophagus may be distended with air, swallowed by the eclosing adult to increase the force exerted by its blood against the pupal cuticle to facilitate ecdysis (Snodgrass 1930). The rectum is greatly enlarged and, in the newly emerged moth, filled with the reddish brown waste material that accumulated during the pupal stage. The moth empties the rectum of this meconium soon after it emerges.

Figure 4.1. Male eastern tent caterpillar, *Malacosoma americanum*. The oblique parallel bars on the forewings are diagnostic characteristics of the genus.

Reproduction

Adult Genitalia

The adult female is sexually mature at the time of eclosion, and the contents of her abdomen are largely devoted to the process of reproduction (Fig. 4.5). The ovaries take up most of the abdomen and account for up to two-thirds of the biomass of the female (Ruggiero and Merchant 1986). Each ovary has four ovarioles filled with mature oocytes. The material that is used to fix the eggs to the substrate and, in some

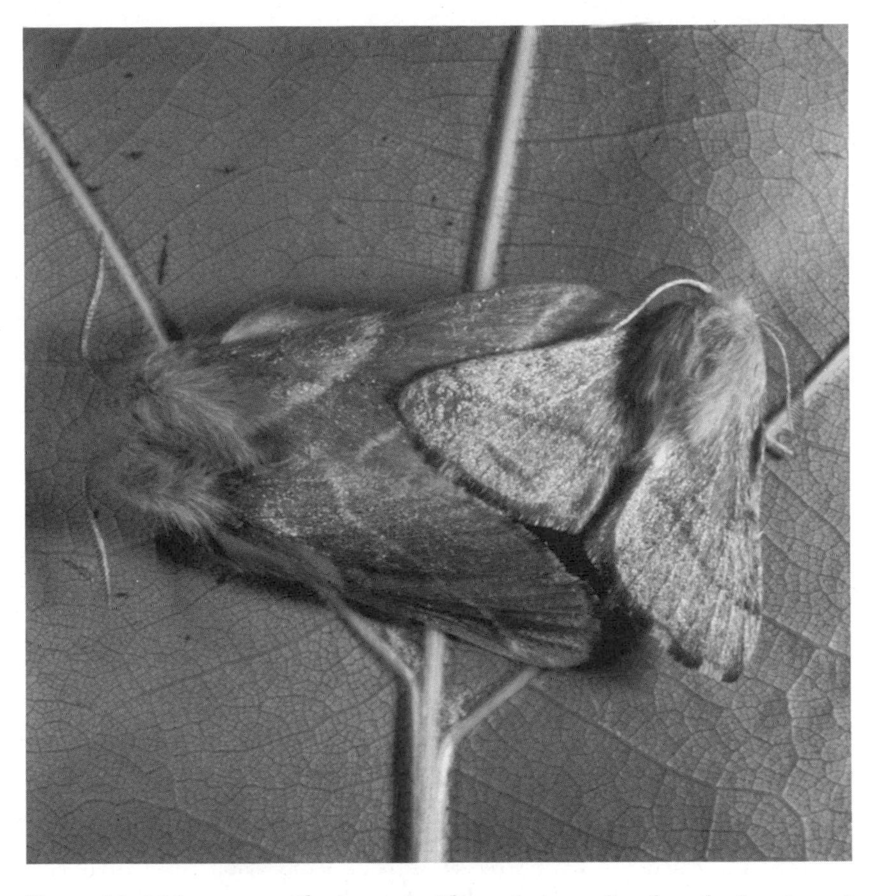

Figure 4.2. *Malacosoma* moths *in copula*. The male is smaller than the female and has more prominent branching of the antennae.

species, to cover them with a frothy coating is stored in the large reservoirs of the colleterial glands.

The bursa copulatrix is the copulatory organ of the female (Williams 1939). The male's ejaculate passes through a narrow duct from the bursa to a storage pouch, or spermatheca, which arises from the wall of the vagina. When the female ovulates, sperm move from the spermatheca, intercepting and fertilizing the eggs as they pass through the vagina to the outside. The tip of the abdomen surrounding the opening of the vagina serves as the ovipositor.

The abdomen of the male moth is richly endowed with fat that fuels

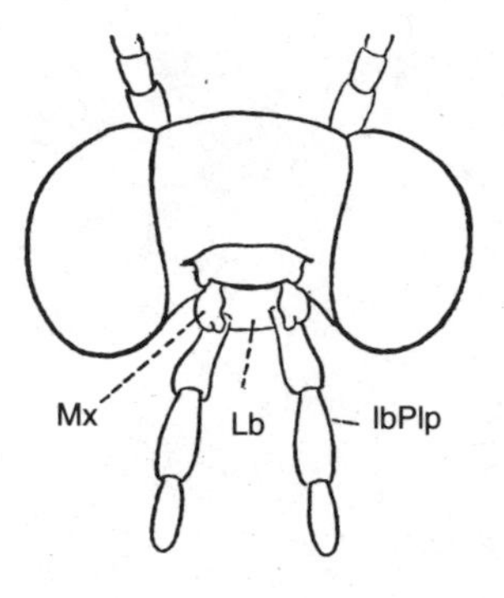

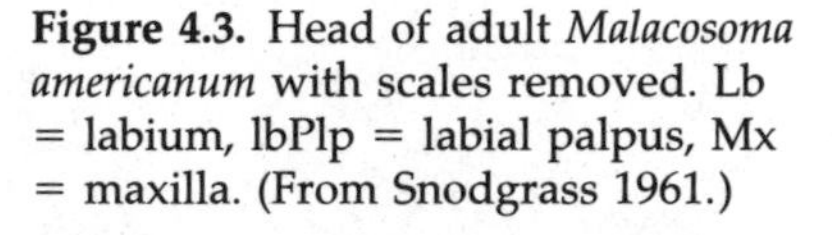
Figure 4.3. Head of adult *Malacosoma americanum* with scales removed. Lb = labium, lbPlp = labial palpus, Mx = maxilla. (From Snodgrass 1961.)

its energy-demanding flights in search of mates. The external genitalia are complex, and the shapes of the component parts vary among species (Fig. 4.6). The genitalia serve as the primary basis for distinguishing species. Prominent opposable claspers that apparently hold the female during copulation cover the sides and top of the genitalia. The intromittent organ, or aedeagus, is sharply pointed and saber-shaped and bears the opening of the ejaculatory duct, the gonopore.

The internal reproductive organs of male *M. neustrium* were described by Williams (1940; Fig. 4.6). The testes lie within a common testicular sac. The vasa deferentia are expanded near their distal ends to form a seminal vesicle in which sperm are stored. The ends of the vasa deferentia are inserted into the duplices. A duct arising from each duplex terminates in an accessory gland, and the two glands are bound together by connective tissue. The glands secrete fluids that mix with the sperm to form the ejaculate, which passes through the ejaculatory duct and gonopore to the outside during the act of mating.

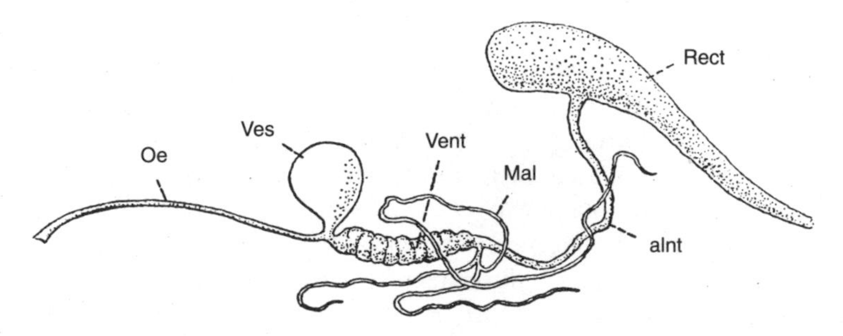

Figure 4.4. Alimentary tract of adult *Malacosoma americanum*. aInt = anterior intestine, Mal = Malpighian tubules, Oe = esophagus, Rect = rectum, Vent = ventriculus, Ves = vesicle. (From Snodgrass 1961.)

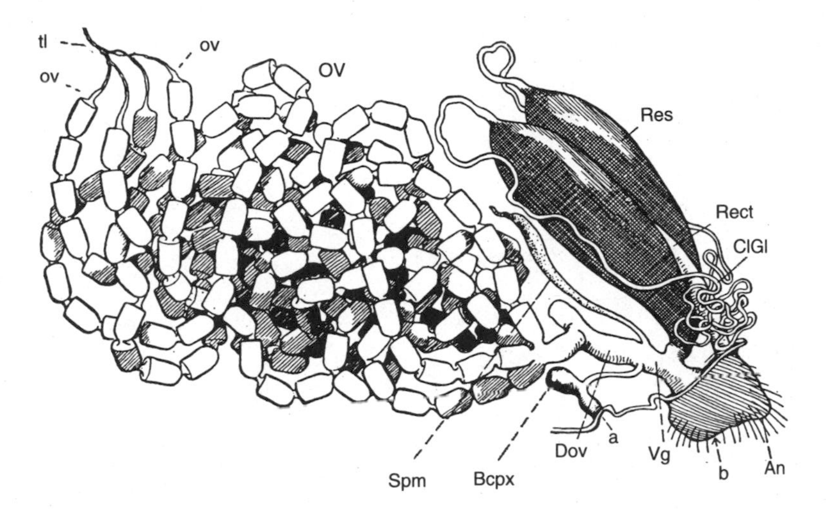

Figure 4.5. Reproductive tract of female *Malacosoma americanum.* a = opening of bursa copulatrix, An = anus, b = opening of vagina, Bcpx = bursa copulatrix, ClGl = colleterial gland, Dov = oviduct, ov = ovariole, Ov = ovary, Rect = rectum, Res = reservoir of colleterial gland, Spm = spermatheca, tl = terminal ligament, Vg = vagina. (Relabeled from R. E. Snodgrass, *Principles of Insect Morphology.* Copyright © 1993 by Cornell University. Used by permission of the publisher, Cornell University Press.)

Mating Behavior

Tent caterpillars eclose and mate in late spring or early summer, but the exact date varies from location to location (see Fig. 3.20). Adults may eclose as early as May at the southern limits of the distribution of the genus but not until mid to late July in northern regions and at higher elevations in more southern regions. It has been commonly observed that the males eclose and fly before the females and that they occasionally copulate with females whose wings have not yet fully expanded (Hodson 1941, Stelzer 1968, Bieman and Witter, 1981). Mating is by apposition: copulating moths are joined end to end and face in opposite directions (see Fig. 4.2).

Much of our knowledge of the mating behavior of *Malacosoma* is derived from studies of field populations of *M. disstria* carried out in western Ontario (Shepherd 1979) and in Michigan (Bieman 1980, Bieman and Witter 1983). Females emerge in late afternoon and early evening but stay at or near the cocoon until twilight. Males apparently eclose earlier in the day and initiate flight in search of females as early as midafternoon. Mating in *Malacosoma* is facilitated by sex pheromones that are secreted by the female. Synthesis of the pheromone is completed before adult eclosion, and the material is stored in an intersegmental gland

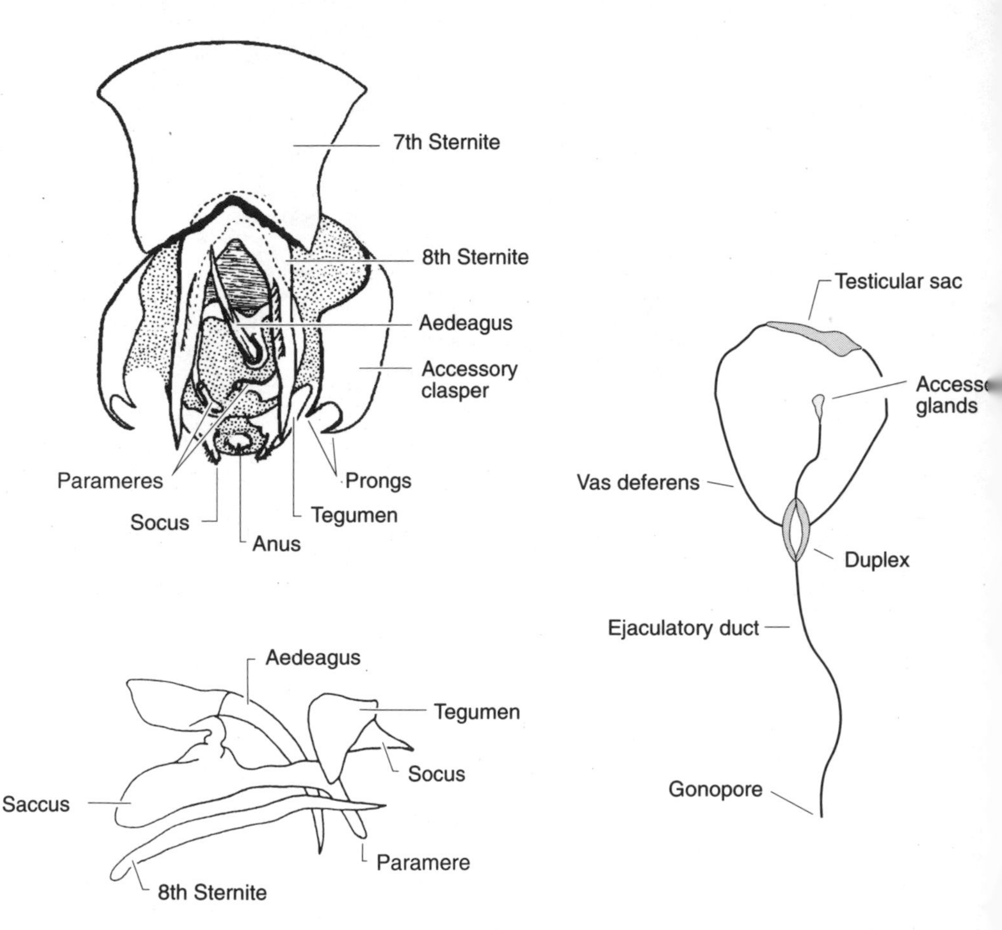

Figure 4.6. Internal and external genitalia of male *Malacosoma americanum.* (From Stehr and Cook 1968, Williams 1940.)

located in the membrane between the eighth and ninth abdominal segments. The gland completely encircles the abdomen and appears as a saddle-shaped area on the dorsal surface of the intersegmental membrane (Percy and Weatherston 1971, Coroiu et al. 1986).

Several straight-chained, 12-carbon aldehydes, acetates, and alcohols have been recovered from the glands of various species. Field and laboratory bioassays have so far led to the identification of four of these compounds as active components of the sex attractants of tent caterpillars (Fig. 4.7, Table 4.1). The effectiveness and specificity of these compounds depend not only on their presence but also on the ratios of the pheromonal components. For the forest tent caterpillar, aldehyde-alcohol

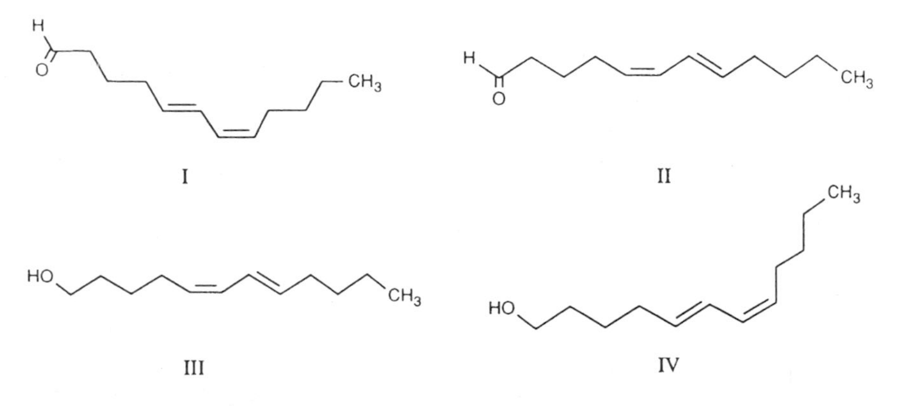

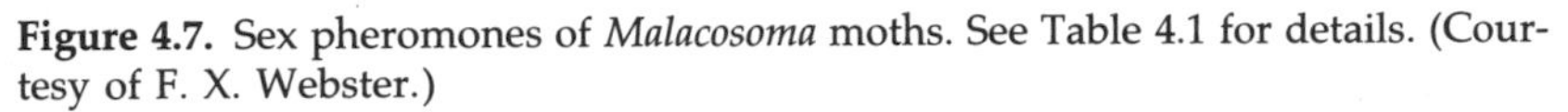

Figure 4.7. Sex pheromones of *Malacosoma* moths. See Table 4.1 for details. (Courtesy of F. X. Webster.)

ratios of from 1:10 to 1:3 are most effective in attracting males under field conditions (Chisholm et al. 1980). An aldehyde-alcohol ratio of 3:1 reportedly gives the best field results for *M. neustrium* (Konyukhov and Kovalev 1988). Although our knowledge of the pheromonal systems of the species listed in Table 4.1 is not yet complete, sympatric species appear to respond optimally to different combinations of the four compounds.

The sex pheromone of *M. disstria* is secreted during the first half-hour after emergence. Females call males by hanging with abdominal segments 8–10 protruded and their wings slightly spread (Percy and Weatherston 1971). Females that fail to attract mates on the evening of

Table 4.1. Active components of the pheromone gland secretions of *Malacosoma* moths

Component	*M. disstria*	*M. americanum*	*M. californicum*	*M. neustrium*
I. (*E*)-5,(Z)-7-dodecadienal		X	X	X
II. (Z)-5,(*E*)-7-dodecadienal	X			
III. (Z)-5,(*E*)-7-dodecadien-1-ol	X			
IV. (*E*)-5,(Z)-7-dodecadienl-1-ol		X		X
Reference:	Chisholm et al. 1980, 1981, 1982	W. Roelofs, pers. comm., 1992	Underhill et al. 1980, Chisholm et al. 1981	Konyukhov & Kovalev 1988

their emergence may call again at dawn. Virgin females were observed to call and mate for as long as five days after emergence (Struble 1970) and were as successful in attracting males after five days as they were on the date of emergence (Shepherd 1979).

Forest tent caterpillar males that enter the plume of a calling female adopt a characteristic zigzagging flight pattern as they move in and out of the odor plume, but they also appear to rely on visual cues to locate females. They are attracted to cocoons and to small brown objects that resemble moths (Bieman 1980, Palaniswamy et al. 1983). The males investigate such objects by hovering near them or by tactile probing. Visual orientation is relatively nondiscriminatory; males have been observed hovering near empty cocoons, cocoons containing parasitoid flies, and even cocoons containing male pupae.

In both the Ontario and Michigan studies (Shepherd 1979, Bieman 1980, Bieman and Witter 1983) of the forest tent caterpillar, females called until about one hour after dark, and in Michigan most matings were initiated during the period from 2.5 hours before sunset until approximately an hour after sunset (mean time of mating = 1939). Although light-trap catches in West Virginia indicate that *M. disstria* moths fly throughout the night (Sample 1992), the moths are not active for long after sunset in northern parts of the species' range. In western Ontario, males were rarely caught at traps baited with virgin females after 2100. Although females were observed to call at temperatures as low as 9°C, ambient temperature typically fell below the threshold flight temperature of 11°C soon after dark.

In Michigan, matings were initiated earlier when there were many moths flying than when population density was low. The duration of copulation also varied significantly with population density. These differences in the time of onset and the duration of mating appear to be related to the density of males and to reflect intrasexual competition for mates. During periods when moths were abundant, males were often observed to interfere with mated pairs. The intruder bashed into the copulating moths and in some instances inflicted abdominal wounds in either the male or female of the pair as it probed the couple with its sharply pointed aedeagus. In one case, Bieman (1982) observed that a copulating female was injured to the extent that an ovary and accessory gland protruded from her wound. Although the investigators observed an average of nearly six encounters between intruders and copulating pairs at high population densities, intruders enjoyed little success in dislodging the copulating males.

Female *M. disstria,* whether mated or not, fly at dusk and signal their intention to initiate flight by vibrating their wings. This behavior serves to discourage males that are still copulating, and they soon disengage the female. But when male density is high, copulating males are reluctant to terminate the union and may walk and even fly in tandem with the female. Thus, in the Michigan study, copulation lasted an average of 85 minutes when population density was low but as long as 187 minutes when male density was high (Bieman 1980).

Although male *Malacosoma* moths may mate more than once (Stehr and Cook 1968), Beiman (1980) observed only a single instance of multiple matings for *M. disstria* females during his field studies (he also observed a female *M. americanum* mate twice in the laboratory). It is difficult, however, to ascertain by observation alone the exact proportion of females that may mate more than once under field conditions. More accurate information on mating patterns may be had by electrophoretic studies of egg masses. Such a study by Costa and Ross (1994) provided strong evidence that 19% of the *M. americanum* females in their field study population mated more than once, but other considerations suggest that the figure may be nearly twice as great. The investigators also ascertained that the mean degree of relatedness among caterpillars hatched from single egg masses that were products of multiply mated females was very close to that of full sibs ($r = 0.466 \pm 0.042$). Thus, there is apparently low sperm penetrance by all but one male so that, despite multiple copulations, the effective number of matings per female is very close to one. The tendency of the female moths to disperse and oviposit shortly after copulating markedly limits their opportunities for multiple mating. All the female forest tent caterpillar moths observed to mate in the Michigan study commenced ovipositing within two hours of initiating wing vibration preparatory to flight and, on average, had completed laying their eggs by 2318.

The temporal pattern of mating and oviposition of the eastern tent caterpillar differs from that of the forest tent caterpillar. Both males and females emerge in the late afternoon and early evening (Williams 1939, Bieman 1980). The females disperse from their hidden pupation sites before calling males. When ambient temperatures permit, both sexes are active throughout the night and may mate at any time until dawn. Twenty-two matings observed by Bieman (1980) had a mean initiation time of 2335. Females oviposit soon after mating, but if mating is not terminated until daybreak they may delay the process until the following dusk. Under laboratory conditions, females were reported to take from 55 to 105 minutes to deposit a complete egg mass (Williams 1939).

The temporal pattern of the mating behavior of *M. neustrium* appears to be similar to that of *M. americanum*. Coroiu et al. (1984) maintained moths under a laboratory light regime in which the scotophase occurred

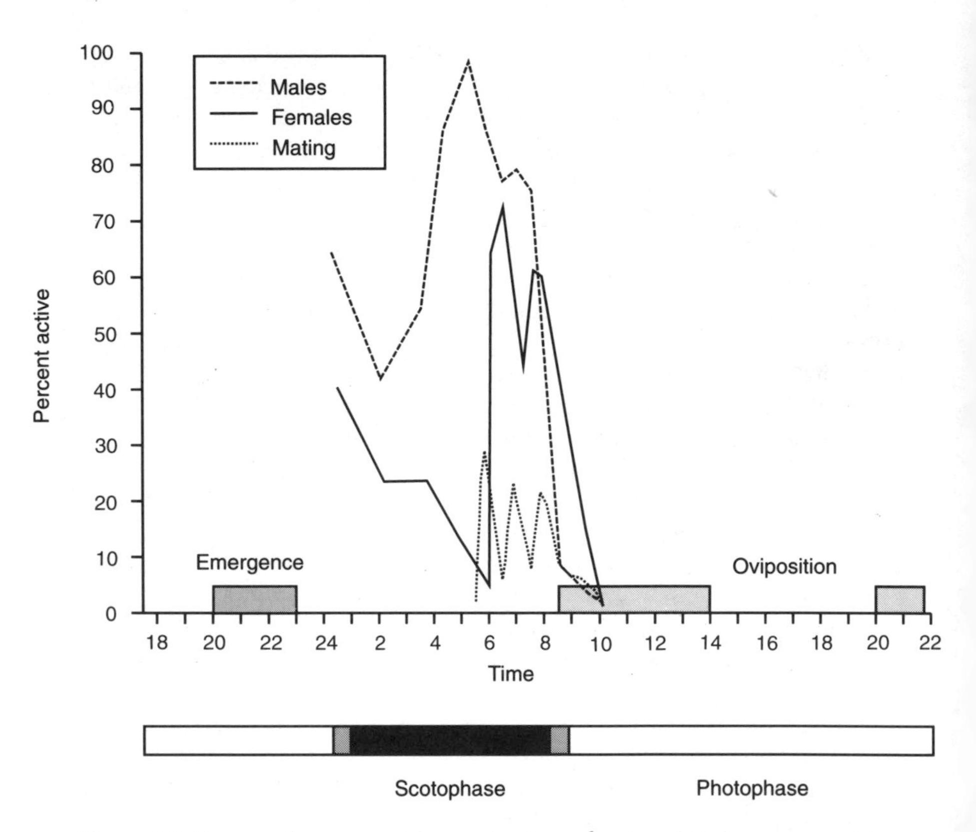

Figure 4.8. Temporal sequence of reproductive behavior of *Malacosoma neustrium* moths maintained under an artificial light regimen in the laboratory. (After Coroiu et al. 1984.)

from approximately 2400 to 0800 (Fig. 4.8). Emergence occurred late in the photophase, corroborating the observations of Williams (1940) that moths emerged in the late afternoon and early evening. Moths flew throughout the scotophase but were most active between 0400 and 0800. Mating occurred most often between 0530 and 0800. The moths oviposited in daylight, at either the beginning or the end of the photophase. Four matings observed by Williams (1940) lasted 45–63 minutes, and females deposited their egg masses within 43–60 minutes (Coroiu et al. 1984). Under laboratory conditions, adult *M. neustrium* females that had mated lived an average of 3.8 days, and males lived 6.3 days (Coroiu et al. 1984). Hodson (1941) stated that both the males and females of *M. disstria* live for approximately five days. The longevity of free-living *Malacosoma* adults is unknown for any species.

Table 4.2. Morphometrics and hovering-flight metabolism of male *Malacosoma americanum* moths

Body mass (mg)	88.8 ± 21.8 (SD)
Total wing mass (mg)	5.0 ± 4.7
Total wing area (cm^2)	2.36 ± 0.46
Wing load	0.396 N/m^2
Total metabolism during hovering-flight	125 ml $O_2 \cdot g^{-1} \cdot h^{-1}$
Energy expanded during hovering-flight	698 mW·g

Source: Reprinted from T. M. Casey, *Physiological* Zoology 54(1984): 362–371, © 1981 by the University of Chicago, by permission of the University of Chicago Press.

Flight Energetics

Both male and female moths of *Malacosoma* fly either before mating or before ovipositing or at both times. Tent caterpillar moths have relatively small wings compared with other species of comparable weight and must work proportionately harder to achieve flight. A gypsy moth (*Lymantria dispar*) with body mass similar to that of a *M. americanum* moth has 2.3 times the wing area of the latter. The larger wings of the gypsy moth produce greater lift and, on a per-unit-area basis, they carry only half the load of those of the tent caterpillar moth. In addition, although the energy required to fly is approximately the same per stroke for the two species, the eastern tent caterpillar must generate a higher wing-beat frequency to sustain flight and consequently uses 2.6 times as much energy per unit of flight time (Casey 1981; Table 4.2).

The eastern tent caterpillar is an ideal subject for the study of flight energetics because it will maintain sustained flight in a small respirometer for as long as 15 minutes. It can be tethered with fine wires implanted at critical sites to measure body temperature and wing-beat frequency. Studies show that the moth cannot fly until it achieves a wing-beat frequency of approximately 57 strokes per seconds (Fig. 4.9A; Casey 1981). The frequency with which flight muscle contracts is strongly temperature-dependent, and the muscle cannot produce this stroke rate until the thoracic temperature (T_{th}) has been elevated to 37–39°C. Like most moths, the tent caterpillar flies when the radiant heat of the sun is largely diminished, or completely absent, and ambient temperature (T_a) is lower than the required T_{th}. Thus, the thorax of the tent caterpillar must be heated through an endogenous process during a preflight warm-up.

To produce the required T_{th}, the moth undergoes a period of "shivering" before initiating flight by simultaneously contracting the antagonistic dorsal-ventral and longitudinal flight muscles. Because muscle

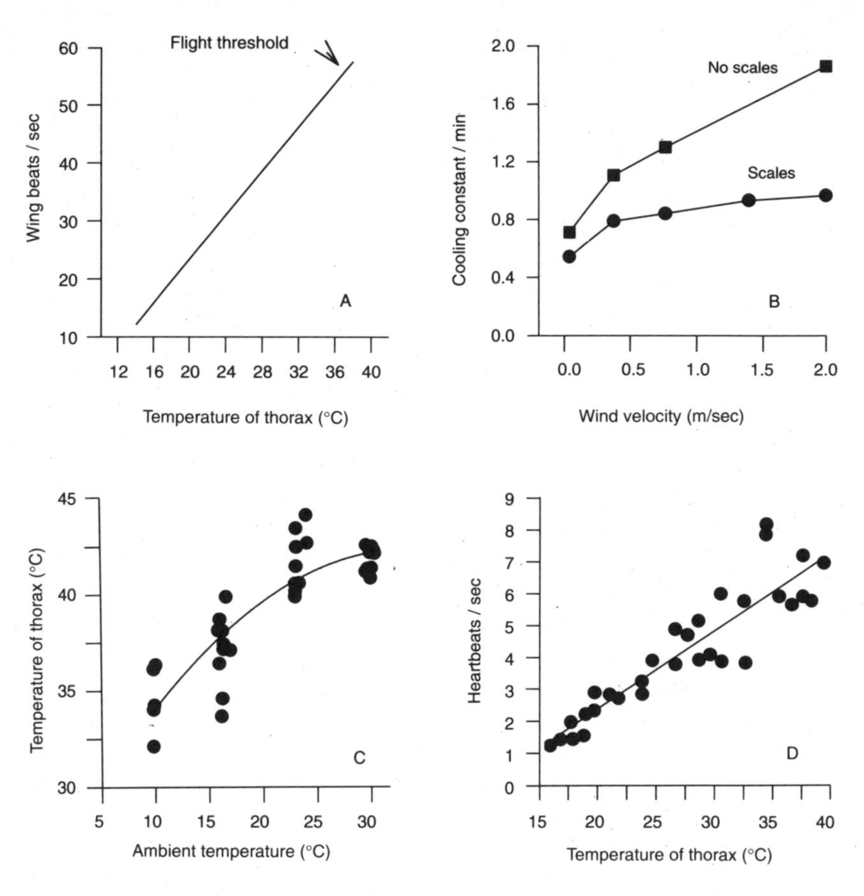

Figure 4.9. (A) Relationship between thoracic temperature and wing-beat frequency of *Malacosoma americanum* moths during preflight warm-up. (B) Rate of cooling in still and moving air of *M. americanum* moths with scales intact and removed. Rate of cooling increases with increasing values of the cooling constant. (C) Relationship between air temperature and thoracic temperature during preflight warm-up and hovering flight of *M. americanum* moths. (D) Relationship between thoracic temperature and pulse rate of the dorsal vessel for *M. americanum* moths. Increased blood flow to the head occurs as the moth warms to flight temperature. (B, C based on T. M. Casey, *Physiological Zoology* 54[1981]:362–371, © 1981 by the University of Chicago, by permission of the University of Chicago Press. D based on Casey et al. 1981, *Journal of Experimental Biology* 94[1981]:119–135, by permission of the Company of Biologists Ltd.)

is relatively inefficient, from 80 to 90% of the energy expended during this process is degraded to heat (Casey 1988). The thick pile of the thorax insulates against heat loss, and moths that have the pile removed are only about 70% as effective in retaining heat as intact moths (Casey

1981). The pile is most effective in insulating the moth against heat loss at high flight speeds when losses due to forced convection would otherwise be substantial (Fig. 4.9B).

The rate of thoracic warming is strongly dependent on ambient temperature, increasing from 3.3°C/min at a T_a of 14°C to 12.8°C/min at 28°C. In turn, wing-beat frequency increases linearly with thoracic temperature, rising from approximately 13 strokes per second at a T_{th} of 14°C to 59 strokes per second at 39°C (Fig. 4.9A). Thus, at $T_a = 22$°C it takes a moth about three minutes to elevate its thoracic temperature to the 37–39°C required to achieve the lift-off stroke frequency. Once flight is initiated, the tent caterpillar moth is able to regulate T_{th} and concomitant stroke frequency. At T_a's of 10–30°C, the moth regulates T_{th} between 35 and 42°C, though it is not entirely clear how this is done (Fig. 4.9C). At T_a's below 10°C the moth is too cold to initiate preflight warm-up.

Because of adaptive changes in the normal circulatory pattern of the moth, little of the heat generated during preflight warm-up is transferred to the abdomen. The head temperature, however, is similar to that of the thorax, and it has been suggested that the brain needs to be warmed to the thoracic temperature to facilitate the finely coordinated movements that occur during flight (Casey 1981). The rate of the heartbeat is directly related to T_{th} (Fig. 4.9D).

Low overnight temperatures may inhibit flight. Eastern tent caterpillar moths can fly at temperatures as low as 10°C, but they require a long period of preflight warm-up and a correspondingly large energy expenditure (Casey and Hegel-Little 1987). Once flight is initiated, however, the cost of sustaining it is largely independent of T_a (Casey 1981). In marked contrast, the winter moth *Operophtera bruceata* can initiate flight at temperatures as low as 0°C, with a T_{th} only slightly above T_a. It can do so because this species has a large wing surface area relative to its body mass, and it flies at a wing-beat frequency of only a few strokes per second (Heinrich and Mommsen 1985).

Dispersal

Little is known of the extent to which *Malacosoma* moths disperse during the flight period. There have been no attempts to release and recapture marked moths. But the fact that new infestations often originate in areas far from known infestations indicates that the females may disperse over moderate distances before ovipositing. Hodson (1941), for example, found that adults of *M. disstria* were attracted to lights at distances two to three miles from known infestations. Moreover, electrophoretic studies of *M. americanum* that indicate that their widely sep-

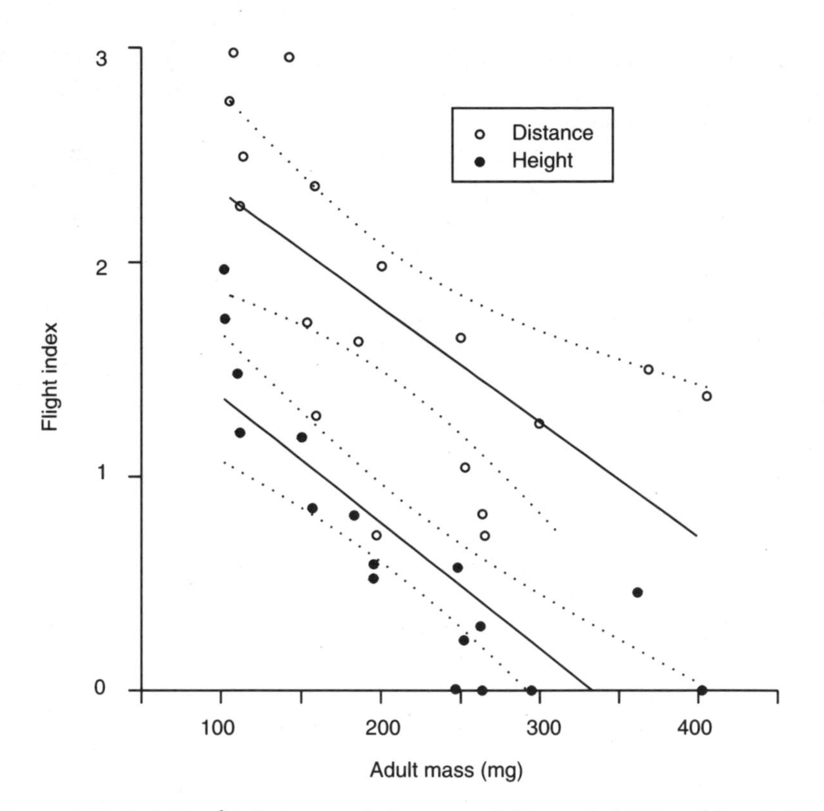

Figure 4.10. Relationship between adult mass and dispersal abilities of female *Malacosoma neustrium testacum* moths. Flight index is based on laboratory measurements of mean distance and height of flight of repeatedly tested individuals. The arbitrary scale indicates relative performance with high values assigned to moths that flew the highest and longest. Dotted lines show 95% confidence limits. (After Shiga 1977.)

arated populations have high genetic homogeneity suggesting that there are high rates of dispersal among populations (Costa and Ross 1994).

In his studies of *M. neustrium testacum*, Shiga (1977) found that the extent of preovipositional flight dispersal by the female was dependent on the size of the adult, with small adults having greater dispersal capabilities than larger ones (Fig. 4.10). Small size, in turn, was attributable to intense larval competition for food. Thus, dispersal of small preovipositional adults from densely populated areas may enable them to enjoy higher fitness by establishing broods at sites remote from established infestations.

Dispersal of moths may also be aided by wind, allowing the insects

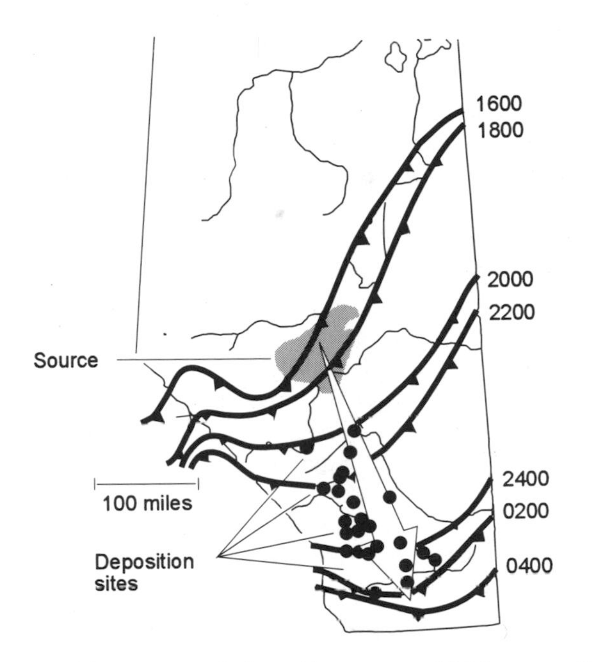

Figure 4.11. Probable source and deposition areas of *Malacosoma disstria* moths carried by cold front. Position of front is shown at various times. (Redrawn, by permission of the Entomological Society of Canada, from Brown 1965.)

to travel remarkably long distances in short spans of time. During the summer of 1964, large numbers of forest tent caterpillar moths suddenly appeared in several areas in southern Alberta, Canada, despite the absence of any known infestations nearby (Brown 1965). Subsequent investigation indicated that the moths originated at an outbreak site in central Alberta up to 300 miles north of the regions where the moths were alighting (Fig. 4.11). Circumstantial evidence indicates that the moths were carried by a cold front that passed through the outbreak site in the late afternoon of the previous day. On the basis of climatological maps, the moths appear to have been carried by the front at an average rate of about 25 miles an hour, covering the 300-mile distance in as little as 12 hours.

The Egg Mass and the Pharate Larva

Although the egg of the tent caterpillar may appear to be the most durable of the insect's life forms, it is the small larva that develops within the egg shell that is the most persistent. Embryogenesis produces a fully formed pharate larva within several weeks of oviposition, and the pharate larvae remain encased and nearly motionless through the remainder of the year. In northern regions, eggs deposited by tent caterpillars in July hatch in April or May of the following year.

Ovipositional Patterns

Tent caterpillars lay all their eggs in one batch. Such cluster oviposition by social caterpillars appears to have evolved in response to selective pressures to maximize female fecundity and to facilitate larval aggregation (Fitzgerald 1993a). As discussed in Chapters 2, 6, and 7, caterpillars benefit from aggregation in several ways, and field and laboratory studies show that small aggregates are less successful than larger groups.

The egg mass of the tent caterpillar is tough and durable. The structure protects the encapsulated pharate larvae from predators, parasitoids, and the elements of the weather for up to 10 months. The egg masses of *M. castrensis,* laid on salt marsh grasses, are regularly inundated by salt water with little apparent harm to their occupants (Tutt 1900).

Tent caterpillar eggs are glued to each other and to the substrate with a secretion produced in the colleterial gland. The resultant egg mass may either completely encircle a branch or enclose only part of it (Fig. 4.12; Stehr and Cook 1968). The forest tent caterpillar, which produces an egg mass that completely encircles a branch, deposits successive rows of four to six eggs aligned at an angle of approximately 45° to the axis of the twig. The rows are laid in a continuous band, which may wrap around the twig several times. As she nears the end of her egg supply, the female adds fewer eggs to each row so that the finishing edge of the egg mass has an even appearance. The eggs in adjoining sections of the band are aligned so that they appear to be neatly arranged in long diagonal rows (Fig. 4.12).

Egg masses of species that only partially enclose the branch consist of successive long rows of eggs laid more or less parallel to the axis of the stem. The female *M. americanum* lays these rows of eggs by placing her

Figure 4.12. (Above) Clasping egg masses of *Malacosoma americanum* and helical egg masses of *M. disstria* and *M. tigris*. Egg masses of *M. americanum* and *M. disstria* are shown with spumaline removed. *M. tigris* is the only American species that does not coat its egg mass with spumaline. (Below) The egg placement sequence of *M. disstria* that gives rise to the helical pattern.

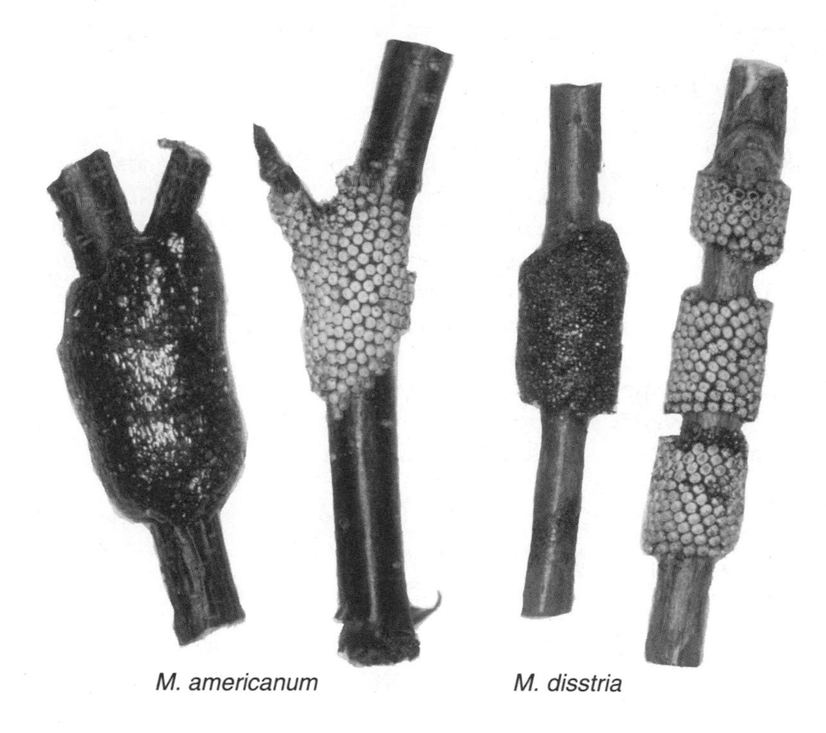

M. americanum *M. disstria*

M. tigris

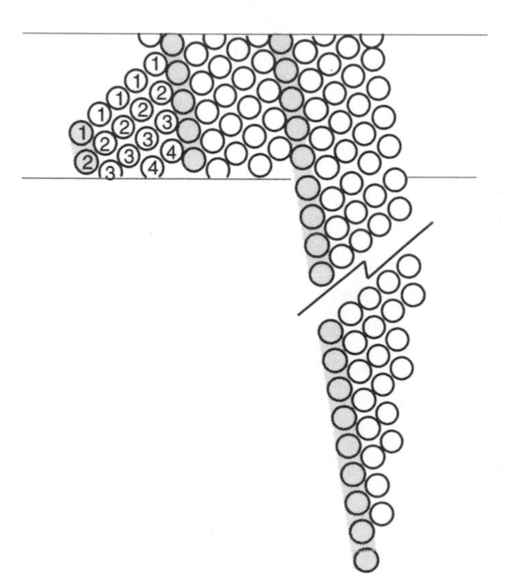

body at a right angle to the axis of the stem. She then extends the tip of her abdomen (the ovipositor) as far as possible to one side of her body to establish the point on the twig where the row is to be initiated. As each successive egg is deposited, the ovipositor is slowly swung through a shallow arc until it reaches the extreme position on the opposite side of her body. Thus, the eggs are deposited in a long, slightly curved row. The female may then move ahead to repeat the process. She usually exhausts her supply of eggs before completely encircling the branch. The fully laden ovarioles occupy much of the abdominal cavity, and the female appears to be completely spent after completing the egg mass (Fig. 4.13).

Encircling egg masses have been referred to as "helical" masses and nonencircling as "clasping" (Stehr and Cook 1968). Of the American species, *M. disstria*, *M. constrictum*, and *M. tigris* lay helical masses, and all other species deposit clasping masses. Some of the European and Asian species such as *M. neustrium*, *M. alpicolum*, and *M. castrensis* deposit eggs in the helical pattern, and others such as *M. parallelum* and *M. indicum* deposit clasping masses. Both types of egg masses are typically deposited on branches of small diameter, but the clasping pattern also allows the female to lay her eggs on larger branches. Indeed, nearly flat masses have occasionally been deposited by *M. americanum* near the bases of small trees with main stem diameters of several centimeters (Stehr and Cook 1968), but this is not a common habit for this species.

Egg masses of tent caterpillars, stripped of their spumaline coating, typically measure 4–7 mm in diameter by 7–15 mm in length. The egg masses of Nearctic species contain approximately 150–250 eggs (Stehr and Cook 1968). The number of eggs per egg mass is positively correlated with the mass of the ovipositing female (Fig. 4.14) and is typically highly variable within a species. Thus, the egg masses of the eastern tent caterpillar in Arkansas were reported in two studies to contain from 75 to nearly 600 eggs (mean = 299) (Baerg 1935, Stacey et al. 1975). Hodson (1941) found similar variability among the egg masses of the forest tent caterpillar attacking aspen in Minnesota. Of the over 10,000 egg masses observed, some contained as few as 15 eggs and others had as many as 327. Averaged over longer periods, the number of eggs per egg mass for populations of this species in Minnesota is approximately 150–170 (Hodson 1941, Witter et al. 1975).

Unusually small egg masses are likely to be the result of interrupted bouts of oviposition or female infertility. Egg masses of above-average size may reflect low levels of intraspecific competition, high-quality food, and ideal weather conditions during larval development. Thus, the average number of eggs per egg mass of adults of the forest tent caterpillar

Figure 4.13. An exhausted female *Malacosoma americanum* after depositing her egg mass.

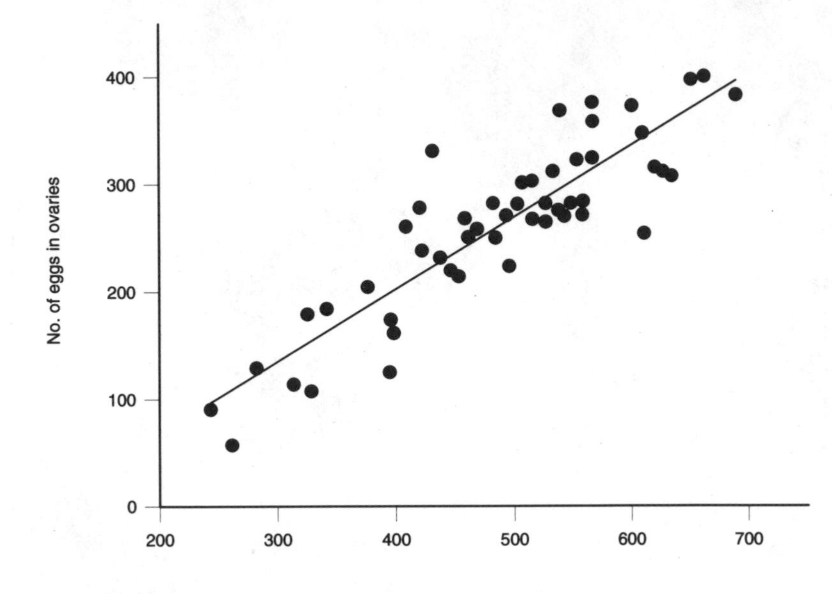

Figure 4.14. Relationship between mass of female pupa and size of egg complement of *Malacosoma neustrium.* (Data from Shiga 1977.)

derived from caterpillars fed on water tupelo (*Nyssa aquatica*) in Louisiana (approximately 300–400; Smith and Goyer 1986, Goyer et al. 1987) is markedly higher than the number of eggs per egg mass produced by adults of the same species reared on aspen in Minnesota. In contrast, the average number of eggs per egg mass produced by adults derived from larvae fed on either water oak (*Quercus nigra*) or flowering dogwood (*Cornus florida*) in North Carolina (185 and 188, respectively) was only slightly greater than that of the Minnesota populations. The differences among these southern populations appears attributable to variation in the nutritional quality of the three host tree species (Smith et al. 1986).

Most European and Asian species of *Malacosoma* produce about the same number of eggs per egg mass as North American species (Shinga 1977, Schwenke 1978, Bhandari and Singh 1991). Egg masses of *M. alpicolum* and *M. castrensis* are the exception. The eggs of these species are laid on the stems of herbaceous plants that are as small as 0.9 mm in diameter. The egg masses are approximately 4 mm in diameter and up to 32 mm in length and may contain over 750 eggs (Martin and Serrano 1984; Fig. 4.15).

The otherwise tedious process of determining the number of eggs per egg mass can be facilitated through the use of formulas that relate one

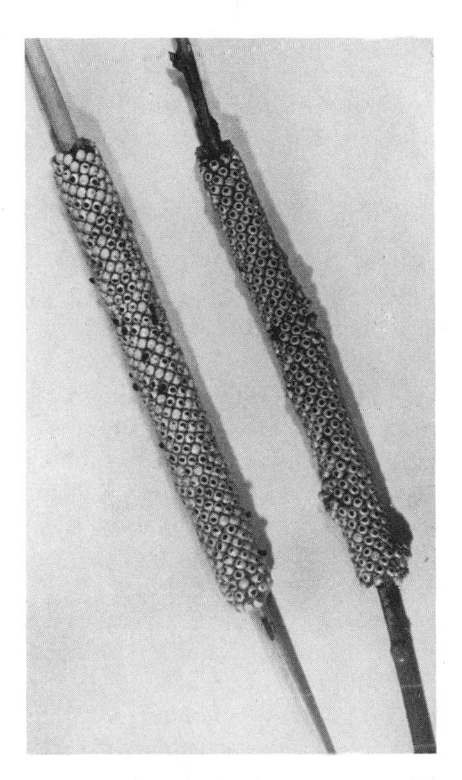

Figure 4.15. Egg masses of *Malacosoma alpicolum* (left) and *M. castrensis* (right). (Reprinted, by permission of Consejo Superior de Investigaciónes Cientificas, Museo Nacional de Ciencas Naturales, from Martin and Serrano 1984.)

or more dimensions of the egg mass to the number of eggs it contains. Thus, for southern populations of the forest tent caterpillar, estimates of the approximate number of eggs per egg mass can be obtained by multiplying the length of the bare egg mass in centimeters by the constant 404.3 for water tupelo populations, 294.4 for water oak populations, and 273.9 for flowering dogwood populations (Goyer et al. 1987). Witter and Kulman (1969) developed a more complex formula to estimate the number of eggs in egg masses of northern populations of the forest tent caterpillar. They obtained good estimates with the formula $N = \pi dln$, where d = diameter of the bare egg mass, l = length of the bare egg mass, and n = number of eggs per square millimeter, and they provide tables of variables based on this formula.

The basis for ovipositional host plant selection is unknown for any species of *Malacosoma*. Since females typically fly before ovipositing, airborne host plant volatiles are likely to be involved, but the female may also use contact chemoreception after alighting. The female must also assess branch dimensions before ovipositing. How this assessment is accomplished is unknown, but oviposition typically occurs on branches of similar size throughout the crown. Egg masses of *M. californicum* are typically laid on branches of *Populus tremuloides* that average 3–4 mm in diameter, though they are sometimes found on branches with diameters up to 14 mm (Schmid et al. 1981). On apple and black cherry trees, *M. americanum* moths oviposit at a mean distance of approximately 10 cm from tips of branches that average approximately 3–4 mm in diameter (Fitzgerald and Willer 1983). There is no evidence that females avoid branches visited by other females, and egg masses are often deposited close to others even when population densities are low. Although the egg masses may be widely dispersed around the circumference of the crown of the trees, studies of *M. americanum* (Fitzgerald and Willer 1983), *M. californicum pluviale* (Moore et al. 1988) and the lasiocampid *Eriogaster lanestris* (Balfour-Browne 1933) indicate that oviposition sites are commonly clustered on the south side of the tree.

Oviposition sites selected by eastern tent caterpillars in old-field successions indicate that the moths do not assess the size of the resource. Eggs are commonly found on trees much too small to support a colony to maturation (Fig. 4.16). Moreover, Robison (1993) found that, although the survival of larval forest tent caterpillars varied widely among different clones of poplar, the adults did not discriminate among clones when ovipositing. The clones are newly developed varieties of poplar, however, and there has not been adequate time to determine if the moths would eventually evolve selective ovipositional behavior. It is unknown whether adult tent caterpillars oviposit discriminately among native hosts, some of which are also known to vary in the extent to which they support larval growth and survival.

Spumaline

The eggs of many species of *Malacosoma* are covered with a coating termed "spumaline" by Hodson and Weinman (1945). The material is stored in the large reservoirs of the colleterial glands (Fig. 4.5) and is secreted at the time of oviposition. In *M. tigris* (Stehr and Cook 1968) and the Old World species *M. neustrium, M. alpicolum,* and *M. castrensis* (Martin and Serrano 1984) the accessory gland secretion serves mainly as an adhesive and is not elaborated into a frothy covering. In all other North American species and in the Asian species *M. parallelum* (Li 1989) and *M. indicum* (Joshi and Agarwal 1987, Bhandari and Singh 1991), the

Figure 4.16. *Malacosoma americanum* moths apparently do not assess the size of the food resource, and small trees are often completely defoliated before the larvae mature, forcing the caterpillars to disperse in search of food.

secretion is mixed with air to create an expanded foamy coating over the egg mass (Fig. 4.12). The material is secreted over each egg after it is positioned in the egg mass and flows together to form a continuous coating (Liu 1926). The superficial properties of the spumaline coating of the egg masses of North American species vary from species to species to the extent that spumaline has taxonomic value (Stehr and Cook 1968).

The function of spumaline has been the subject of several inquiries. Hodson and Weinman (1945) and Carmona and Barbosa (1983) showed that the material is hygroscopic and serves as a water reservoir. Spumaline picks up moisture from the air most effectively when there is a large saturation deficit, and it becomes saturated during rainy weather. This property appears to ameliorate the potential for the eggs to desiccate; laboratory studies have shown that significant mortality occurs when the covering is not allowed to recharge. When held in desiccators over calcium chloride, the postdiapause eggs of *M. disstria* dry rapidly, and, after only eight days, survival is reduced to approximately one-third of that which occurs after a single day's exposure to the regimen (Hodson and Weinman 1945). In the short term, hardening of the chorion, which is also hygroscopic, may contribute to the mortality by making it difficult or impossible for the larva to chew its way out, but desiccation eventually leads to the insect's death.

Spumaline traps infrared radiation, accounting for the fact that egg masses are often warmer than the ambient temperature on clear, sunny days in the winter (Wellington 1950, Carmona and Barbosa 1983). Excesses of 1.6–6.0°C ($T_{egg} - T_{air}$) have been measured when ambient temperatures ranged from +9 to −20°C, with the largest excesses occurring at the coldest temperatures. This feature is likely to be a side effect of properties of spumaline selected for other reasons. Although temperature amelioration could become significant when ambient temperatures approach the supercooling point, such extremes are most likely to occur in the early morning before sunrise. The effect of spumaline on the temperature of egg masses during the summer has not been reported. Eggs may be exposed to direct sunlight for extended periods, and heat absorption and heat trapping by spumaline could, potentially, have deleterious consequences, particularly at high air temperatures. Also uninvestigated is the possibility that evaporation of water from the spumaline reservoir may act to cool the eggs on hot, dry days. It has been proposed that spumaline may also serve as a mechanical barrier to egg parasitoids, but the evidence is circumstantial and equivocal (Chapter 8).

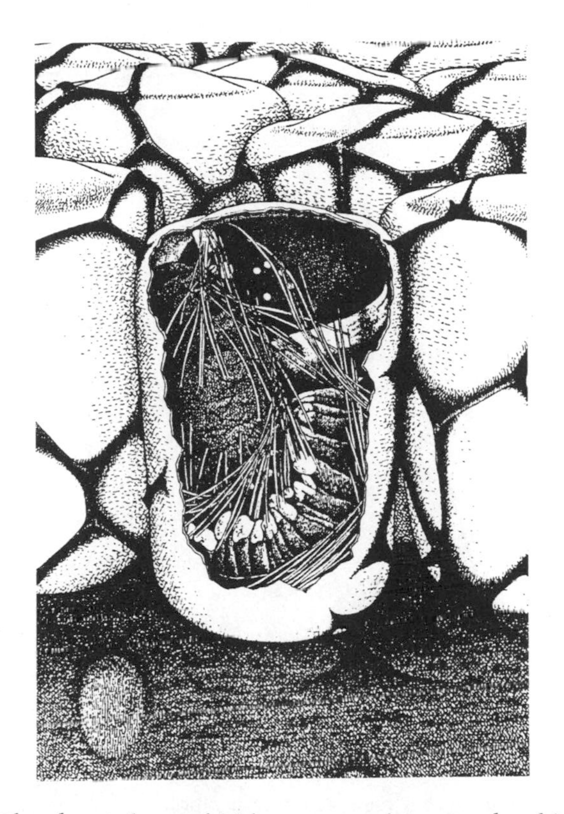

Figure 4.17. The pharate larva of *Malacosoma americanum* enclosed in the chorion of its egg. (Reprinted from *The World of the Tent Makers: A Natural History of the Eastern Tent Caterpillar,* by Vincent Dethier. Drawing by Abigail Rorer. Amherst: University of Massachusetts Press, 1980. Copyright © 1980 by The University of Massachusetts Press.)

Embryogenesis and Cold Hardiness

Studies of forest and eastern tent caterpillars indicate that embryogenesis is initiated soon after the eggs are deposited in early summer and is completed in about three weeks (Fig. 4.17; Hodson and Weinman 1945, Mansingh 1974). The pharate caterpillars then undergo an obligatory diapause. Histological studies show little change in the condition of the tiny caterpillars that lie within the chorions until November, at which time they begin to assimilate the yolk. Cold weather slows the process, and the yolk is not completely absorbed until just before hatching in the spring.

Nothing is yet known of the factors responsible for the onset of dia-

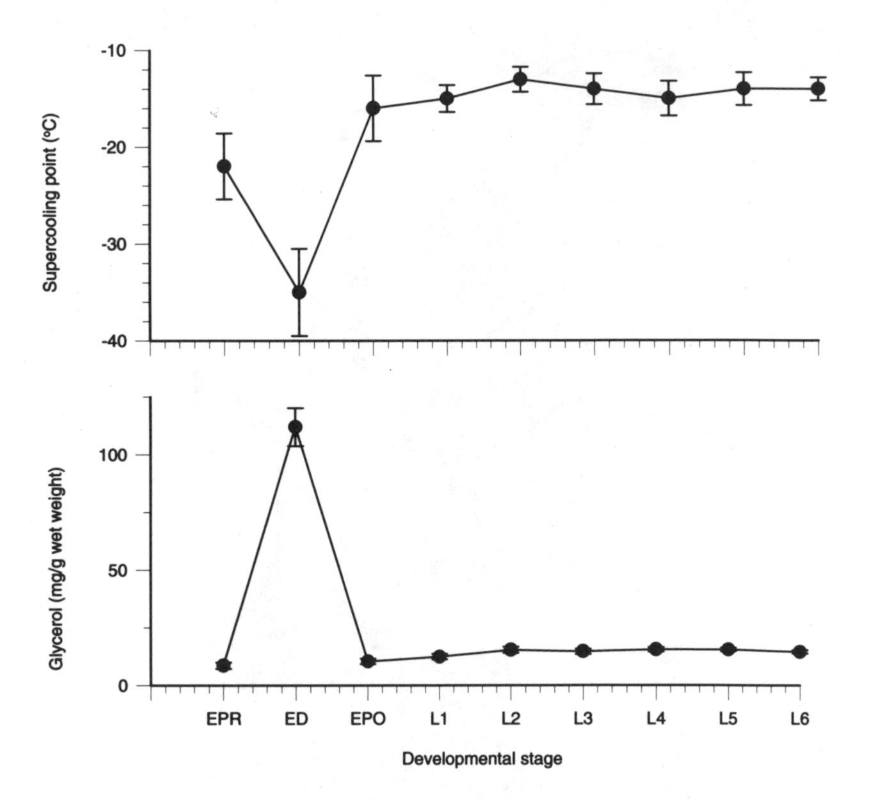

Figure 4.18. Relationship between supercooling point of various stages of *Malacosoma americanum* collected in Ontario, Canada and the glycerol content of the stage. Ed = diapausing pharate larva, Epo = postdiapause pharate larva, Epr = prediapause egg, L*x* = larval instar. (After Mansingh 1974.)

pause, but its termination is temperature-dependent (Flemion and Hartzell 1936, Hodson and Weinman 1945, Hanec 1966, Mansingh 1974). If the eggs are collected in August and held indoors at 25°C, yolk absorption proceeds rapidly and is completed within three months. But because they have not been exposed to cold temperatures, the larvae die within the chorions. In contrast, when eggs are held at 2°C for three months, to simulate overwintering, then moved to 25°C to hatch, over 90% of the caterpillars eclose within eight days of transfer. The caterpillars will also eclose after being transferred to temperatures as low as 10°C following cold treatment at 2°C, but it takes over a month to achieve a 90% hatch rate. Thus, the pharate larvae are sensitive to ambient temperature, and eclosion can be delayed or accelerated, allowing the larvae to eclose under appropriate conditions in the spring.

Before the initiation of diapause in the summer, the pharate larvae are

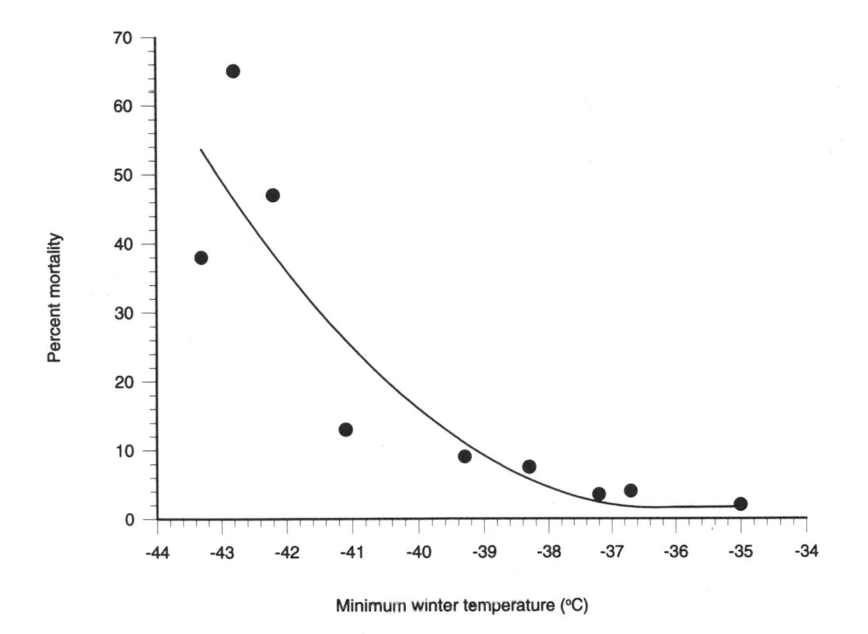

Figure 4.19. Overwintering mortality of pharate larvae of *Malacosoma disstria* in Minnesota as related to the minimum winter temperature to which the caterpillars were exposed. (After Witter et al., *The Canadian Entomologist* 107[1975]:845, by permission of the Entomological Society of Canada.)

cold-sensitive and begin to die if they are held at 5°C for four weeks. During diapause, the pharate larvae are highly resistant to cold temperatures and remain so throughout the winter and spring, until just before they eclose. Paralleling the rise and fall of cold tolerance is change in the glycerol content of the caterpillars. Although other cryoprotectants are also involved, glycerol appears to be a major component of the "antifreeze" responsible for the cold tolerance of the insect. The temperature at which both the forest tent caterpillar and the eastern tent caterpillar freeze (supercooling point) closely follows changes in their glycerol content (Fig. 4.18).

Unlike some freeze-tolerant insects, such as the goldenrod gallmaker, *Eurosta solidaginis,* that can survive being cooled far below their supercooling point (Hanec 1966), tent caterpillars die if their body fluids freeze. Indeed, the active stages of the insect cannot tolerate for long cold temperatures well above their supercooling point. Thus, newly hatched and unfed larvae of the forest tent caterpillar survive for a mean maximum of 19 days when held at −1°C, even though their supercooling point is −13.4°C (Hanec 1966). In a laboratory experiment, approximately 25% of pharate forest tent caterpillars were killed when exposed to a temperature of −15°C for 72 hours, several days before they would have hatched (Wetzel et al. 1973). In another study, 100% of fed or unfed first-instar forest tent caterpillars died when held at −18°C for only 48 hours, but over 80% survived when they were held at −12°C for up to 168 hours (Raske 1975).

Clearly, tent caterpillars are physiologically adapted to survive for only brief periods at temperatures approaching their supercooling points. Under most circumstances this limited ability provides adequate protection, but unusually cold winters can result in significant mortality of the pharate larva (Fig. 4.19). Gorham (1923) noted that all the pharate larvae of the forest tent caterpillar found within 5.6 km of the Salmon River in northwestern Brunswick were killed when the temperature dropped to −43°C in the river basin. The supercooling point for diapausing larvae of this species collected in Manitoba, Canada, is −40.8°C (Hanec 1966).

5

Interactions between Tent Caterpillars and Their Host Trees

Tent caterpillars are the first folivores to appear in spring, and their seasonal development is closely tied to that of the host plant. The caterpillars typically eclose just as the new leaves of their natal plants begin to grow. But in some years the eggs hatch even before the buds of the host tree have opened, and the small caterpillars must survive for days with little or no food. The caterpillars also must endure the vicissitudes of early spring weather, and, in the colder parts of their ranges, whole colonies have been known to perish when late-season frosts destroy their food supply. Why don't the caterpillars remain within the relative security of their eggs until the leaves are well formed and the weather is more hospitable? Although the selective forces that pushed back the time of larval eclosion to the earliest possible moment may include depredation by vespid wasps and songbirds, whose demands for food increase as the spring progresses, it is clear that tent caterpillars are now adapted to feed within a narrow phenological window. Numerous studies show that seasonal patterns of eclosion, feeding, and quiescence of tent caterpillars are constrained to no small degree by the chemical and physical characteristics of early spring foliage.

Feeding Efficiency

When properly synchronized with their host trees, tent caterpillars are efficient herbivores. The larvae grow exponentially, fueling their growth with exponentially increasing quantities of food (Fig. 5.1). Such exponential growth accounts for the common observation that during out-

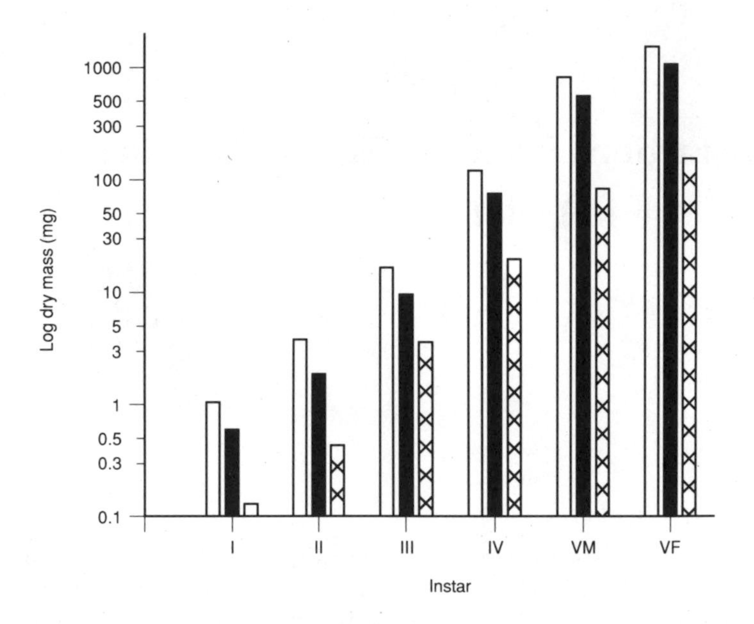

Figure 5.1. Mass of leaves consumed (white), fecal pellets produced (black) and mass gained (hatched) by an average larva of *Malacosoma neustrium testacum.* Data for the first four instars are for males and females combined, separate values are given for the fifth instar. (Data from Shiga 1976.)

breaks, hordes of caterpillars appear, seemingly from out of nowhere, and proceed to rapidly defoliate trees. Hodson (1941) determined that while the first through fourth instars accounted for 0.23%, 0.78%, 3.4%, and 13.3%, respectively, of the aspen foliage consumed during the lifetime of a forest tent caterpillar, consumption by the last instar accounted for 82.2% of the total. Thus, during the first three stadia a caterpillar consumes approximately half a leaf, during the fourth stadium, one leaf, and during the last stadium, seven leaves. Slightly more than 1000 fifth-instar larvae, a small company compared with the numbers that occur during outbreaks, can consume all of the approximately 7000 leaves that occur on an average-sized aspen sapling (Hodson 1941).

Caterpillars must process large quantities of food because they are able to capture only a small fraction of the total energy contained in leaves. A large part of the leaf consists of nondigestible components such as cellulose and lignin. The material passes through the digestive tract without being absorbed and is eventually egested as fecal pellets, becoming part of the detritivore food chain. Furthermore, part of the energy content of the fraction of the leaf that is assimilated is used for

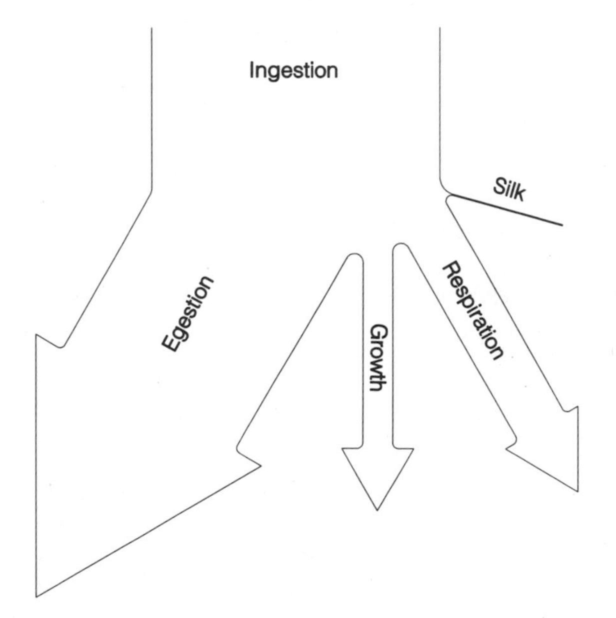

Figure 5.2. Partitioning of energy in the last- (5th-) instar larva of *Malacosoma neustrium testacum* reared on Japanese pear. Growth accounts for approximately 10% of the total energy budget. (Data from Shiga 1976a.)

maintenance (respiration) and is lost in the process. Only the remaining energy is stored as caterpillar body or made available for reproduction.

Of the energy contained in the leaves that the caterpillars ingest (I_e), some fraction (A_e) is assimilated. Of the assimilated energy some fraction (G_e) is used for growth. The balance of the ingested energy is egested as fecal pellets (F_e) or used in respiration. A small portion of the energy is also lost to detritivores in the form of exuviae and silk (Fig. 5.2). There are three measures of the efficiency of this process of energy transfer.

1. ECI, the *efficiency of conversion of ingested material,* measures the fraction of energy in the ingested leaf that is utilized for growth. It is calculated as follows:

$$\mathrm{ECI} = \frac{G_e}{I_e}$$

The value typically is small, because a relatively large fraction of I_e is contained in nondigestible material or used in respiration.

2. AD, the *approximate digestibility* of the leaf, is a measure of assimilation and is calculated as follows:

$$\mathrm{AD} = \frac{(I_e = F_e)}{I_e}$$

The proportion of the energy contained in nondigestible material is equal to 1 − AD.

3. ECD, the *efficiency of conversion of digested (assimilated) energy*, is calculated as follows:

$$\mathrm{ECD} = \frac{G_e}{(I_e - F_e)}$$

The proportion of ingested energy used in respiration is equal to 1 − ECD.

In practice, these measurements are often based on ratios of oven-dry weight rather than energy. Because the energy contents of dry leaves, larvae, silk, and frass are roughly equivalent (approximately 22–23 J/mg; Ruggiero and Merchant 1986), calorimetric and gravimetric measurements yield similar results.

The efficiency of energy transfer reported for tent caterpillars varies widely. For the eastern tent caterpillar reared on spring leaves, mean ECI values range from 17 to 31%, AD values from 40 to 54%, and ECD values from 35 to 72% (Schroeder and Malmer 1980, Futuyma and Wasserman 1981, Ruggiero and Merchant 1986, Schroeder 1986, Peterson 1987). Thus, eastern tent caterpillars void as fecal pellets nearly half of the energy they ingest, and they use up to 65% of their assimilated energy in respiration. Approximately 2% of the assimilated energy is lost as exuviae, and at least 3% as silk (Ruggiero and Merchant 1986). Nonetheless, compared with other species of caterpillars that feed on the leaves of black cherry, the eastern tent caterpillar is one of the most efficient.

In a study of 34 species of lepidopterous caterpillars that feed on black cherry, the eastern tent caterpillar had the highest ECI and ECD values and the third highest AD value (Schroeder and Malmer 1980). The efficiency with which this seasonal succession of caterpillars digests and utilizes food for growth decreases with time, reflecting a marked deterioration in food quality as the season advances. By late summer, the leaves are dry and fibrous. The fall webworm, which feeds on such leaves from late July to September, is 25 to 30% less efficient than the eastern tent caterpillar in assimilating food and in converting digested food into caterpillar body. Its overall ability to convert whole-leaf tissue into caterpillar body is only about half that of the eastern tent caterpillar.

The efficiency of energy transfer has also been calculated for other species of tent caterpillars. When reared on Japanese pear or willow, *Malacosoma neustrium* had AD, ECI, and ECD values ranging from 34 to 38.4%, 14 to 16.8%, and 41 to 43.7%, respectively (A. C. Evans 1939, Shiga 1976a). Both AD and ECD are lower for the last instar, probably because it feeds on a more aged food supply. *M. disstria* reared on black cherry or tupelo gum (*Nyssa aquatica*) had values of 30–42% for AD, 7–17% for ECI, and 25–47% for ECD (Futuyma and Wasserman 1981, Smith et al. 1986). In all cases, the lower values were for caterpillars fed on tupelo.

Wide variation in the indices of energy transfer reported for tent caterpillars reflects not only the varying quality of the foliage of different host species but also the high sensitivity of these insects to within-tree differences in the age of foliage. In almost all of these studies, the age of the leaves fed to the larva was not stated, so it is difficult to assess the extent to which the values approximate the efficiencies of freely foraging caterpillars. When larvae are fed exclusively on the stage of leaf they prefer, values even higher than these may be expected. In one study, eastern tent caterpillars given only leaves less than 10 days old had an ECD of 75% (Peterson 1987). In contrast, when caterpillars are fed exclusively on mature leaves, they grow more slowly and produce smaller larvae and pupae than caterpillars fed young leaves.

Schroeder (1986) compared the growth performance of the last two instars of eastern tent caterpillars fed spring leaves, summer leaves, and fall leaves of black cherry. Mean conversion rates ranged from 27 to 11% for ECI, 72 to 37% for ECD, and 44 to 27% for AD. In all cases the high value was for spring leaves, and the low for fall leaves. Larvae took 45% longer to achieve maturity when fed summer leaves rather than spring leaves and had a final dry mass only 58% of that of caterpillars fed spring leaves. Larvae fed fall leaves exclusively weighed only one-third as much as larvae fed spring leaves at maturity. The variable nutritional qualities of the leaves are also reflected in the foraging behavior of the larvae. In one study, 74% of the diet of larvae that were allowed to choose between young and aged leaves consisted of the younger leaves (Peterson 1987).

Differences in the quality of leaves of the same tree species grown under different conditions may also affect the efficiency of energy transfer. Black cherry trees that grow in the shade of the forest are rarely attacked by the eastern tent caterpillar, and studies have shown that this insect is not well adapted to the leaves of such trees. Caterpillars reared exclusively on shade leaves in laboratory experiments produced female pupae that weighed only 67% as much as the pupae of caterpillars reared on sun leaves from open-grown trees (Futuyma and Saks 1981).

In addition to the spatial and temporal variability that occurs among leaves within trees, the more polyphagous species of tent caterpillars, such as *M. disstria,* must deal with interspecies variability as well. It might, therefore, be hypothesized that these caterpillars would be less well adapted to any one of their host trees than would a specialist species. But in one test of this hypothesis, it was reported that eastern tent caterpillars, which specialize largely on rosaceous tree species, were no better at utilizing the foliage of their preferred host than were the larvae of the forest tent caterpillar. Both species were collected in the egg stage from ovipositional sites on black cherry, then reared on that host in the laboratory. The indices AD, ECI, and ECD did not differ significantly between the species during the fourth and fifth stadia, indicating that the specialist was no better at utilizing host leaves than was the generalist (Futuyma and Wasserman 1981). But the investigators did not control for leaf age or measure growth efficiency during the earlier stadia. The eastern tent caterpillar prefers the youngest leaves found at the tips of branches, and its diet may consist largely of these during the first several instars under field conditions when little else may exist. It remains to be determined if the forest tent caterpillar is equally adapted to these young, often only partially expanded, leaves, particularly during the first several stadia. As discussed below, such leaves are highly cyanogenic.

Plant Chemistry and Foraging Success

Tent caterpillars attack trees just as the season's new foliage is forming (Fig. 5.3), and they are capable of causing considerable damage to their hosts. Indeed, the caterpillars commonly defoliate their natal trees and during prolonged outbreaks may eventually kill their hosts. It may appear that trees can do little more than tolerate such assaults, but it is clear that they are far from passive victims. It is of particular significance for tent caterpillars that leaves are highly unstable in the early spring. Leaves grow rapidly and in the process undergo profound physical and chemical changes that render them increasingly less acceptable to the caterpillars. Such rapid maturation prevents tent caterpillars from tracking their hosts through time, restricting them to a single generation each year wherever they occur. Moreover, their window of opportunity is

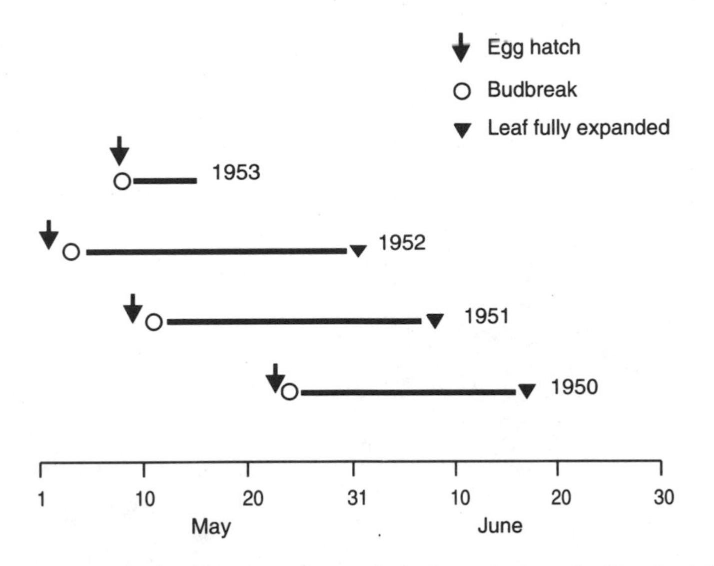

Figure 5.3. Relationship between the median date of aspen budbreak, full leaf expansion, and hatching of the eggs of *Malacosoma disstria* during four successive field seasons. Egg hatching of the various species of tent caterpillars occurs early in the spring and is typically tightly synchronized with budbreak. In 1953, a spring freeze killed the host leaves. (Reprinted, by permission of the Society of American Foresters, from Rose 1958.)

particularly narrow. Slow growth, and even population collapse, may result if caterpillars fail to synchronize their development with that of the host tree.

Seasonal Changes in Nutritional Value of Leaves

The water and nitrogen content of leaves are among the most important of the seasonally variable characteristics of plants known to influence their palatability. When black cherry leaves are experimentally allowed to vary only in their water content, caterpillars confined to a diet of drier leaves eat more but grow more slowly than caterpillars given leaves with higher water content. This is true because when water is limiting the caterpillars are less efficient at utilizing both nitrogen and leaf energy and in assimilating food (Scriber 1977). Like water, the nitrogen content of leaves has a marked influence on the rate at which caterpillars grow. When nitrogen is in short supply, caterpillars may have to process large quantities of food to meet their needs for protein.

Leaf toughness also impedes herbivory. Sclerophyllous tissue forma-

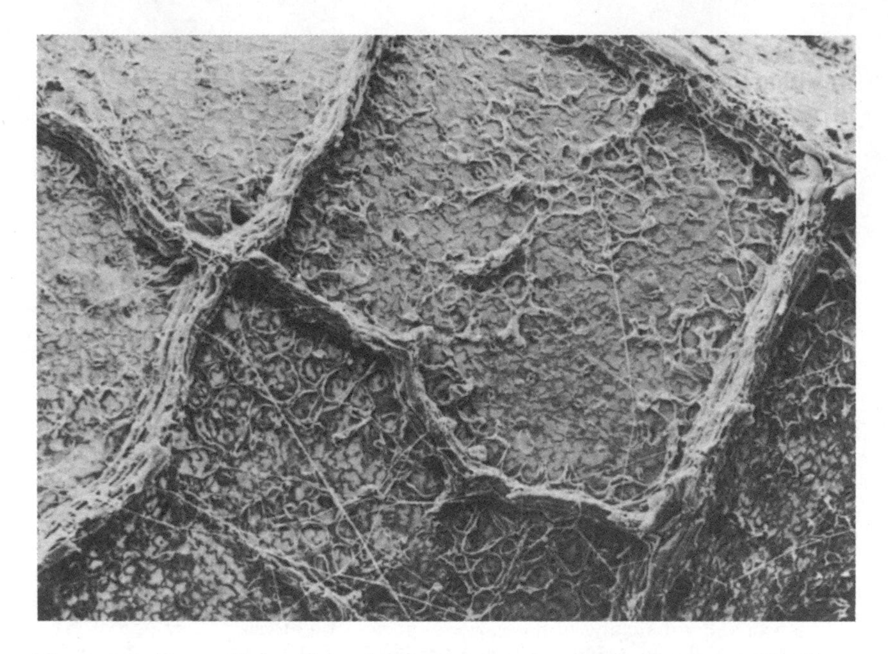

Figure 5.4. Upper leaf surface of sugar maple, *Acer saccharum*. Caterpillars have removed the soft tissue, leaving behind a network of inedible vascular bundles. (Reprinted, by permission of Munksgaard, from Hagen and Chabot 1986.)

tion proceeds rapidly in the leaves of many trees and renders them increasingly less palatable with time. Small larvae have difficulty penetrating such tissue, and their digestive systems may be taxed to process large quantities of nonnutritive fiber. It has been suggested that although vascular tissue exists primarily to translocate water within the tree, its fortification with thickened bundle sheath cells and extensions may be an adaptation to thwart small herbivorous insects (Hagen and Chabot 1986; Fig. 5.4). Small caterpillars may be restricted to islands of soft tissue defined by impenetrable borders of vascular tissue. This restriction greatly impedes the rate at which they process food.

Chemical assays can be conducted to determine the nitrogen content of leaves; water content is measured by comparing the weight of leaves before and after oven drying. The relative toughness of leaves is determined by measuring the force required to puncture a leaf with a metal rod. The results of a series of such measurements made on the leaves of black cherry and sugar maple during the period when tent caterpillars feed upon them is shown in Figure 5.5. Both water content and nitrogen

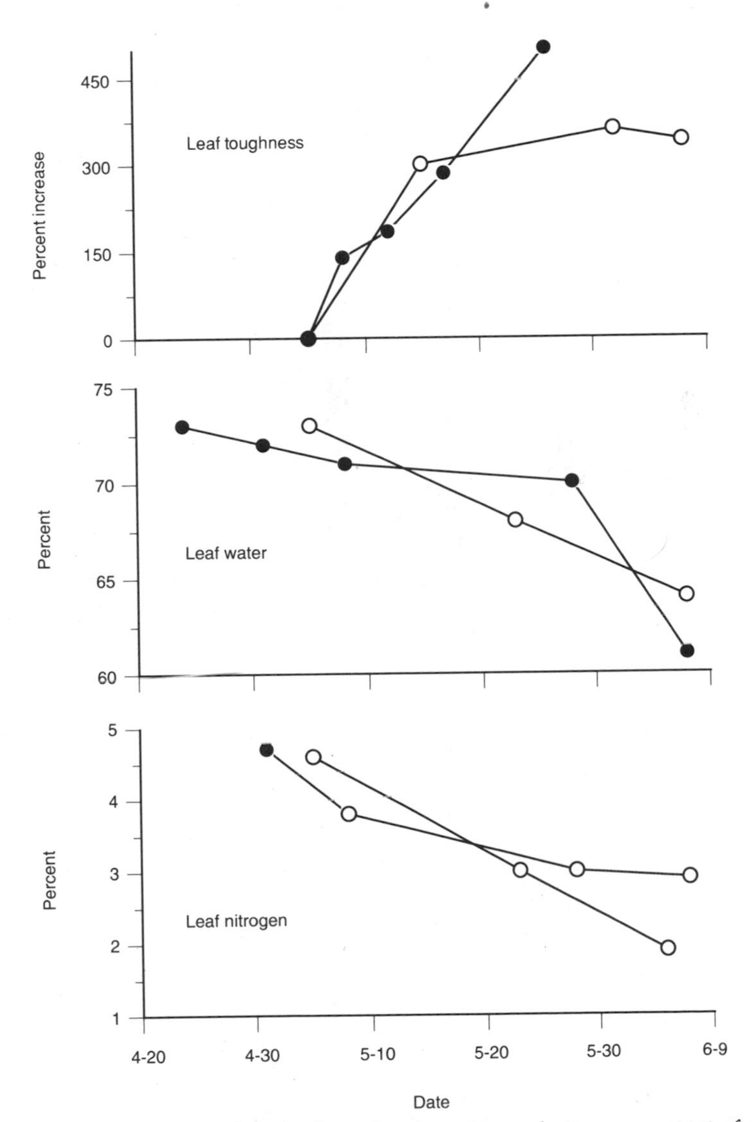

Figure 5.5. Seasonal trends in leaf toughness, water, and nitrogen content of sugar maple (open circles) and black cherry (filled circles) in the northeastern United States during the period when tent caterpillars are active. (Based on data from J. C. Schultz et al. 1982, Segarra-Carmona and Barbosa 1983, Peterson 1985.)

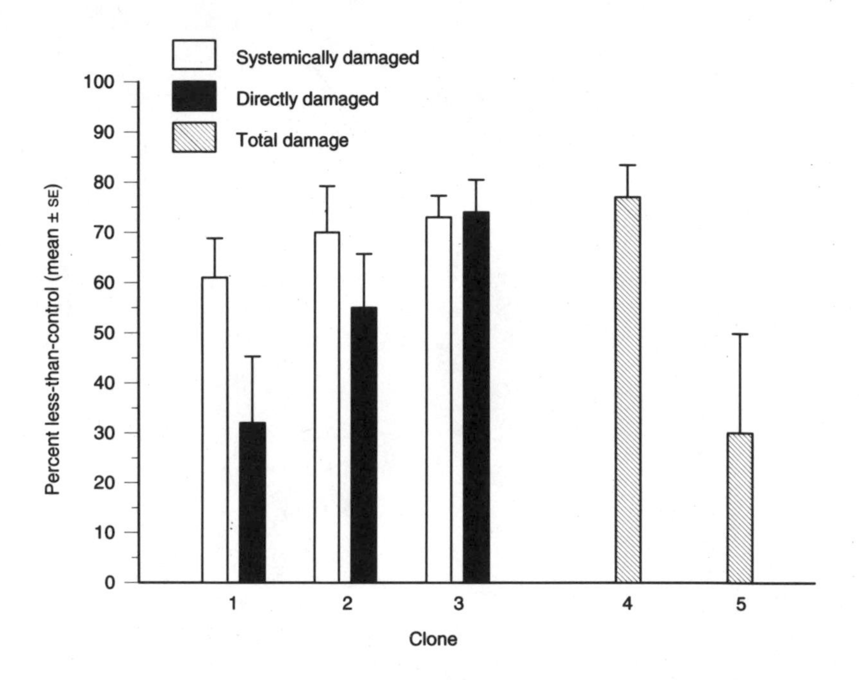

Figure 5.6. Consumption of leaf discs of hybrid poplar clones by forest tent caterpillars. For feeding studies involving clones 1–3, leaf discs fed to the caterpillars were cut from leaves already partially eaten (directly damaged), nearby leaves (systemically damaged), or control trees. For studies involving clones 4–5, caterpillars were allowed to move freely between previously damaged and control trees. (Data from Robison 1993.)

content of leaves decline, and leaf toughness increases markedly during the period when tent caterpillars are active.

The effect of increasing leaf toughness on the growth of caterpillars is particularly apparent. Attempts to rear newly hatched larvae of *M. americanum, M. neustrium,* and *M. disstria* on leaves of their preferred host trees—black cherry, oak, and quaking aspen, respectively—only several weeks older than foliage they normally feed on fail because the caterpillars have great difficulty penetrating the leaf and grow little or not at all (Hodson 1941, van der Linde 1968, T. D. Fitzgerald, pers. observ.). The feeding activity of tent caterpillars may in itself also result in deterioration of leaf quality. Robison (1993) found that caterpillars fed leaf discs from damaged and control hybrid poplar clones or allowed to forage freely on damaged and control trees tied together, consumed significantly more of the control foliage (Fig. 5.6). The remains of leaves partially eaten by caterpillars (directly damaged leaves) and leaves near damaged leaves (systemically damaged leaves) had a lower percentage of nitrogen and water and more fiber and toughness than leaves from control trees, but other unassessed chemical changes may have been involved.

In the field, unusually cold spring weather may interfere with the ability of the caterpillars to forage or to process food, but temperatures may be sufficient to allow the leaves of the tree to continue to develop. If inclement weather continues long enough, the leaves may toughen to a point at which the caterpillars can no longer bite into them or process them efficiently. Blais et al. (1955) reported one such incident involving populations of forest tent caterpillars in Manitoba during a particularly cold spring. For a period of nearly a month after the eclosion of the caterpillars, the temperature rarely exceeded 15°C. Persistent low temperatures prevented the caterpillars from growing but had a lesser effect on the growth of the foliage of the host tree, trembling aspen. When the temperature eventually rose sufficiently to allow the caterpillars to feed, the aspen leaves were fully expanded and tough, and the caterpillars were able to feed on only the softer tissue between the veins. In another study, forest tent caterpillars established experimentally as newly emerged first instars on sugar maple trees in mid-May grew slowly because of particularly cool and cloudy weather (Fitzgerald and Costa 1986). Daytime temperature at the study site averaged only 11.3°C, and the caterpillars spent an average of 27 days, an abnormally long period, completing the first two larval stadia. When warm weather returned, the leaves had matured considerably, and the caterpillars moved repeatedly from site to site, only partially consuming leaves before abandoning them. By mid-June, when the caterpillars normally mature, they had barely achieved the fourth instar. Feeding became more sporadic, and the colonies were extremely restless. Shortly thereafter the caterpillars abandoned their host trees and disappeared from the study site (Fitzgerald and Costa 1986). The cause of this delayed schedule of development was not determined, but leaf toughness, reduced leaf water content, or increased leaf tannin levels (see Phenols, below) were likely causes.

Such asynchrony of host and caterpillar development may be less severe when caterpillars feed on host trees, such as black cherry, that continue to produce new leaves during the period when tent caterpillars are active. By searching extensively, eastern tent caterpillars may meet their demand for young cherry leaves while ignoring the less palatable mature ones (Fig. 5.7). One of the primary host trees of the western tent caterpillar *M. californicum pluviale,* red alder (*Alnus rubra),* also has indeterminate growth and may continue to produce new leaves into late

Figure 5.7. Selective feeding on the youngest leaves of black cherry (*Prunus serotina*) by eastern tent caterpillars.

summer or fall in favorable years. Adams (1989) found that caterpillars would survive and grow when placed on trees even in July or August, although it was necessary to confine them to mesh bags to keep them from leaving the tree.

Secondary Chemicals

The production of so-called allelochemicals by host trees may also affect the foraging behavior of tent caterpillars. Allelochemicals are secondary plant chemicals that do not appear to play a role in the primary metabolic processes of the organism but are deterrents to herbivores. Although they may be incidental by-products of metabolism, in some instances they appear to have originated in direct response to selection pressure from their enemies. Naturally occurring poisons such as nicotine from tobacco and pyrethrum from the *Chrysanthemum* flower are well-known examples. Not all such compounds are so overtly toxic, however. Phenols, especially tannins, may deter herbivores by rendering leaves distasteful, by inhibiting enzyme systems of the herbivore, and by reducing the digestibility of leaf protein. Still other secondary compounds mimic the activity of insect hormones, interfering with growth or development.

Cyanogenic Glycosides

One particularly ubiquitous secondary chemical found in plants is cyanide. More than 60 families of the flowering plants contain species that

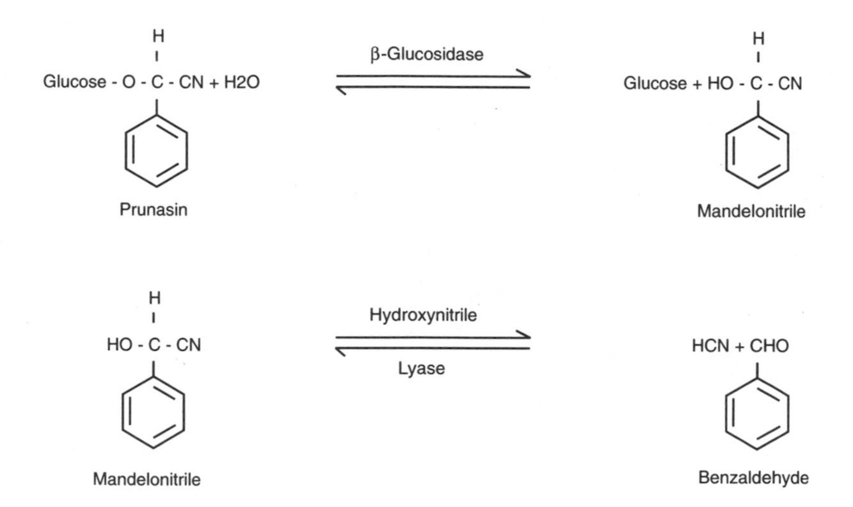

Figure 5.8. Steps in the production of HCN from prunasin in black cherry. (Redrawn from Peterson 1985.)

are cyanogenic (Jones 1972). The cyanide (HCN) produced by these plants in response to injury is a potent toxin that affects both vertebrate and invertebrate herbivores. When freshly picked, young black cherry leaves contain up to 2000 mg of HCN per kilogram of leaves (Morse and Howard 1898, Peterson et al. 1987), some 30–40 times the average fatal dose for humans. Sodium nitrate and sodium thiosulfate are antidotes for accidental cyanide poisoning, but they must be administered in a timely manner; a concentration of as little as 300 ppm can result in death to a human in only a few minutes. Incidences of cattle poisoning have been reported, and many insects are also highly susceptible to cyanide. Indeed, the HCN liberated from crushed leaves can be used as a substitute for potassium cyanide in insect-killing jars (Hall et al. 1969).

In the intact black cherry leaf, cyanide occurs as a component of the cyanogenic glucoside prunasin. When a leaf is eaten by a caterpillar, enzymes convert prunasin to glucose and the cyanohydrin, mandelonitrile. The latter compound then disassociates to form cyanide gas (HCN) and the distinctly odorous benzaldehyde (Peterson 1985, Peterson et al. 1987; Fig. 5.8).

Of the three common cherries that occur in the ranges of the forest and eastern tent caterpillars, black cherry has the greatest HCN potential, followed in order by choke cherry and pin cherry (Morse and Howard 1898). The concentration of cyanide varies from leaf to leaf on a single tree and also seasonally. Leaves that contain as much as 2500 ppm of HCN in April contain less than 50 ppm by late October (Smeathers et al. 1973; Fig. 5.9).

In addition to the eastern tent caterpillar, more than 200 species of lepidopteran caterpillars attack black cherry during the season, but the majority of these species feed during midsummer (Fig. 5.9). Although other factors such as predation pressure from spring nesting birds, inclement early season weather, or even the threat of competing with colonies of tent caterpillars are likely to be involved, the high cyanide potential of spring leaves may be a potent deterrent to most herbivores that otherwise might attack the succulent young growth (Schroeder 1978). In contrast, the eastern tent caterpillar is clearly immune to the ingested toxin and feeds on host leaves at their peak HCN potential. It remains to be determined how the tent caterpillar detoxifies the cyanide, but some other insects, including the southern army worm, are known to produce the cyanide sulphurtransferase rhodanese, which catalyzes a reaction that renders the compound harmless (Brattsten 1979).

The eastern tent caterpillar is not only unaffected by cyanide, but one study suggests that the caterpillar may actually use the compound to its advantage. Eastern tent caterpillar larvae commonly regurgitate defensively when attacked by predators or parasitoids. Regurgitated juices collected from caterpillars that have just eaten contain both benzaldehyde and HCN at the same concentration in which they occur in the leaves. Laboratory experiments have shown that predatory ants are repelled by concentrations of benzaldehyde far below those found in caterpillar regurgitations, but they are insensitive to HCN (Peterson 1986a). Indeed, when food items are treated with HCN alone, ants continue to feed until they collapse and may be irreversibly poisoned. It has been suggested that the eastern tent caterpillar may arm itself intentionally by feeding preferentially on the youngest host leaves, which have the highest benzaldehyde potential (Peterson et al. 1987). The effectiveness of this tactic has not been assessed under field conditions, and it remains to be determined for how long after a meal the regurgitations contain repellent and toxic components.

Phenols

Another important class of secondary compounds shown to deter herbivores are the phenols and the polyphenols, particularly tannins. Larvae of the winter moth, *Operophtera brumata*, like tent caterpillars, are early spring feeders and have a single generation each year. Attempts to rear the larvae on leaves of English oak more than six weeks old fail because the increased tannin content of the aged leaves inhibits larval feeding

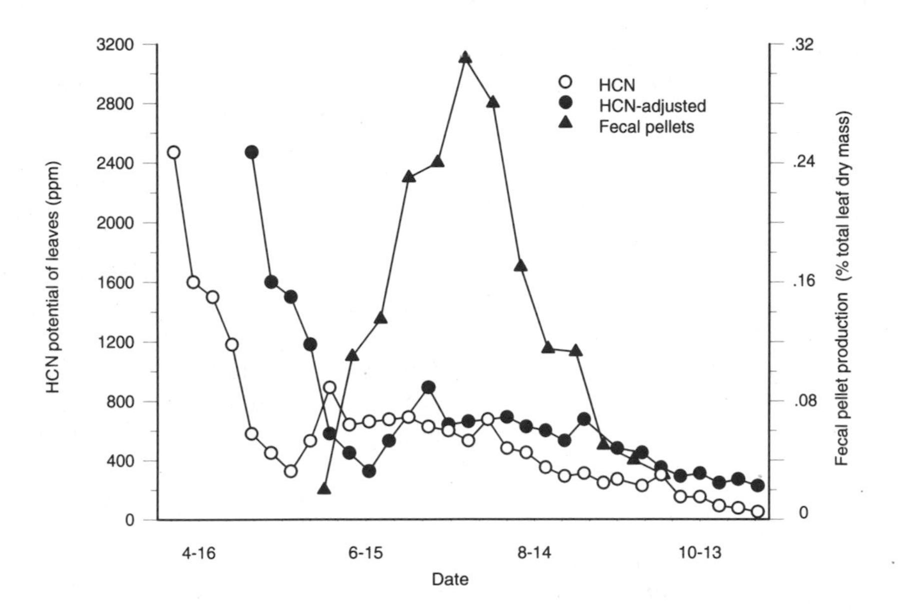

Figure 5.9. Seasonal variation in the HCN potential of black cherry leaves and the abundance of folivores as indicated by fecal pellet production. The HCN curve, based on measurements made in Kentucky, is shifted four weeks to the right to approximate the growing season in Ohio, where the fecal pellet data were collected. Although the eastern tent caterpillar initiates spring feeding when the HCN potential is at its peak, most folivores feed when the potential is markedly lower. (Redrawn from Smeathers et al. 1973, Schroeder 1978.)

(Feeny 1970). Tannins may not only affect the growth of caterpillars by deterring ingestion, but they may also inhibit enzymes, reduce the digestibility of leaf protein, or act as a toxin.

In sugar maple, a principal host of the forest tent caterpillar, the tannin content of leaves rises gradually, reaching a plateau by mid-June, where it remains until the leaves abscise. The leaves of this tree develop rapidly, so that by June all the leaves that are to be formed that season are already fully grown. Tannin may constitute as much as 30% of the dry weight of these mature leaves (J. C. Schultz 1983). The tannins are widely dispersed, occurring in the vacuoles of the mesophyll, epidermis, and vascular cells, and cannot be altogether avoided by herbivores (Hagen and Chabot 1986). There is, however, considerable heterogeneity among leaves on the same tree; tannin contents of adjacent leaves may vary by as much as 300%. Herbivores that are sensitive to tannin must, therefore, forage selectively among the host leaves. The larvae of the saddled prominent have been reported to sample dozens of leaves during feeding bouts, rejecting as many as two-thirds of those tasted (J. C. Schultz 1983). The addition of tannic acid to artificial diets at levels as low as 0.5% (dry weight) was also shown to reduce survival and growth rates of forest tent caterpillars and to cause lethal deformations in the pupae (Karowe 1989). The chemical was found not to affect the digestibility of the diet but to act as a toxin and feeding deterrent.

Studies show that adding phenolic glycosides to synthetic diets also suppresses the growth of forest tent caterpillars. The allelochemical agents salicortin and tremulacin reduce both growth rates and survivorship when incorporated into artificial diets at levels characteristic of those found in the leaves of aspen (Lindroth and Bloomer 1991). The effect is particularly pronounced when the diets have relatively low levels of protein. Reduced growth of caterpillars on such low-protein diets appears to be attributable to small changes in both consumption rates and ECD because the allelochemic agents do not affect the digestibility of the diet. Moreover, when first-instar forest tent caterpillars are fed a mixture of these phenolic glycosides and the esterase inhibitor *S,S,S*-tributylphosphorotrithioate mortality increases markedly over that which occurred when the chemicals were fed separately, indicating that the caterpillar is normally able to enzymatically detoxify the glycosides.

The levels of both salicortin and tremulacin increase in leaves that are torn to simulate caterpillar attack (T. P. Clausen et al. 1989). Furthermore, when aspen leaves are crushed to simulate the state of the material in the gut of the caterpillar, leaf enzymes are suspected of cleaving the glycosides to form 6-hydroxycyclohexenone (6-HCH). Under the alkaline conditions of the caterpillar gut, 6-HCH may then be converted to catechol. Both 6-HCH and catechol were found to reduce the growth of the aspen tortrix (*Choristoneura conflictana*) when added to artificial diets, but their potential effect on tent caterpillars has not been assessed. Although studies of both phenolic glycosides and tannins demonstrate the negative effect that these allelochemicals have on laboratory populations of forest tent caterpillars, amended synthetic diets may not be realistic models for the caterpillar's natural food supply. Such diets isolate the allelochemicals from other leaf chemicals that might have an ameliorating effect.

Unlike cyanide, phenols accumulate in leaves as the season progresses, and their potential impact may be minimized by a regimen of early spring feeding. When properly synchronized with its host trees, the forest tent caterpillar may be little affected by normal seasonal increases in defensive leaf chemistry. Indeed, when compared with other polyphagous species—species whose host range includes plants in at least 11 plant families—the forest tent caterpillar has below-average microsomal mixed-function oxidase (MFO) activity (Krieger et al. 1971). The measurement of midgut MFO activity provides a rough index of the ability and need for caterpillars to detoxify secondary plant compounds. Midgut MFO activity of the forest tent caterpillar is only slightly greater than that of the eastern tent caterpillar, a species that, at least during the majority of its larval life, specializes on plants in the family Rosaceae.

In addition to the genetically programmed seasonal increase in leaf phenols, herbivory may induce increased synthesis of allelochemicals in both damaged and undamaged leaves of the host tree. Such a response has been demonstrated from three hosts of the forest tent caterpillar: sugar maple, hybrid poplar, and red oak. Increases of 50% to over 100% in phenol content occur in maple and poplar, respectively, after attack by herbivores. On red oak, gypsy moth larvae can induce a rapid increase of up to 300% in the tannin content of unattacked leaves. Laboratory populations of gypsy moth caterpillars reared on leaves with these high tannin contents grow slowly, produce small pupae, and have low fecundity (J. C. Schultz and Baldwin 1982).

Adams (1989) determined that the grazing history of a tree also affected the growth of the western tent caterpillar *M. californicum pluviale.* He found that red alder leaves on branches attacked by the larvae of the western tent caterpillar had increased concentrations of phenolics and decreased nitrogen levels. Laboratory populations of *M. californicum pluviale* caterpillars fed leaves from previously grazed branches had significantly lower average weight gains than did caterpillars fed on leaves from uninfested trees. Because nitrogen and leaf phenolics covaried, it was not possible to determine from the results of this study whether the reduced growth was attributable to lower nitrogen or to allelochemicals. In addition, the feeding study was conducted for a period of only nine days, so the long-term effects of the feeding regime are not known.

Other studies of the effects of allelochemicals on the growth and development of the western tent caterpillar have yielded contradictory results. Myers and Williams (1984, 1987) found that western tent caterpillars fed leaves of red alder collected from trees under heavy attack from tent caterpillars grew as well as caterpillars fed on trees experiencing low attack or none at all. In one experiment (Myers and Williams 1987), the investigators tore leaves in half in situ; the next day they took the remaining half-leaf and an adjacent leaf from the tree and fed them to caterpillars. Growth of caterpillars on these leaves was not significantly different from that of caterpillars fed intact leaves collected from sites remote from the damage. Similar results were obtained when caterpillars were fed unattacked leaves immediately adjacent to leaves eaten by field populations or leaves from unattacked areas of the tree.

Moreover, the researchers found no evidence to support the hypothesis that trees under attack release a volatile, pheromone-like chemical (see Baldwin and Schultz 1983; Rhoades 1983, 1985; Fowler and Lawton 1985) that causes changes in the defensive chemistry of unattacked, near-neighbors sufficient to affect the growth of caterpillars.

Plant-Ant Symbiosis

In addition to chemical and physical defense mechanisms, plants may also defend themselves against herbivores by forming associations with predators of their enemies. Such symbiotic relationships have been described from tropical plants. Typically, in such associations, plants secrete a sugary fluid from extrafloral nectaries to attract nectivorous insects, usually ants. The ants, in turn, attack and dispatch small phytophages and even the large vertebrate herbivores they encounter while foraging for nectar. In one of best known of these associations, tropical *Acacia* trees not only secrete nectar from glands located near the base of the compound leaf but also provide their ants with proteinaceous bodies, which are harvested from the tips of leaflets. In addition, the hollow thorns of the tree are used by the ants for shelter. That the ants do indeed repay their host was demonstrated by fumigating an experimental group of trees to rid them of their ant colonies. Trees so treated were promptly attacked by herbivores, while nearby trees, still in possession of their ant symbionts, remained relatively unscathed (Janzen 1966).

The phenomenon of plant-ant symbiosis is less well known among species from more temperate latitudes, but the presence of extrafloral nectaries on a diversity of plants suggests that it may be more common than is presently known. Four genera of trees that serve as hosts for tent caterpillars—*Prunus, Crateagus, Quercus,* and *Populus*—have extrafloral nectaries (Wheeler 1910). Ants have been reported to be predaceous on forest tent caterpillars feeding on *Populus tremuloides* and on eastern tent caterpillars feeding on *Prunus virginiana* (Green and Sullivan 1950, Ayre and Hitchon 1968). In neither case, however, has it been demonstrated that ants are attracted to the trees because of their nectar production. But Tilman (1978) presented compelling evidence that the eastern tent caterpillar is both the raison d'être and the victim of an association between black cherry and the red-headed ant *Formica obscuripes.*

Foliar nectaries occur on the marginal teeth of the unfolding leaflets of black cherry. As many as 60–90 of the nectaries are active on each leaf soon after budbreak, but only 2–5 are still active after a week. Each

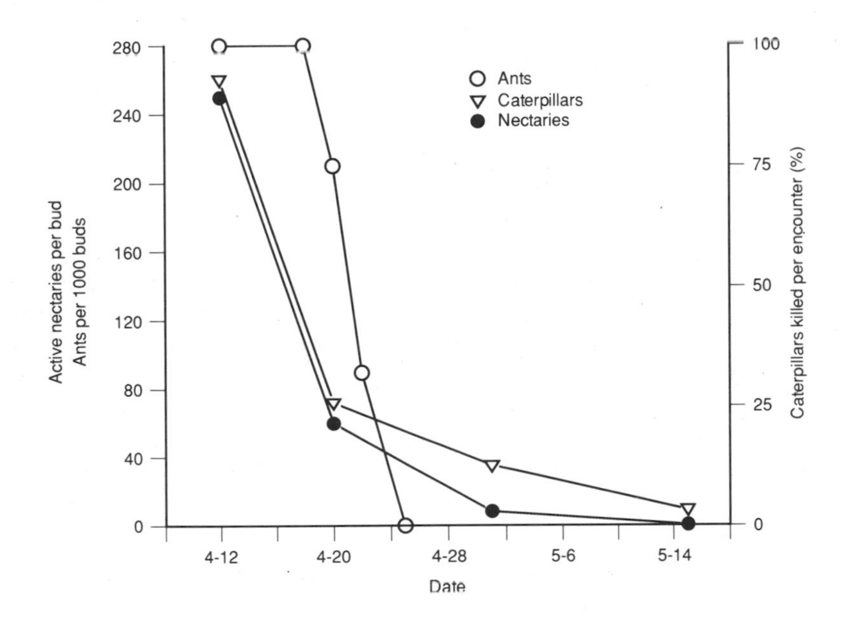

Figure 5.10. Relationship between the average activity of foliar nectaries of black cherry trees, the number of *Formica obscuripes* foraging on a tree located 1.3 m from their colony, and the proportion of encountered eastern tent caterpillars that are killed. After April 25, the caterpillars are too big to capture. (Modified from Tilman 1978.)

newly unfolding leaf follows a similar schedule, so that, for the tree as a whole, the number of active nectaries decreases gradually to less than 5% of the initial level in three weeks. Red-headed ants from nearby colonies are attracted to the nectaries and start to forage on the leaves soon after budbreak. The ants are predaceous on tent caterpillars and attack and kill the larvae they encounter. Because the amount of nectar secreted is small, an ant spends only a minute or so on a leaf before moving to the next one, and a colony may thoroughly scour the leaf surface of a small tree. When the caterpillars exceed about two times the size of the ants they become too formidable a prey, and ants no longer attempt to capture them. It takes approximately three weeks for the larvae to reach that size, so there is a remarkable correlation between the activity of plant nectaries and the susceptibility of the plant's major spring herbivore to ant predation (Fig. 5.10). It is unclear how effectively the ants serve as population-regulating agents, but Tilman found that survivorship of tent caterpillar colonies within 5 m of an ant nest was close to zero.

Host Range

Tent caterpillars are highly successful folivores, attacking a wide range of trees and shrubs (Tables 5.1 and 5.2). There are no monophagous tent caterpillars among the North American species, but *M. constrictum constrictum, M. constrictum austrinum,* and *M. tigris* oviposit only on various species of oak (*Quercus*). All other American *Malacosoma* have multiple ovipositional hosts, including at least one tree species in the family Rosaceae. Trees in the genus *Prunus* are the most common of all ovipositional hosts, and willow and poplar are also widely used (Stehr and Cook 1968; Table 5.1).

Most tent caterpillars, particularly in the later instars, feed on a much broader range of host species than those they select as ovipositional hosts (Stehr and Cook 1968): the caterpillars frequently defoliate their natal tree before they are fully grown, so they are forced to disperse and search elsewhere for food. Hodson (1941) noted, for example, that quaking aspen is the principal ovipositional host of the forest tent caterpillar in northern Minnesota, yet the dispersed caterpillars also attack, in order of preference, sugar maple, basswood, red oak, bur oak, paper birch, and American elm. One tree not attacked by the forest tent caterpillar is the red maple (*Acer rubrum*), for an unknown reason (Fig. 5.11). During outbreaks, the caterpillars may attack larch and even such garden crops as strawberries, cabbage, peas, beets, and potatoes. Fyre and Ramse (1975) also reported a broad feeding range for dispersed *M. disstria* populations in North Dakota. The tree species most heavily defoliated, *Tilia americanum* (basswood), *Fraxinus pennsylvanica* (green ash), and *Betula papyrifera* (paper birch), are not known to be ovipositional hosts of this species.

The eastern tent caterpillar, a species that oviposits almost exclusively on trees in the family Rosaceae, has been reported to feed on more than 50 species after dispersing from the natal tree (Tietz 1972). Fitzgerald and Peterson (1983) also reported that the newly hatched caterpillars of *M. americanum* will accept the young, partially unfolded leaves of some nonovipositional hosts, including *Vaccinium, Populus tremuloides,* and *Amelanchier arborea.*

One instance of host specificity that has puzzled researchers is the marked reluctance of populations of the eastern tent caterpillar, found in the northeastern United States, to oviposit on pin cherry, *Prunus pensylvanica.* In a survey of trees in infested mixed-species stands in Massachusetts, 58.4% of 471 black cherry trees had one or more egg masses of the eastern tent caterpillar, but only 2.3% of 686 pin cherry trees had egg masses. Several experiments have shown that the tree is capable of

Table 5.1. Documented ovipositional hosts of American species of *Malacosoma*

Ovipositional host	*M. disstria*	*M. constrictum constrictum*	*M. constrictum austrinum*	*M. tigris*	*M. americanum*	*M. californicum*	*M. c. californicum*	*M. c. ambisimile*	*M. c. recenseo*	*M. c. pluviale*	*M. c. lutescens*	*M. c. fragile*	*M. incurvum incurvum*	*M. i. discoloratum*	*M. i. aztecum*
Salicaceae															
Populus alba														X	
Populus angustifolia														X	
Populus fremontii													X	X	
Populus tremuloides	X									X					X
Populus grandidentata	X[a]														
Salix spp.	X[a]				X[b]		X	X		X	X		X	X	X
Salix lasiolepis															
Corylaceae															
Alnus sp.										X					
Alnus oregona										X					
Alnus rubra										X[c]					
Betulaceae															
Betula glandulosa										X					
Betula pumila										X					
Betula papyrifera										X[d]					
Fagaceae															
Quercus spp.	X	X	X	X			X								
Quercus borealis	X[a]														
Quercus nigra	X[e]														
Saxifragaceae															
Ribes spp.										X	X				
Ribes aureum											X				
Ribes cereum												X[f]			
Ribes velutinum												X[f]			
Hamamelidaceae															
Liquidambar styraciflua	X														
Rosaceae															
Amelanchier sp.										X	X				
Amelanchier utahensis												X[f]			
Cercocarpus ledifolius												X[f]			
Chamaebatiaria millefolium												X[f]			
Chaenomeles lagenaria					X[b]										
Cowania stansburiana												X[f]			
Crategus spp.					X										
Holodiscus spp.															X
Holodiscus dumosus												X[f]			
Malus spp.	X				X		X	X		X					
Paraphyllum ramosissimum												X[f]			

Table 5.1—*cont.*

Ovipositional host	*M. disstria*	*M. constrictum constrictum*	*M. constrictum austrinum*	*M. tigris*	*M. americanum*	*M. californicum*	*M. c. californicum*	*M. c. ambisimile*	*M. c. recenseo*	*M. c. pluviale*	*M. c. lutescens*	*M. c. fragile*	*M. incurvum incurvum*	*M. i. discoloratum*	*M. i. aztecum*
Prunus spp.	X				X		X	X		X	X	X	X	X	
Prunus andersonii												X			
Prunus capulis															X
Prunus emarginata										X					
Prunus fasciculata												X			
Prunus pensylvanica	X[a]				X					X[d]					
Prunus serotina[h]	X				X										
Prunus virens													X		
Prunus virginiana					X				X	X	X				
Purshia glandulosa												X			
Purshia tridentata									X	X		X			
Rosa spp.							X		X	X	X				
Rosa woodsii												X[f]			
Sorbus americana					X[b]										
Aceraceae															
Acer saccharum	X[g]														
Rhamnaceae															
Ceanothus cordulatus									X						
Ceanothus cuneatus										X					
Ceanothus incanus								X							
Ceanothus integerrimus									X						
Ceanothus thyrsiflorus								X							
Ceanothus velutinus									X						
Cornaceae															
Comus florida	X[e]														
Nyssaceae															
Nyssa aquatica	X														
Nyssa sylvatica	X														

Sources: Primarily from Stehr and Cook 1968. Additional sources: [a]Sippel 1957; [b]Fitzgerald, unpubl. observ. (only one or a few instances observed on each tree; [c]Adams 1989; [d]Atwood 1943; [e]Goyer et al. 1987; [f]Weaver 1986; [g]numerous references.

Note: The host range of most species is not limited to ovipositional hosts.

[h]Although primarily eastern in distribution, *P. serotina* is reported to occur in the mountains of Mexico (Little 1971), where it could serve as an ovipositional host for species whose ranges extend into those regions.

Table 5.2. Partial list of food plants of some European and Asian tent caterpillars

Species	Host plants	Reference
Malacosoma alpicolum	Polyphagous on herbaceous and woody plants including *Quercus, Salix, Rubrus, Rosa pimpinellifolia, Alchemilla, Polygonum bistorta*	de Freina & Witt 1987, Martin & Serrano 1984
M. castrensis	Polyphagous on herbaceous and woody plants including *Quercus, Betula, Salix, Euphorbia cyparissias, Limonium vulgare, Artemisia maritima*	Tutt 1900, de Freina & Witt 1987, Martin & Serrano 1984, Dreisig 1987
M. franconicum	Polyphagous on herbaceous plants, especially *Artemisia, Achillea, Plantago, Rumex*	de Freina & Witt 1987
M. indicum	Polyphagous, food plants include *Malus pumila, Prunus* spp., *Pyrus* spp., *Juglans regia, Populus, Salix, Rosa, Berberis, Cotoneaster, Quercus, Betula utilis*	Joshi & Agarwal 1987, Bhandari & Singh 1991
M. laurae	*Limonastrium monopetalum*	de Freina & Witt 1987
M. luteum	Polyphagous on herbaceous plants	de Freina & Witt 1987
M. neustrium spp.	Highly polyphagous on almost all trees and shrubs especially *Prunus, Quercus, Betula, Populus, Tilia, Pyrus*	Tutt 1900, Shiga 1976, de Freina & Witt 1987
M. paralellum	Polyphagous; principal food plants include *Prunus* spp., *Malus, Salix*	Li 1989

Note: The food plants of most of the Asian species have not yet been documented. The food plant range given here is likely to exceed the ovipositional host range.

supporting the growth of tent caterpillars, but the caterpillars grow more slowly than they do when reared on the leaves of a preferred host, black cherry (Segarra-Carmona and Barbosa 1983, Wagge and Bergelson 1985). Moreover, larvae allowed to choose between the two species clearly prefer black cherry over pin cherry, even when reared on pin cherry through the first two stadia. Thus, the leaves appear to lack appropriate feeding stimulants or have unknown deterrent properties. The foliar nitrogen levels of pin cherry are markedly lower than those of black cherry during the period when the caterpillars are active (Segarra-Carmona and Barbosa 1983). Despite pin cherry's marginal status as a host tree in the southern part of its range, the caterpillar appears to have adapted to the tree in Canada, where it is a common ovipositional host of not only *M.*

Figure 5.11. A stand of sugar maple (*Acer saccharum*) defoliated by an outbreak population of the forest tent caterpillar. The trees that still have leaves are red maples (*A. rubrum*), a species which is not attacked by tent caterpillars.

americanum but also *M. californicum pluviale* (Atwood 1943; Prentice 1963). One possible reason may be that more favorable host species are relatively rare in the northern regions where *P. pensylvanica* is both widely distributed and abundant.

Although the nutritional value of leaves is of undoubted importance in the establishment of host preferences, the ability of tent caterpillars to feed on a broad range of distantly related plant species suggests that primary, or ovipositional, hosts are not selected solely because they have favorable nutritional qualities. A necessary feature of the natal tree is that it produce leaves that are acceptable to the newly eclosed caterpillars, regardless of its ability to support the growth of caterpillars later on. The newly unfolded leaves of pin cherry, for example, are particularly sticky and impede the feeding of young caterpillars (Wagge and Bergelson 1985). Hodson (1941) reported a similar situation, noting that although paper birch, *Betula papyrifera*, may be completely defoliated later in the season by forest tent caterpillars, the young larvae cannot move about on the sticky surface of the young leaves, and the tree does not serve as an ovipositional host. Robison (1993) found that the stickiness of the buds and new leaves of hybrid poplars varied markedly among clones. Larvae that he placed on the expanding buds of particularly sticky clones were unable to free themselves during a 24-hour observation period.

Adult *Malacosoma* also appear to select as ovipositional hosts trees that are among the earliest to produce leaves in the spring. Thus, even though the forest tent caterpillar feeds on both poplar and oak in northern Minnesota, it prefers poplar over oak as an ovipositional tree, possibly because the leaves of the latter do not appear until a week or more after the flushing of aspen (Hodson 1941). Sippell (1957) noted that although big-toothed aspen, *Populus grandidentata,* can serve as an ovipositional host for the forest tent caterpillar, it typically does not leaf out until several weeks after the caterpillars eclose in synchrony with their preferred host *P. tremuloides.* During the 1952 season, caterpillars feeding on *P. tremuloides* were in their third instar when *P. grandidentata* began to produce its first leaves. Eclosion of the eastern tent caterpillar in central New York is also synchronized with the host species that is the earliest to produce leaves. In fields where black cherry and apple co-occur, eclosion corresponds to the phenology of the cherry tree, even though the flushing of cherry leaves sometimes precedes the flushing of apple leaves by as much as two weeks (Fitzgerald and Willer 1983).

On apple, prematurely eclosed colonies are highly active and often lay down an extensive trail system in a futile search for leaves (Fig. 5.12). They are able to obtain a small quantity of nutrient from partially expanded buds, and they imbibe water. They usually survive in this fashion until the host leafs out, though some mortality is likely. Variability in the date of leaf-flush among the many varieties of apple may explain the retention of this species as an ovipositional host. Hawthorn (*Crataegus*) may also serve as a host for the tent caterpillar in these same areas, but egg masses are rarely found on this species. Hawthorn is typically among the latest of the trees to produce leaves, lagging considerably behind both cherry and apple.

Caterpillars that begin their larval life on a tree species that is among the earliest to produce leaves in a given locale maximize the probability of finding young, palatable leaves on other tree species later on if they are forced to disperse from the natal tree. But the relationship between the phenology of leafing among tree species and the average date of caterpillar eclosion in any given area is known for only a few species of

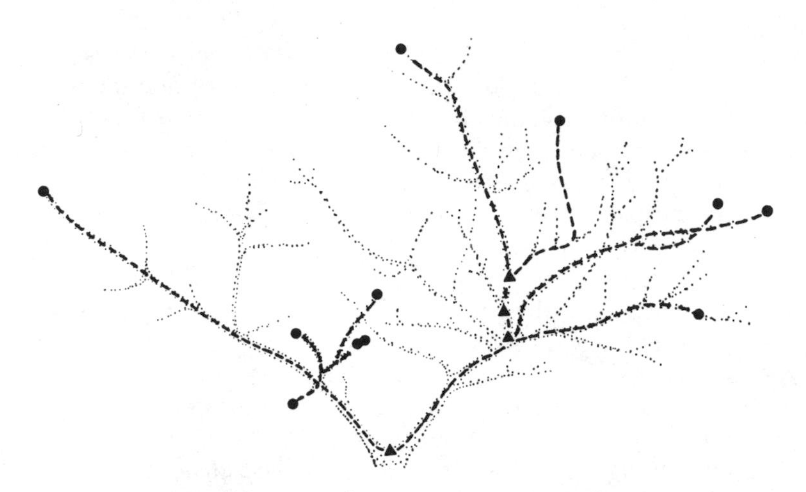

Figure 5.12. Trunk trails (heavy dashed lines) laid down in the crown of an apple tree by 11 colonies of unfed *Malacosoma americanum* that eclosed 10 days before leaf flush. Filled circles = egg masses, filled triangles = sites where tents were eventually constructed. (Reprinted, by permission of the Kansas Entomological Society, from Fitzgerald and Willer 1983.)

tent caterpillar, so much remains to be learned about this aspect of host selection.

It is not known how tent caterpillars synchronize eclosion to the phenology of their ovipositional hosts. Like their host trees, they may measure degree-days to assure synchrony. There is currently no indication that the caterpillars are able to monitor the state of the host from within the egg. Caterpillars, however, may spend several days chewing through the chorion before eclosing, and conceivably they could sample airborne volatiles emanating from the newly unfolding leaves during that period, delaying or accelerating their emergence from the egg accordingly.

6

Aggregation and Foraging Behavior

Of all the characteristics of an animal's behavior, those associated with reproduction and feeding contribute most directly to its fitness. When measured against the duration of a life span, reproductive acts typically take little time, but animals are preoccupied with feeding and they have evolved an endless variety of foraging behaviors to meet their continual demand for energy. The foraging patterns of caterpillars are characteristically species-specific and stereotypic and are shaped most often by adult ovipositional patterns, the temporal and spatial characteristics of the food supply, pressure from predators and parasitoids, and daily and seasonal temperature trends. These selective forces have led tent caterpillars to establish temporary or permanent bases from which they launch cooperative forays in search of food. Like ants and termites, some species have been shown to use trail-based systems of elective recruitment communication to direct hungry tentmates to profitable feeding sites.

Foraging Patterns

Although the details of their foraging behaviors vary considerably, social caterpillars may be broadly classified as patch-restricted foragers, nomadic foragers, or central-place (fixed-base) foragers (Fitzgerald and Peterson 1988, Fitzgerald 1993a). The basic differences among these foraging patterns are illustrated in Figure 6.1 and discussed below.

Colonies of patch-restricted foragers obtain most of the food required for their larval development from the leaves found in a single contiguous

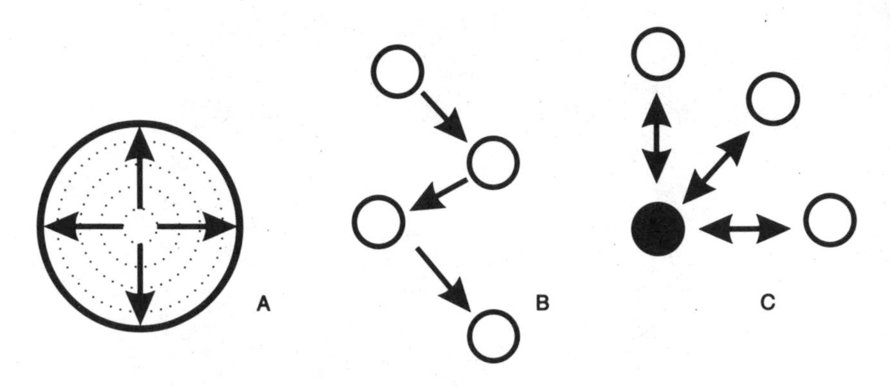

Figure 6.1. (A) Patch-restricted, (B) nomadic, and (C) central-place foraging patterns of social caterpillars. Arrows show how the larvae move (A) within patches, (B) between patches, and (C) between the resting site (solid) and feeding sites. (After Fitzgerald and Peterson 1988.)

patch. The caterpillars typically envelop leaves or whole branches of the host tree in silk and feed from within the envelope. Because they are in constant contact with their food supply, they are likely to require only rudimentary forms of communication to facilitate aggregation and group foraging. Chemical and tactile cues associated with the silk envelope may suffice to define the foraging arena and to hold the colony together.

Patch-restricted foraging is the most common feeding pattern found among other social caterpillars (Fitzgerald 1993a), but it has yet to be described from the tent caterpillars. Details of the foraging patterns of most species of *Malacosoma* are sketchy, or altogether lacking, but of those whose behavior is known, none has been reported to limit its feeding to a single contiguous patch. Demand by colonies of tent caterpillars for large quantities of high-quality food is more likely to favor colony mobility and the evolution of nomadic or central-place foraging patterns.

The only tent caterpillar known to utilize a nomadic foraging strategy is the forest tent caterpillar, *M. disstria*. The larvae of this species are unrestrained by tents or webs and move frequently to new and distant locations, often after only partially depleting a feeding site (Table 6.1, Fig. 6.2). Although they are the most economically important of the North American tent caterpillars and have been studied extensively, their foraging ecology has been documented in only a few instances, and the adaptive significance of their nomadic behavior is not clear. Their nomadic habits may be a response to marked temporal and spatial variation in the quantity and quality of their food supply, to the patchiness

Table 6.1. Nomadic foraging pattern of the first four instars of nine field colonies of *Malacosoma disstria* feeding on sugar maple

Instar	No. of feeding sites	Distance between feeding sites ± SE (cm)	No. of leaves attacked	Mean percentage of leaf consumed ± SE
1	3.3 ± 0.3	45.2 ± 10.7	9.5 ± 0.9	38.4 ± 3.3
2	4.7 ± 0.8	115.1 ± 20.2	15.3 ± 2.7	46.7 ± 3.4
3	6.8 ± 1.3	130.0 ± 38.2	20.7 ± 4.1	48.0 ± 3.1
4	—	145.1 ± 22.4	—	52.0 ± 4.6

Source: From Fitzgerald and Costa 1986, with permission of the Entomological Society of America.
Note: — = Not measured.

of sunlight in the forest habitat, or to contagious diseases that fester in debilitated siblings, but other as yet undocumented factors may also be involved.

Central-place foraging appears to be the most common form of foraging pattern utilized by tent caterpillars. With the exception of *M. disstria,* all American species of *Malacosoma* are, to a greater or lesser extent, central-place foragers. The Eurasian species *M. neustrium, M. indicum, M. parallelum, M. castrensis,* and *M. alpicolum* also launch forays from a permanent or semipermanent gathering site, to which they return between successive bouts of feeding. Some species of central-place foragers, such as *M. americanum,* typically construct a single tent that is used as a foraging center for the life of the colony. Others such as *M. tigris, M. constrictum, M. castrensis,* and *M. alpicolum* periodically abandon their tents, establishing a series of new tents during the lifetime of the colony. The full extent and adaptive significance of variants of the central-place mode remain largely unknown.

The caterpillars of some genera of the Lasiocampidae other than *Malacosoma* exhibit various degrees of gregariousness and group foraging. Colonies of substantial size have been reported for caterpillars in the genera *Gloveria* and *Eutachyptera* among the American species and *Eriogaster* among the Old World species. Little is known of the extent to which these species forage cooperatively or of the role of communication in the organization of caterpillar behavior. Some species in these genera are central-place foragers and may have systems of recruitment communication similar to those of the tent caterpillars. Our knowledge of these species is rudimentary, and at the present time *Eriogaster lanestris* is the only lasiocampid other than the tent caterpillars that has been shown to utilize a chemical trail system (Weyh and Maschwitz 1978).

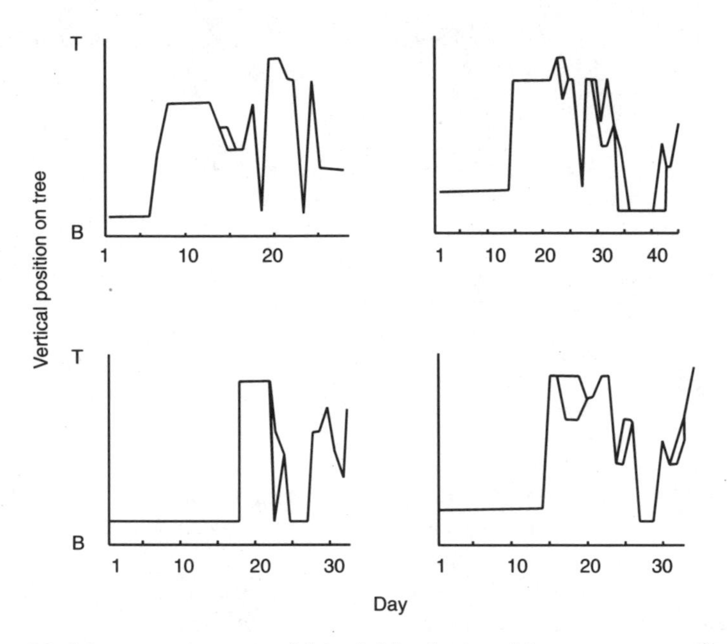

Figure 6.2. Movement patterns of four field colonies of forest tent caterpillars on 4–8 m tall maple trees. Lines that diverge or converge at a common point indicate colony fragmentation and regrouping, respectively. During periods of colony fragmentation, groups of caterpillars were separated by a mean distance of 136 cm. An interconnecting trail system enabled the fragments to regroup even after several days of separation. T, B = top and bottom of tree, respectively. (Redrawn, by permission of the Entomological Society of America, from Fitzgerald and Costa 1986.)

Group Size and Foraging Success

Studies of several species of caterpillars have shown that both the survivorship of caterpillars in colonies and the growth rate of individuals are influenced by the size of the colony (Fitzgerald 1993a). The size of a sibling cohort is ultimately limited by the number of oocytes that a single female can bring to maturation, though larger, multicolony aggregates are not uncommon among caterpillars. As E. O. Wilson (1975) pointed out, the modal number of insects in colonies is a compromise value determined by the various selective forces that shape the genetic structure of the population. For tent caterpillars, the factors that bear on survivorship and growth that would appear to favor larger colony size include the need for caterpillars to search cooperatively for food and, possibly, when small, to overcome tough plant surfaces; the need to spin sufficient silk to provide adequate footing on smooth plant surfaces; and

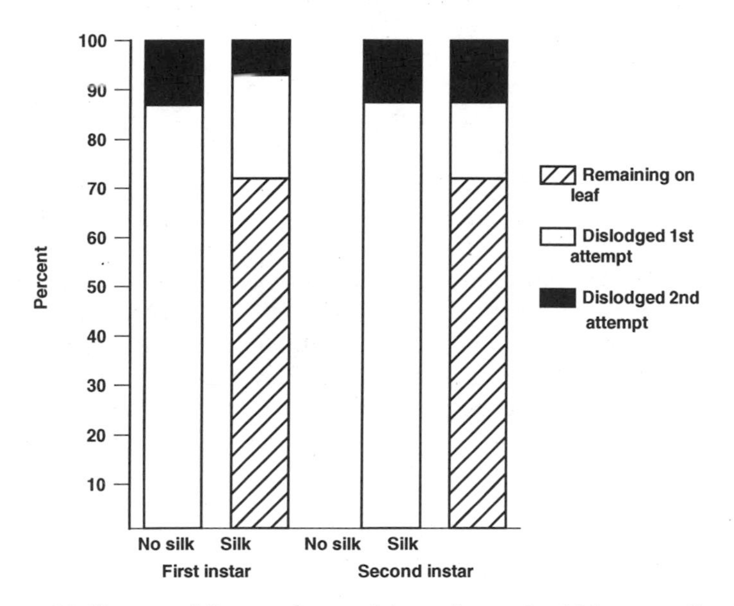

Figure 6.3. Percent of first- and second-instar larvae by *Malacosoma disstria* dislodged from silked and unsilked leaves of poplar by tapping the stem. (Redrawn, by permission of D. J. Robison, from Robison 1993.)

the need to construct a tent or mat to facilitate molting, thermoregulation, and defense against predators and parasitoids. Factors that may favor females that produce smaller broods include the greater conspicuousness of large colonies, demand for energy that may exceed the food supply provided by the natal tree, and the increased likelihood of infection and transmisson of disease in large colonies.

Although studies to assess the effect of group size on the survival and growth of tent caterpillars have all suffered to some extent from small sample sizes, there is general agreement that larger aggregates of tent caterpillars do better than smaller groups. Studies of *M. neustrium* by Sedivy (1978) and Shiga (1979) and of *M. disstria* by Robison (1993) indicate that under laboratory conditions first-instar caterpillars kept in groups of fewer than five individuals feed less, grow more slowly, and have lower survival rates than groups of five or more caterpillars. The likelihood of survival is particularly low for caterpillars maintained in isolation. Robison attributed much of the early mortality of isolated individuals to their inability to spin adequate silk to provide steadfastness on the plant (Fig. 6.3). He observed that solitary individuals placed on leaves that had been presilked had higher survival rates and grew faster than caterpillars placed on unsilked leaves. Moreover, my own observations of eastern tent caterpillars separated from their colony show that the isolated caterpillars are prone to wander extensively, in an apparent attempt to reestablish contact with siblings, and are more likely than their aggregated siblings to fall from the plant.

The number of caterpillars needed to assure colony success appears to be much higher under field conditions than in the laboratory. On the basis of both studies of colonies of *M. neustrium* established under field conditions and studies of naturally occurring colonies, Shiga (1976b, 1979) concluded that caterpillars in colonies with fewer than 200 individuals had lower survival and growth rates than caterpillars in larger colonies (Fig. 6.4). Smaller colonies constructed less adequate tents and trail systems and appeared to be more vulnerable to predators. Moreover, small groups of first-instar caterpillars appeared to have difficulty feeding on the swelling buds of the tree, which provided the only source of energy at the time of eclosion.

Exploration and Recruitment Trail Marking

Studies of *M. disstria, M. americanum,* and *M. neustrium* indicate that the foraging behavior of tent caterpillars is largely governed by trail-based systems of chemical communication (Fitzgerald and Edgerly 1979b, Fitzgerald and Costa 1986, Fitzgerald and Peterson 1988, Peterson 1988). The larvae of these species make synchronized, mass movements between resting sites and distant feeding sites and have evolved systems of chemical communication that facilitate group foraging.

Two distinct types of chemical trails are employed by tent caterpillars: exploratory trails and recruitment trails. Hungry larvae searching for food lay down exploratory trails. Exploratory trails allow colony mates to follow each other without physical contact or temporal synchrony. They also enable central-place foragers to find their way back to the resting site after feeding. Studies of *M. disstria* and *M. neustrium* indicate that exploratory trails are marked primarily by caterpillars in the vanguard. A forager is less likely to reinforce a trail that has been recently marked by other caterpillars. Exploratory trails become less effective with time, and the caterpillars readily distinguish old and new deposits that differ in age by only a few hours.

Recruitment trails are laid down by caterpillars immediately after feeding. In *M. americanum* colonies, the first caterpillar to finish feeding overmarks the exploratory trails it follows with a recruitment pheromone as it returns to the tent. Additional caterpillars that have fed at the same food find follow the precise pathway of the first returning caterpillar and rein-

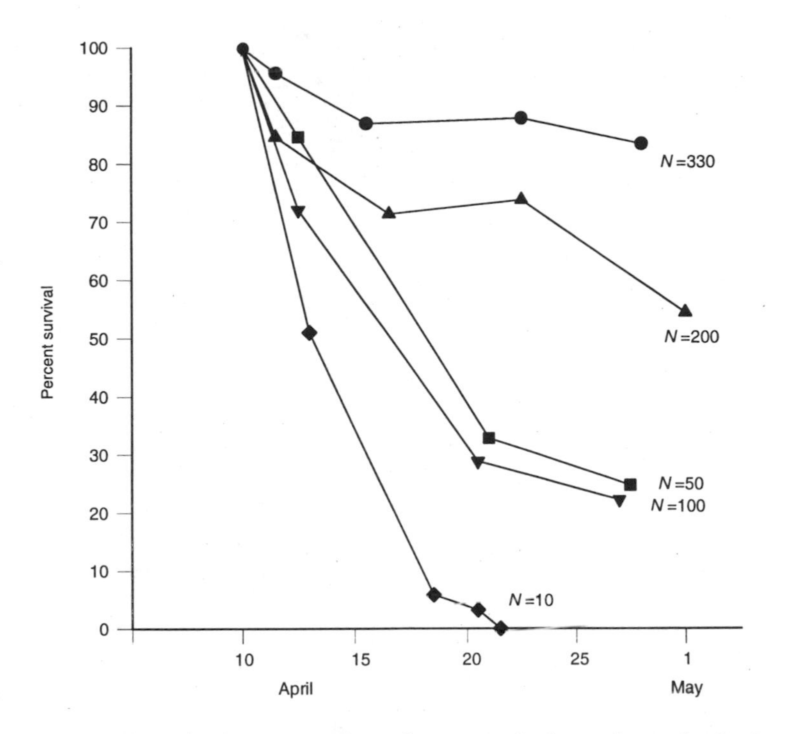

Figure 6.4. Effect of colony size (*N*) on the survival of experimental colonies of *Malacosoma neustrium* maintained in the field. (Data from Shiga 1976b.)

force the trail as they proceed. Thus, these overmarked trails stand out against the background of the exploratory trail system and are followed preferentially by both hungry and sated caterpillars.

Recruitment trails of tent caterpillars have thus far been shown to have two distinct functions (Fig. 6.5). The trails of *M. americanum* and *M. neustrium* serve to lead hungry tentmates to food finds (Fitzgerald 1976, Peterson 1988). The trails of *M. disstria* facilitate aggregation at a new resting site, established immediately after eating at a distance from the feeding site (Fitzgerald and Costa 1986).

Tent caterpillars deposit both exploratory and recruitment trails by lowering and brushing the venter of the last abdominal segment against the substrate (Fig. 6.6; Fitzgerald and Edgerly 1982). The source of the pheromone is unknown, but the material accumulates on the surface of the cuticle and can be collected by wiping the folded edge of a sheet of filter paper across the venter (Fig. 6.7). Scanning electron microscopy of the region has revealed no obvious openings, suggesting that the material may be the product of epidermal secretory cells.

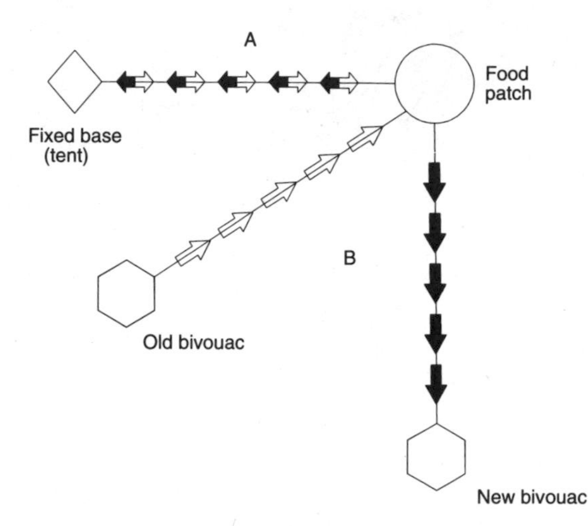

Figure 6.5. Trail marking by (A) *Malacosoma americanum* and (B) *M. disstria*. Unfilled arrows = exploratory trail marking, filled arrows = recruitment trail marking. Recruitment trails of *M. americanum* draw tentmates to food finds. The trails of *M. disstria* lead fed siblings to a new aggregation site. (Based on Fitzgerald and Costa 1986.)

Although silk is a prominent feature of the trails of tent caterpillars, studies indicate that silk in itself does not elicit a trail-following response, though it may enhance the effectiveness of the chemical trail (Fitzgerald and Edgerly 1979a). Thus, the basis of trail marking by tent caterpillars differs from that of larvae of the range caterpillar, *Hemileuca oliviae* (Capinera 1980); the scarce swallowtail, *Iphiclides podalirius* (Weyh and Maschwitz 1982); the ermine moth *Yponomeuta cagnagellus* (Roessingh 1989, 1990); and the ugly nest caterpillar, *Archips cerasivoranus* (Fitzgerald 1993c)—all species in which the pheromone has been demonstrated to be a component of the silk itself.

Little is yet known of the chemical bases of communication in tent caterpillars, but the steroids 5β-cholestane-3,24-dione and 5β-cholestane-3-one have been shown to be major components of the trail of *M. americanum* (Fig. 6.7). Chemical analysis of the trail pheromone of *M. disstria* indicates that the insect produces only 5β-cholestane-3-one, but also responds to 5β-cholestane-3,24-dione (Fitzgerald and Webster 1993). Although there has been no attempt to analyze the trail pheromone of *M. neustrium*, bioassays show that the caterpillar follows trails prepared from 5β-cholestane-3,24-dione (Crump et al. 1987, Peterson 1988). All three species follow trials prepared from the synthetic compounds dissolved in hexanes at levels as low as 10^{-11} g/mm of trail (Fig. 6.8). The

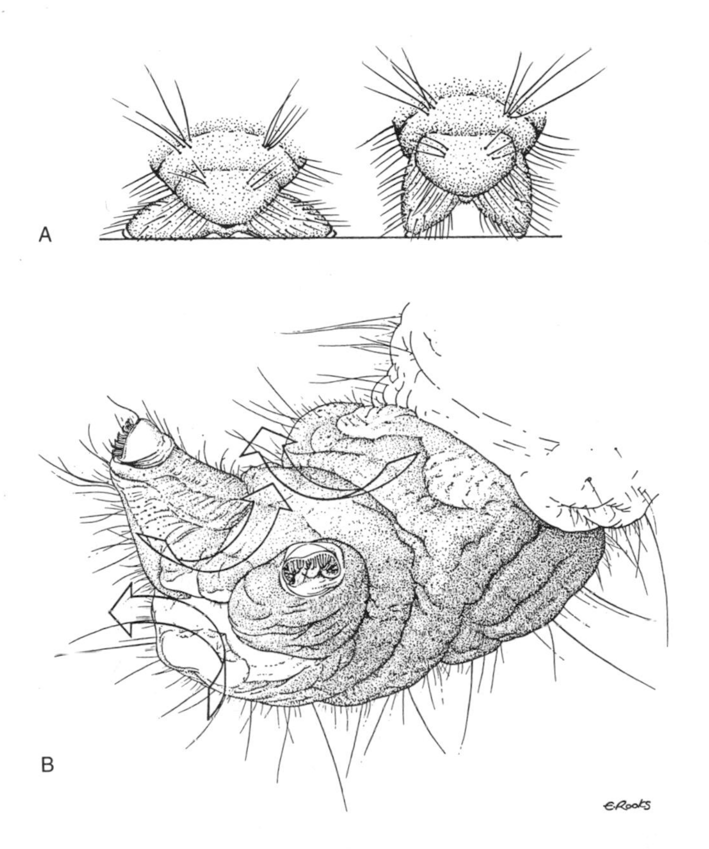

Figure 6.6. (A) Position of the abdomen of the larva of a tent caterpillar when marking trails (left) and not marking (right). (B) Ventral surface of the tip of the abdomen of *Malacosoma americanum*. Trail pheromone can be collected by wiping the folded edge of a sheet of filter paper across the region indicated by the center arrow but not across the regions indicated by the other arrows. (Drawing B by Edward Rooks, used with permission of the American Entomological Association.)

threshold sensitivity of the caterpillars to the pheromone is likely to be at least an order of magnitude lower because, when the marker is applied in a solvent, much of the material is lost below the surface, where it cannot be detected by the contact chemoreceptors of the maxillary palps. These are the only trail pheromone components thus far identified from the Lepidoptera.

Studies of the trail-following behavior of larval *M. americanum* and *M. disstria* indicate that caterpillars choose stronger trails over weaker trails when foraging. The larvae are highly sensitive to the amount of phero-

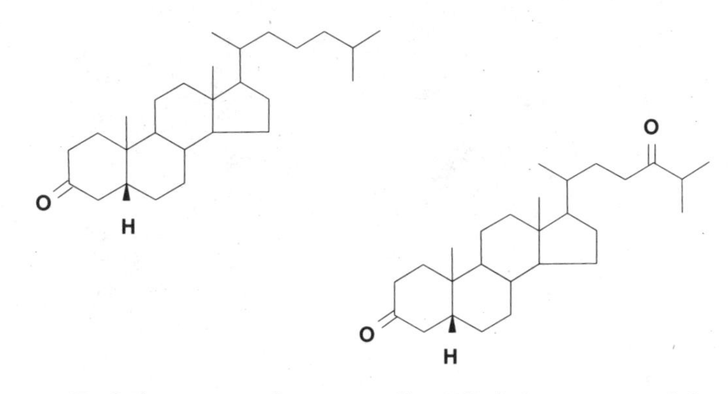

Figure 6.7. Trail pheromones of tent caterpillars, 5β-cholestane-3-one (left) and 5β-cholestane-3,24-dione.

mone present in trails and carefully compare trails by sweeping their heads from trail to trail at branch junctures. The eastern tent caterpillar is able to distinguish artificial pheromone trails differing by only one billionth of a gram per millimeter of trail, and the forest tent caterpillar is only slightly less sensitive (Peterson and Fitzgerald 1991, Fitzgerald 1993b, Fitzgerald and Webster 1993). Caterpillars abandon stronger trails in favor of weaker ones or unmarked substrate only after the stronger trails have been followed repeatedly and proven unprofitable.

Because the trail pheromone of the forest tent caterpillar apparently consists of only the single component 5β-cholestane-3-one, the caterpillars probably use more of the material to mark recruitment trails and less to mark exploratory trails. The significance of there being two active compounds in the trail pheromone of the eastern tent caterpillar remains speculative. The more active compound is 5β-cholestane-3-one, and both laboratory and field studies have shown that it is fully competitive with authentic exploratory and recruitment trails. The caterpillars also readily follow artificial trails prepared from 5β-cholestane-3,24-dione, but they will not move off their authentic recruitment trails onto trails prepared from the chemical unless it is applied at abnormally high levels. It is consistent with what is presently known of the trail-following behavior of the eastern tent caterpillar that the larvae use diketone to mark exploratory trails and monoketone to mark recruitment trails. It is unclear, however, how they do it. It may be that the two chemicals are stored and secreted independently, but it is also plausible that a single compound is secreted and subsequently converted to the other. Although the monoketone does not appear to be capable of autoxidizing to the diketone, enzymatic activity at the secretory site could act to convert a

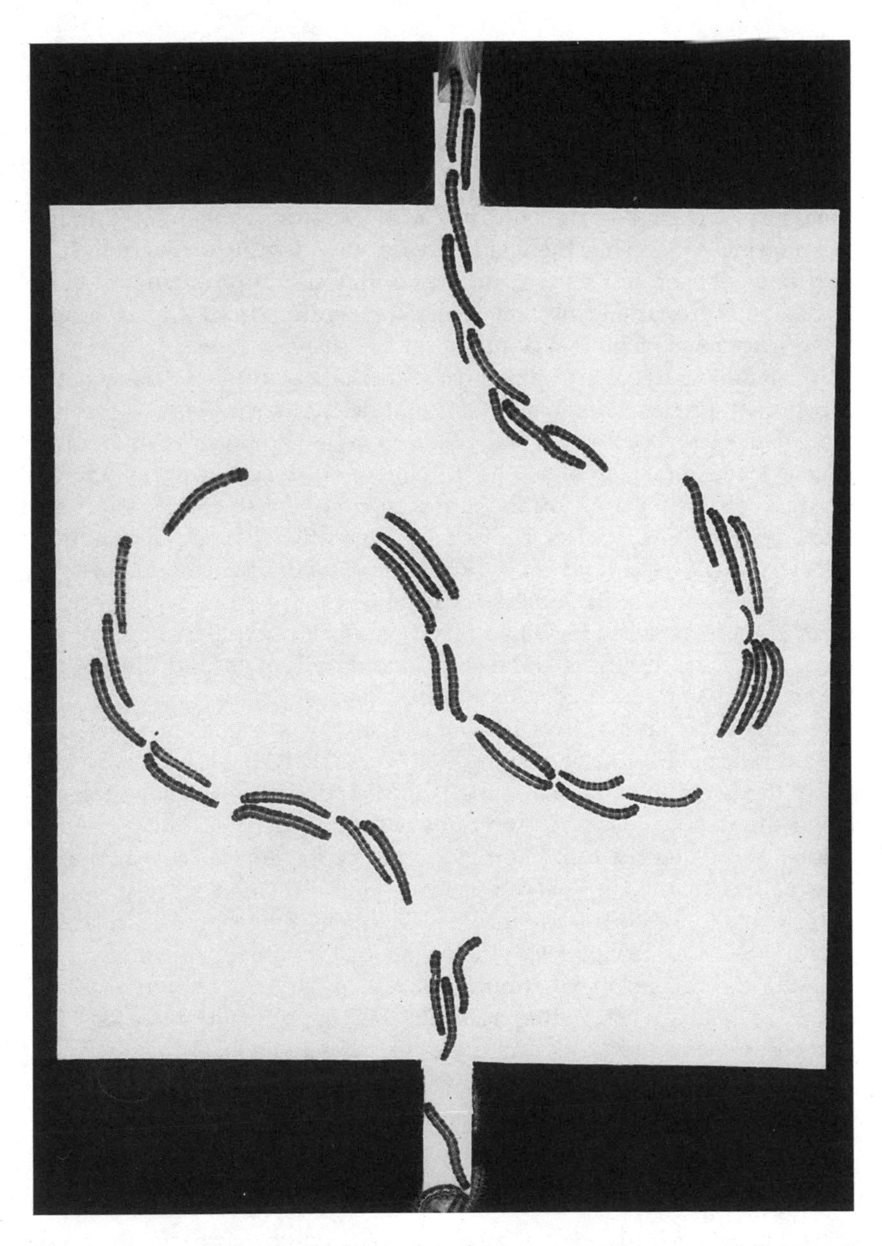

Figure 6.8. Response of *Malacosoma americanum* caterpillars to an artificial trail prepared from 5β-cholestane-3,24-dione laid out on a card positioned between their tent and feeding site.

surface residue of the material to the diketone. There would be adequate time for this conversion to occur; the fed caterpillars typically rest for six hours after secreting the recruitment marker. During their next activity bout the caterpillars could mark exploratory trails by brushing the diketone from the surface of the cuticle and actively secreting monoketone only to mark recruitment trails as they return to the tent. Though this scheme is plausible, there is at present no scientific evidence to support it. Moreover, our current understanding of the physiological basis of bilevel trail marking by tent caterpillars suffers from a total lack of knowledge of the histological nature of the secretory site.

It remains a possibility that host chemicals also play a role in the recruitment process. Hungry eastern tent caterpillars milling on the surface of the tent become excited when volatile components of crushed cherry leaves are wafted at them. Moreover, if an extract of freshly crushed leaves is placed on the surface of the tent, the caterpillars are attracted to the site from short distances. Recently fed caterpillars almost certainly carry fresh residues of host leaf juices on their mouthparts, so host volatiles may play some, as yet undiscovered, role in the recruitment process, possibly serving to excite hungry caterpillars or to attract them over short distances to the recruitment trails of successful foragers. Because the steroidal trail pheromones have very low volatility, they cannot by themselves attract the caterpillars to the trail from a distance.

Tent caterpillars are superb trail followers. They are able to sense precisely the lateral boundaries of their linear trail, and they move rapidly along the pathway with a minimum of lateral displacement. Ablation studies show that the trail pheromone receptor is located on the maxillary palpus (Fig. 3.10; Roessingh et al. 1988). Bilateral ablation of the palps leads to loss of the caterpillars' ability to follow trails. Unilateral ablation leads to pronounced circling toward the intact side, indicating that the caterpillars make simultaneous comparisons of trail strength with both palps when following trails (Fig. 6.9). But the receptors can also be used as a unit, allowing the caterpillars to make sequential comparisons of trail strength at trail junctures by swinging their heads from side to side (Peterson and Fitzgerald 1991).

The foraging behavior of *M. americanum* has been studied more extensively than that of any other tent caterpillar. A flow diagram summarizing major aspects of the trail-based foraging system of this species is shown in Figure 6.10. The colony has discrete bouts of activity, the onset of each bout is marked by the mass emergence of the caterpillars from the tent. The caterpillars rarely leave the tent immediately, but move about the surface laying down strands of silk that serve to expand the structure. This milling about allows time for all the caterpillars to emerge

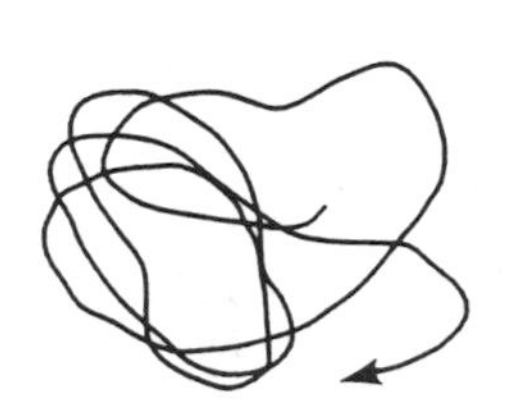

Figure 6.9. Pathway of a larval *Malacosoma americanum* following ablation of the right (above) or the left (below) maxillary palp. The caterpillar turns repeatedly toward the side with the functional palp. (Reprinted, by permission of Plenum Publishing, from Peterson and Fitzgerald 1991.)

and assemble on the surface. It is likely, though undemonstrated, that it is this activity on the tent surface, rather than a coincidence of biological clocks or states of hunger, that brings the siblings to assemble en masse.

The caterpillars at first show marked hesitancy to leave the tent, often venturing only short distances onto the main trunk trails before turning back. This reluctance is due largely to the fact that the trails have not been reinforced with chemical marker since the last bout of activity and are only marginally effective in eliciting trail following. In time, the caterpillars grow restless and show an increased readiness to leave the tent. Indeed, if all pathways are experimentally blocked, trapping the colony on the tent for an hour or so after the onset of an activity bout, the caterpillars immediately stream off the tent when the blocks are removed.

Although the caterpillars may leave the tent en masse, particularly during their evening foraging bout (Plate 2D), a few caterpillars, possibly the hungriest, often leave the tent in advance of their siblings (Fitzgerald 1980). After feeding, the advanced contingent of foragers returns to the structure, laying down recruitment trails as they proceed. The recruitment marker is deposited on the branches of the host tree and is continued onto the surface of the tent. There the trail is encountered by caterpillars that have not yet initiated their bout of foraging or by those that have returned to the tent after having failed to find adequate food. Thus, the tent serves not only as shelter and as a staging area, but also as a communication center, where hungry caterpillars are alerted to the discovery of food.

There is often insufficient foliage to provide a complete meal for all

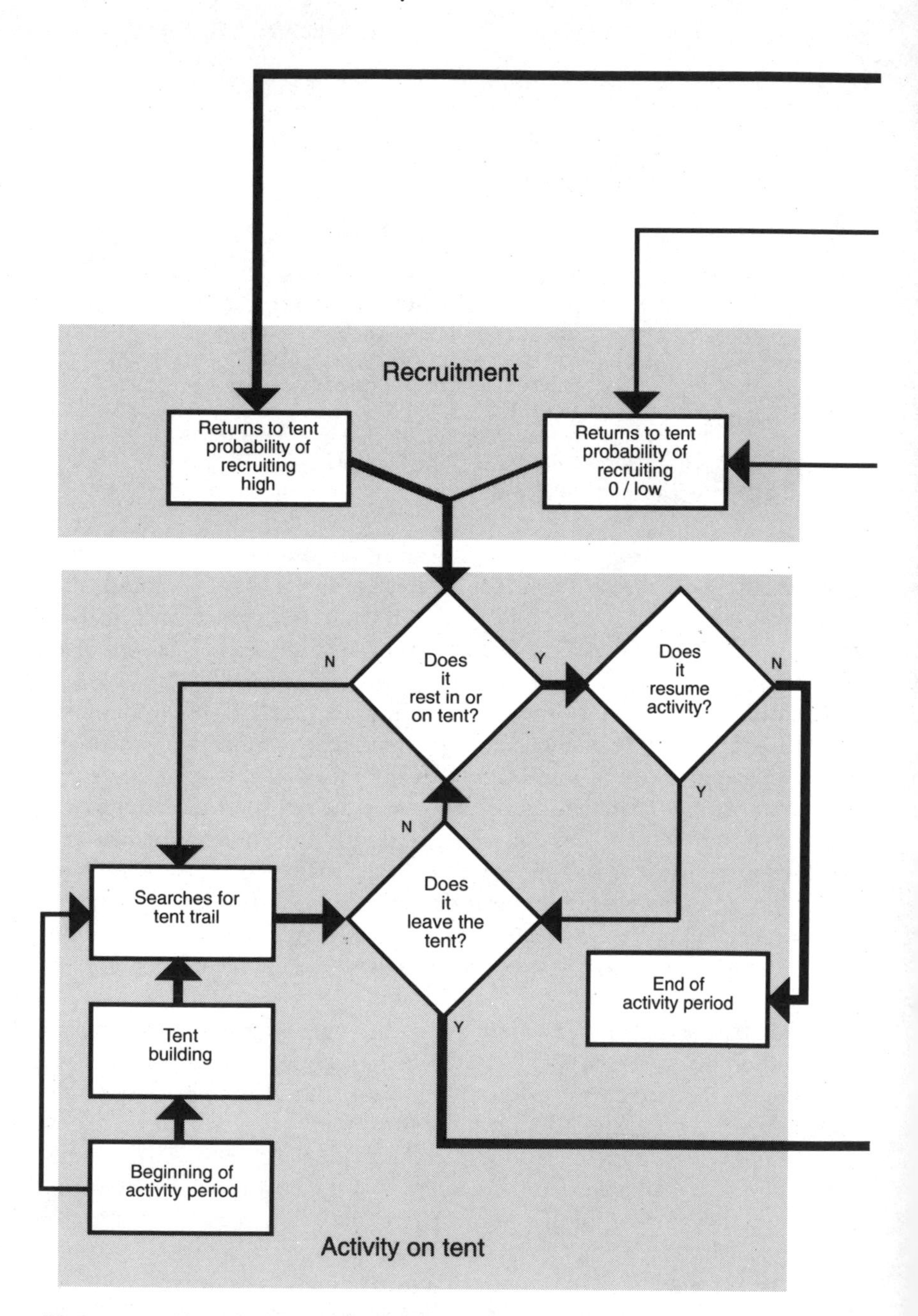

Figure 6.10. Flow diagram of the foraging behavior and trail-based communication system of *Malacosoma americanum*. Heavy lines = most common pathways. (Redrawn, by permission of the American Institute of Biological Sciences, from Fitzgerald and Peterson 1988, © 1988 American Institute of Biological Sciences.)

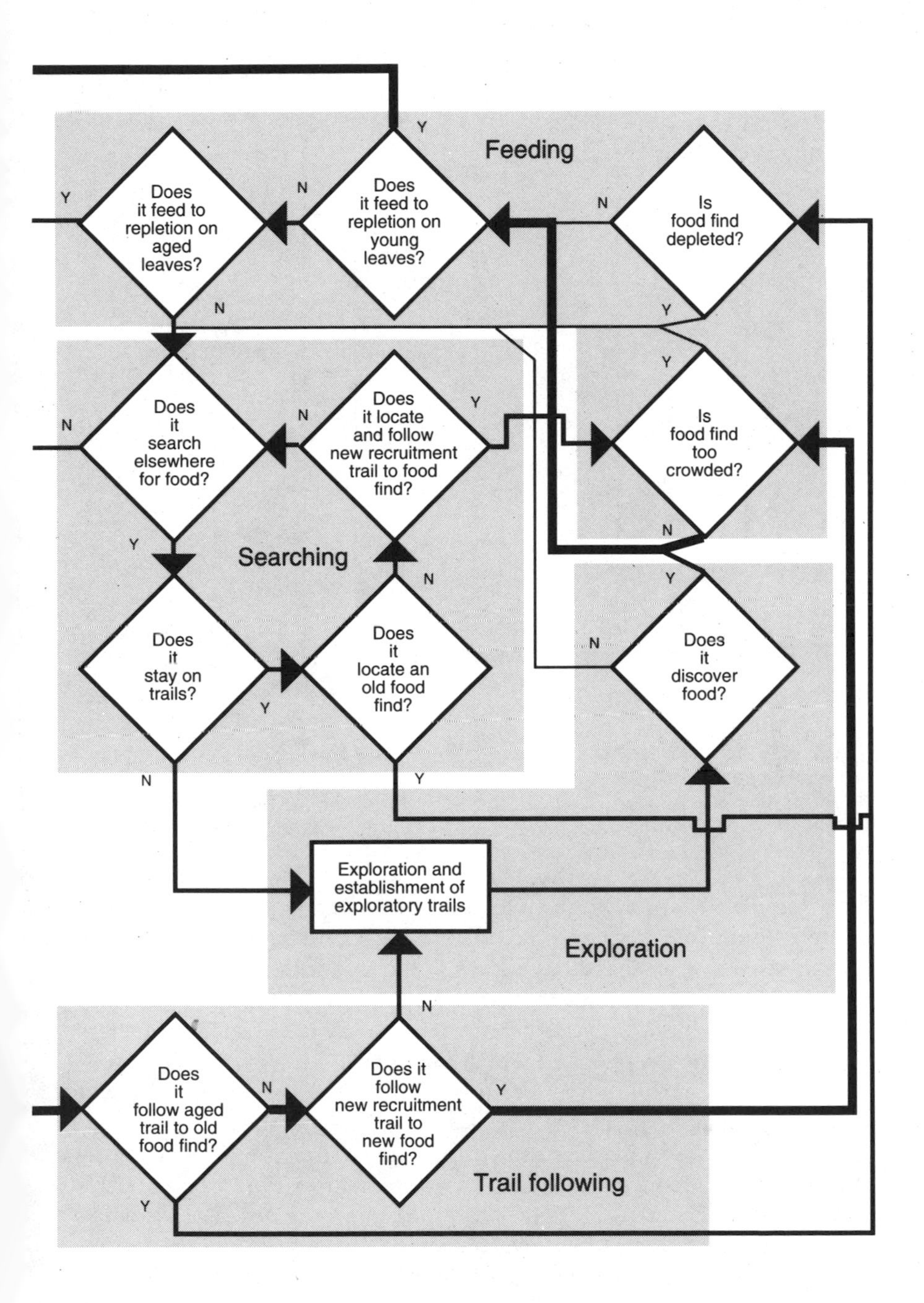

Feeding
Does it feed to repletion on aged leaves?
Does it feed to repletion on young leaves?
Is food find depleted?
Is food find too crowded?
Searching
Does it search elsewhere for food?
Does it locate and follow new recruitment trail to food find?
Does it stay on trails?
Does it locate an old food find?
Does it discover food?
Exploration and establishment of exploratory trails
Exploration
Does it follow aged trail to old food find?
Does it follow new recruitment trail to new food find?
Trail following
Y
N

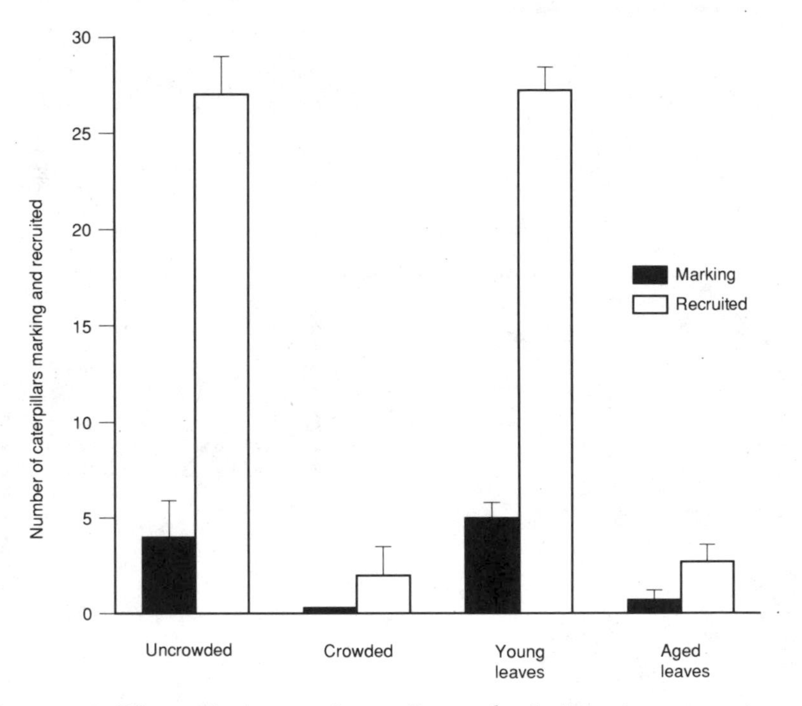

Figure 6.11. Effect of leaf age and crowding at the feeding site on recruitment behavior of larval *Malacosoma americanum.* (After Fitzgerald and Peterson, *Animal Behaviour* 31[1983]:417–442, published for the Association for the Study of Animal Behaviour and the Animal Behaviour Society by Academic Press Ltd.)

the caterpillars that converge on a feeding site during an activity period, and crowding may prevent the caterpillars from settling. Under such conditions, the caterpillars may return to the tent after having eaten only partially; they may continue to search for food by following older trails, or they may explore previously unmarked branches. The last of these possibilities is the main way in which the caterpillars establish new feeding sites. The probability that a caterpillar will strike off on its own over unmarked substrate increases with time since its last meal.

Eastern tent caterpillar recruitment communication is an elective process, contingent upon an individual's assessment of food quality (Fitzgerald and Peterson 1983). Fewer caterpillars are recruited to crowded than to uncrowded sites because caterpillars usually do not feed to repletion at crowded sites and do not mark trails to them. In addition, caterpillars that feed on aged leaves of host trees or on young leaves of nonhost trees recruit fewer tentmates than do caterpillars that feed on the young leaves of host trees (Figs. 6.11, 6.12). This difference occurs

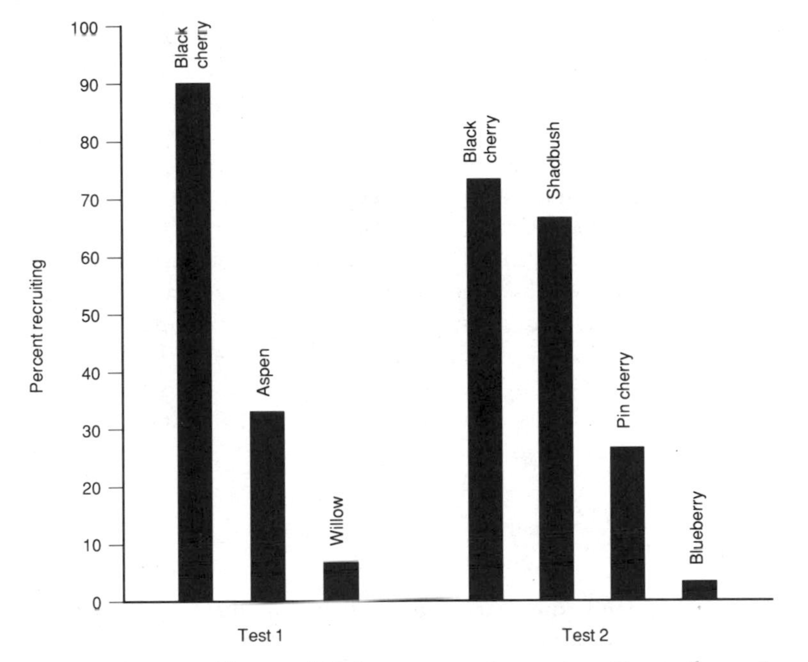

Figure 6.12. Percent of larvae of *Malacosoma americanum* recruiting to the preferred host black cherry, and to the young, partially unfolded leaves of other nonhost tree species, after feeding on the plants. (Redrawn from Fitzgerald and Peterson, *Animal Behaviour* 31 [1983]:417–442, published for the Association for the Study of Animal Behaviour and the Animal Behaviour Society by Academic Press Ltd.)

even when caterpillars have fed to satiation on aged leaves or on the leaves of nonhost species. As discussed in Chapter 5, caterpillars that feed on young leaves of the preferred host grow faster and larger than caterpillars that feed on aged leaves.

The communication system of the eastern tent caterpillar is sufficiently flexible to allow the caterpillars to recruit to suboptimal food sources when preferred food is not available (Fitzgerald and Peterson 1983, Peterson 1986b). When reared on their natural host, black cherry, caterpillars recruit little or not at all after feeding for the first time on either of the nonhost species sweet cherry or scarlet oak. But when caterpillars are reared on either sweet cherry or scarlet oak, they readily lay recruitment trails to these nonhost species. No matter what the previous rearing history, however, caterpillars always recruit preferentially to black cherry when given a choice between it and a nonhost species. These results indicate that caterpillars require a period of deprivation and adjustment before accepting and recruiting to a suboptimal food plant.

Moreover, caterpillars that have been reared on nonhost plants show particularly strong marking behavior after their first encounter with black cherry (Fig. 6.13).

Although the larvae of *M. disstria* occasionally follow trails of sated siblings to food, recruitment to food is clearly not an integral component of the nomadic foraging behavior and concomitant trail-marking system of this species. *M. americanum* larvae efficiently recruit to food because their trail system is partitioned into distinctive recruitment and exploratory components. Selective forces apparently have acted to eliminate spurious marking by unfed caterpillars that might otherwise interfere with the recruitment process. Similar forces apparently have not acted on *M. disstria,* because spurious marking is common in this species and clearly interferes with the potential for the colony mates to recruit to food (Fitzgerald and Costa 1986).

Polyethism and Foraging

The foraging behavior of larval tent caterpillars has been proported to be influenced by individual differences in larval behavior. These differences, it has been argued, have given rise to a sort of primitive division of labor within colonies that involves distinct leaders and followers. Wellington (1957) found that colonies of *M. californicum. pluviae* contained a mixture of dependent and independent larval types. His studies indicated that colonies consisted of from 0 to 38% type I caterpillars, a larval type capable of independent movement. He attributed the success of the colony largely to the presence of these larvae. Colonies also contained sluggish subtypes, which he labeled type II larvae. These caterpillars were largely incapable of independent movement but would move together in tight formation or follow trails established by type I larvae. Wellington (1965) traced the differences in activity of these larval types to the varying amounts of yolk in eggs. Eggs with normal quantities of yolk produced type I individuals. Eggs with less yolk produced type II larvae. Wellington (1957) argued that an adaptive mixture of these larval types resulted in a weak division of labor that benefited the entire colony. Type I larvae were said to stir the colony to activity and to established trails to new food finds. Type II caterpillars were largely responsible for constructing a tightly woven tent and for maintaining colony cohesiveness.

Wellington argued that differences in the activity levels of larvae have a profound effect on the population dynamics of *M. californicum pluviale.* He proposed that active adults produced from type I larvae emigrate to

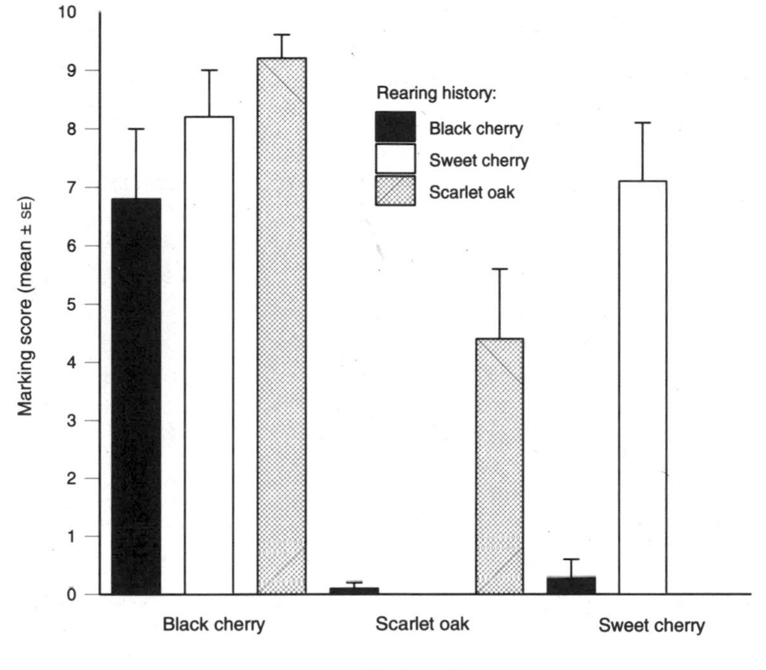

Figure 6.13. Marking scores of *Malacosoma americanum* caterpillars after feeding on leaves of various host and nonhost tree species. The rearing history indicates the plant species on which the caterpillars were reared up to the time of the test. A marking score of 10 indicates that a caterpillar laid down a recruitment trail over the entire length of a demarcated section of a pathway leading from the feeding site to the resting site. (Based on Peterson 1986b.)

new sites, whereas the weaker adults derived from type II caterpillars fail to disperse. Over time, colonies with increasing proportions of sluggish larvae will build up at the site of the original infestation, and eventually the local population will crash.

Laux and Franz (1962) reported that the larvae of *M. neustrium* could also be separated into active and inactive categories. Individual caterpillars of this species, however, were not consistently active or inactive, but shifted from one category to another. Greenblatt and Witter (1976) came to a similar conclusion in their study of *M. disstria*. In his extensive work on *M. neustrium testacum*, Shiga (1979) reported that individual differences in this species were less obvious than those reported for *M. californicum pluviale* and were not an important factor in the population dynamics of this insect.

Wellington (1957) reported that distinct behavioral types also occurred in colonies of *M. americanum,* but provided no supporting data. A preliminary study by Greenblatt (1974), however, indicated that activity levels in this species were not consistent from test to test. Moreover, recent work by other investigators has failed to support the contention that polyethism is a regular feature of colonies of tent caterpillars. A study of *M. americanum* (Edgerly and Fitzgerald 1982) indicated that larval activity levels are normally distributed. The activity level of individuals of this species varies from day to day, and exploration is a group function with no apparent leaders; differences in the rate at which colony mates develop under field conditions are largely the result of parasitism. Myers (1978) also reported that, contrary to previous reports, there is no apparent relationship between the order in which eggs are laid and the activity level of the larvae of *M. californicum pluviale.* Furthermore, a reanalysis of Wellington's (1965) data failed to support the contention that maternal nutrition influences the proportions of larvae that exhibit different activity levels within egg masses (Papaj and Rausher 1983). Thus, although early studies served to stimulate considerable thinking about the role of individual differences in the population dynamics of organisms, more recent studies have failed to provide unequivocal evidence in support of an adaptive polyethism in colonies of tent caterpillars.

Tent Caterpillar versus Eusocial Insect Foraging Behavior

Ants and termites are preeminent trail makers, and an extensive literature exists on the trail-based communication systems of these insects (E. O. Wilson 1971; Hölldobler and Wilson 1990 and references therein). Although some ants have nomadic phases, ants and termites are for the most part central-place foragers, and the insects share with tent caterpillars certain features of their trail-based communication systems.

Eusocial insects lay down pheromone trails to recruit siblings to areas of the nest that need repair, expansion, or defense, but the most common type of trail marking by these insects involves recruitment to food. As do tent caterpillars, individuals that discover food while exploring lay down a recruitment trail as they return to the nest. The recruitment markers of ants and termites serve to attract nestmates from modest distances and to orient them to the food find. Unlike the recruitment chemicals described for ants, the chemical trail marker used by tent caterpillars to recruit siblings is relatively nonvolatile and cannot serve to attract caterpillars to the trail from any significant distance, although, as mentioned above, residues of plant volatiles may function to some extent in that capacity. In any case, it is not necessary to attract tent caterpillars to recruitment trails because the caterpillars are already mobile, and their activity brings them into contact with the trails. The trails, once contacted, serve to stimulate and orient the caterpillars' locomotion, and thus lead the foragers to converge en masse at the food find.

Many species of ants depend on visual cues to maintain contact with their nests and do not deposit exploratory trails. Army ants are an exception. These blind or semiblind nomadic foragers rely on chemically based exploratory trails to relocate their bivouacs and to hold foraging columns together (Chadab and Rettenmeyer 1975, Topoff et al. 1980). Although evidence suggests that the army ants *Eciton* (Chadab and Rettenmeyer 1975) and *Neivamyrmex* (Topoff et al. 1980) use qualitatively different chemicals to mark exploratory and recruitment trails, the chemical basis of these bilevel trail systems is presently unknown. The ant *Leptogenys* (Maschwitz and Mühlenberg 1975) also employs a trail system with features similar to those of tent caterpillars, but it is not understood how it distinguishes long-lived trunk trails from its more ephemeral recruitment trails. Less is known of trail marking by termites, but the insects are blind and depend heavily on chemical cues to maintain contact with the nest. Workers of the subterranean termite *Reticulitermes flavipes,* for example, are known to deposit both exploratory and recruitment trails (Runcie 1987).

Different chemical substances may be responsible for attracting and for orienting both ant and termite foragers during bouts of recruitment. The evidence suggests that the termite *Nasutitermes costalis* uses a volatile chemical to attract nestmates and a longer-lived chemical to orient them along trails (Hall and Traniello 1985). The situation is similar in ants. In the ponerine ant *Cerapachys* sp., a pheromone that serves primarily to orient trail behavior is secreted from the poison gland, but recruitment is attributed largely to a separate chemical secreted from the pygidial gland (Hölldobler and Wilson 1990). Whether the secretion is from a single site or from multiple sites, the evidence suggests that the recruitment pheromones of most ants are likely to consist of mixtures of compounds that may have different effects. In the fire ant, *Solenopsis invicta,* a trail pheromone that is stored in the Dufour's gland and secreted from the sting contains distinct chemical components that control recruitment and motivate orientation (Vander Meer et al. 1988, 1990)

The recruitment process of many species of ants involves the active excitation of workers by the recruiting insect. Excitation may be accomplished by antennation, by the presentation of food samples, by agitated displays by the recruiter, or by the secretion of chemicals that alert nestmates to the presence of a chemical trail that leads to the food find. But the most evolutionarily advanced recruitment systems used by ants involve mass communication. In mass communication, the pheromone itself is the prime, if not the sole, signal. The number of ants recruited is related to the amount of pheromone secreted. Controlling the amount of pheromone enables colonies to regulate the outward flow of workers according to need. The pheromone is adequate to both stimulate and orient trail following; direct physical interaction between individuals is not required (Hölldobler 1978).

The trail systems of tent caterpillars resemble these systems of mass recruitment in that chemical cues alone are adequate to elicit recruitment. Neither motor patterns nor sensory modalities other than chemoreception appear to be involved in the case of the tent caterpillar. Moreover, like the systems of mass communication reported for ants, the flow of tent caterpillars to specific feeding sites depends on the extent to which individuals reinforce trails as they return from the food find. But there appears to be considerably less precision. The recruitment marker of the caterpillar persists long after it is deposited and caterpillars continue to visit the site of food finds for some time after the site has been depleted of leaves.

Despite numerous similarities in the trail-based communication systems of ants, termites, and tent caterpillars, tent caterpillars clearly lack the behavioral flexibility of eusocial insects. Tent caterpillars are only primitively social, and they neither store food nor distribute it to colony mates. Unlike eusocial insects, tent caterpillars follow recruitment trails only when hungry, though the extent to which hunger also influences the readiness of ants and termites to follow trails is not clear. Moreover, individuals analogous to scout ants, which reinforce recruitment trails without carrying food (Jaffe 1980), do not occur in colonies of tent caterpillars.

Perhaps the most significant difference between recruitment communication in ants and termites and in tent caterpillars lies in the manner in which foraging bouts are organized. Caterpillars feed at discrete, predictable intervals. Individual hunger and, possibly, tactile stimulation from milling siblings motivate the caterpillars to initiate mass foraging. During the initial stage of an activity bout, the caterpillars confine their activity to the tent or venture only short distances onto their trunk trails before turning back. During this stage in the activity bout, the caterpillars are highly responsive to recruitment trails deposited on the surface of the tent by the first of the advanced contingent of successful foragers to return. When they encounter such trails their response is typically immediate, and they exhibit excited motor patterns as they follow the

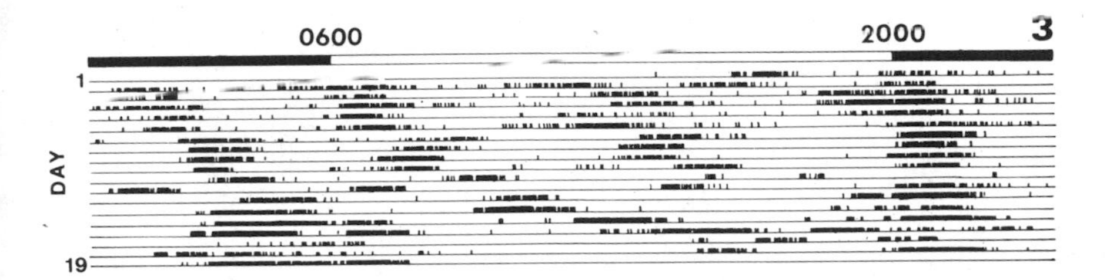

Figure 6.14. Daily foraging patterns of a colony of eastern tent caterpillars recorded in the laboratory over 19 days. (Reprinted, by permission of the Entomological Society of Canada, from Fitzgerald 1980.)

trails to the food find. In marked contrast, the foraging behavior of the nonnomadic ants and termites is not so clearly motivated by individual hunger, nor do the potential foragers periodically and predictably mass in preparation for imminent departure. Rather, workers that are resting or otherwise occupied are stimulated to spontaneous activity when they detect the volatile odors of the recruitment pheromone and are drawn to the odor trail. Moreover, the original concept of recruitment as a process by which workers or soldiers draw siblings to parts of the nest where work is required (E. O. Wilson 1971) is not as neatly applied to the tent caterpillars. Unless the concept of work is understood to include the effort involved in filling one's stomach, it would appear that a less restrictive definition is needed that recognizes a broader range of cooperative foraging strategies, that, though underlain by diverse motivational factors and chemosensory systems, have a similar appearance and proximate effect: the maintenance of the colony's nutritional fitness.

Temporal Patterns of Foraging

Several studies have shown that tent caterpillars have regular bouts of feeding and rest. In the laboratory, under conditions simulating a spring light regimen at approximately 21°C, young colonies of the eastern tent caterpillar foraged four times each day: at dawn, midafternoon, dusk, and in the early morning (Fitzgerald 1980, Casey et al. 1988; Fig. 6.14). Outdoors, there are three peaks each day: at approximately 0600, 1500, and 2000 (Fitzgerald et al. 1988; Fig. 6.15). The omission of the predawn foraging bout under field conditions has been attributed to the

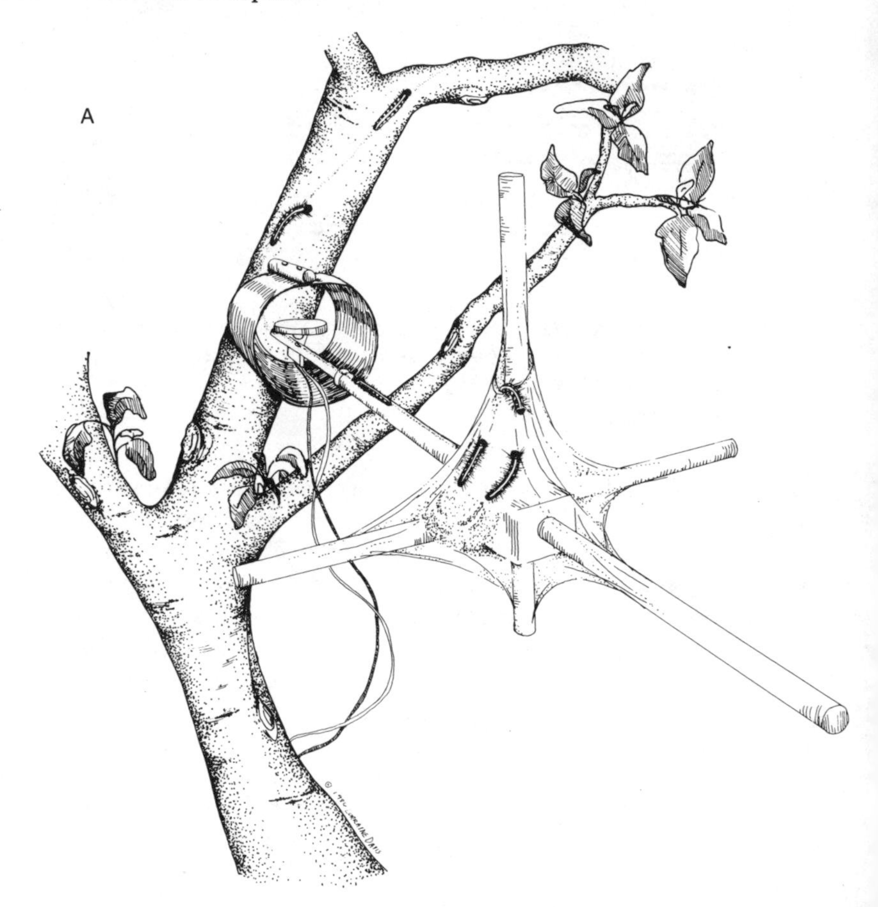

inability of caterpillars to process the food collected during the evening feeding period at the low body temperatures they experience overnight.

Movement to and from food is often tightly synchronized, particularly during the early stadia. Under field conditions, colonies are active for nearly 50% of each day, with approximately 80% of their activity occurring under the cover of darkness at dusk and dawn (Fitzgerald et al. 1988). Morning and evening feeding bouts occur at relatively low temperatures, and whole-colony activity periods in the morning and evening last approximately twice as long as those in the afternoon.

At temperatures greater than about 15°C, the number of daily foraging bouts is temperature-independent. Casey et al. (1988) demonstrated this

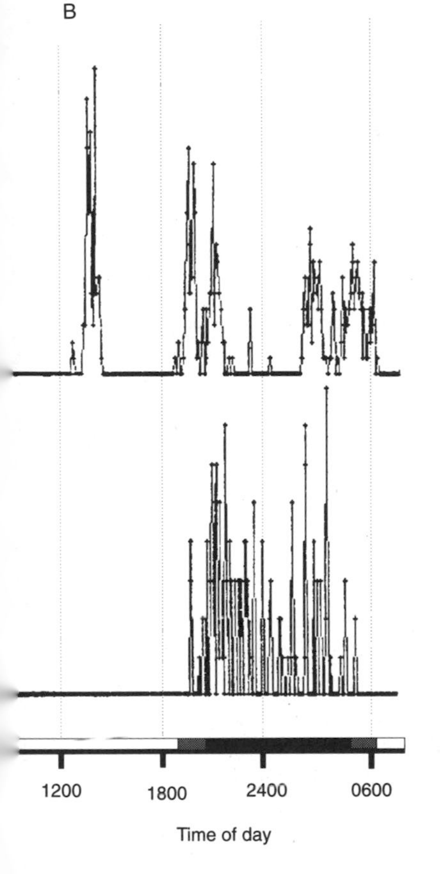

Figure 6.15. (A) Method of recording daily foraging bouts of field colonies of the eastern tent caterpillar. The caterpillars pass over a photosensor as they move from their tent to feeding sites and again as they return. Impulses from the photosensor are transmitted to a data logger and plotted against time. A large metal cylinder shields the sensor from light and weather. (B) Chart recordings of characteristic daily activity patterns of *M. americanum* during the fourth to fifth (above) and the sixth (below) larval stadia, recorded in the field as shown in (A). The height of a trace is proportional to the number of caterpillars passing over the sensor at a particular point in time. Through the fifth stadium, colonies typically eat three times a day, and the caterpillars feed more or less in synchrony, producing bimodal traces as they move to and from feeding sites. In the last stadium, caterpillars forage more independently throughout the hours of darkness. (Reprinted, by permission of Springer-Verlag, Heidelberg, from Fitzgerald et al. 1988.)

relationship by rearing genetically matched colonies under two temperature regimes. One colony was maintained at a constant 20°C , the other at 20°C during the night and at 35°C during the day. Caterpillars exposed to the higher temperature during the day grew significantly faster than those maintained at a constant temperature (see Thermoregulation and Growth, Chapter 7). Despite this difference in growth rate, there was no significant difference in the number of times the two colonies foraged each day or in the duration of activity and rest periods. More rapid growth of caterpillars in the colony exposed to the elevated daytime temperature was attributed to faster food processing. These caterpillars were able to empty their guts completely during their rest periods and to fill them to capacity with new food during their foraging bouts. In contrast, caterpillars held at the lower temperature had only partially

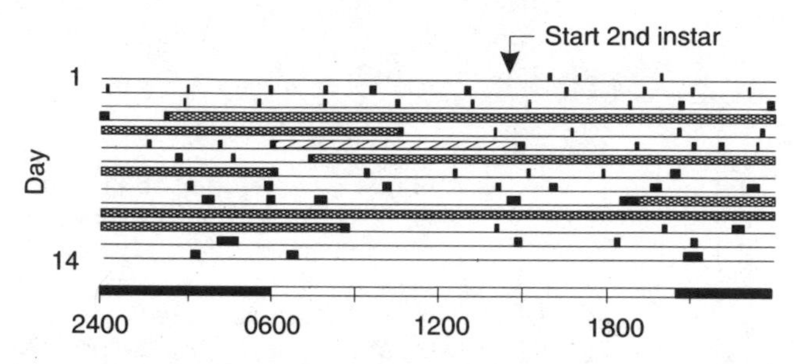

Figure 6.16. Chart recording of the feeding activity of a *Malacosoma americanum* larva reared from the second to the fifth instar in isolation in the laboratory. Cross-hatched bars indicate periods of inactivity preceding ecdysis, diagonally striped bar indicates recorder malfunction. Isolated caterpillars show much higher levels of feeding activity during the early instars than do caterpillars in laboratory colonies. (T. D. Fitzgerald and C. R. Visscher, unpubl. data.)

emptied guts at the outset of foraging bouts and were capable of ingesting significantly less food.

In the last stadium, eastern tent caterpillars are active throughout the hours of darkness, but they rarely forage during the daylight hours. These last instars also forage more independently than the earlier instars and disperse over more of the tree. They do not appear to reinforce trails with either silk or chemical marker, and there is no indication that they continue to recruit siblings to food finds. Unlike the earlier instars, they commonly rest adjacent to their food find and feed intermittently throughout the night and early morning. As long as the host tree supplies the colony with food, last-instar caterpillars may continue to return to the tent shortly after dawn and spend the day sequestered in the structure. The dispersed caterpillars characteristically return to the tent at about the same time, possibly because they respond similarly to increasing ambient light intensity.

The habit of field colonies to forage under the cover of darkness, when ambient temperatures are least favorable for activity and food processing and to rest at the tent during the daylight hours makes it clear that this insect does not follow a schedule of activity designed to maximize its rate of food processing. Indeed, the insect appears to follow a schedule that minimizes its contact with day-active, visually oriented predators and parasitoids. Field observations indicate that, in general, caterpillars may be better able to defend themselves when aggregated than when foraging (Fitzgerald et al. 1988, Fitzgerald 1993a), but there is as yet no hard proof of this assertion for tent caterpillars.

In marked contrast to the tendency of whole colonies of eastern tent

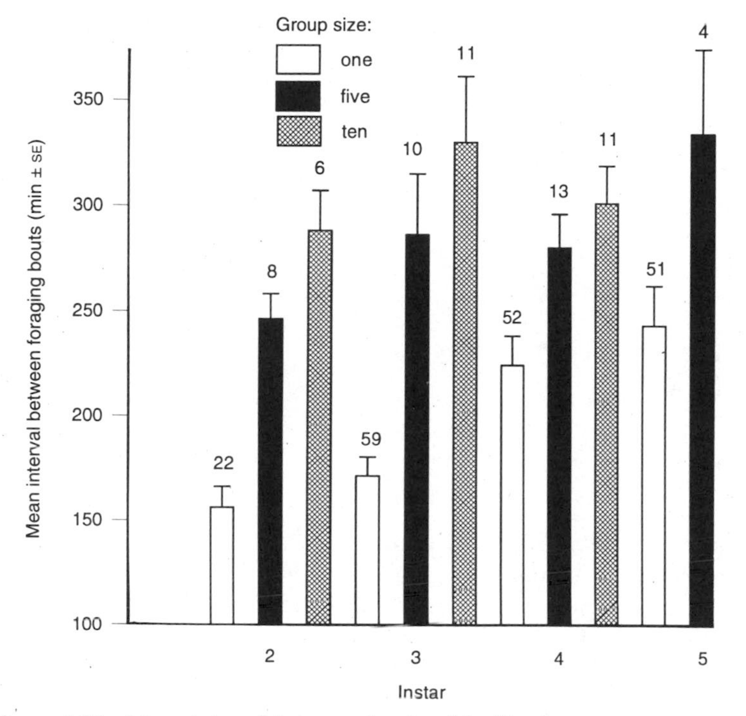

Figure 6.17. Mean interval between bouts of feeding by eastern tent caterpillars reared alone or in groups of 5 or 10. Number of observations indicated above bars. The foraging interval of caterpillars maintained in groups of 5 or 10 approximates that of whole colonies reared under similar conditions, but solitary individuals feed much more often, particularly during the early instars. (T. D. Fitzgerald and C. R. Visscher, unpubl. data.)

caterpillars to feed three or four times a day, caterpillars reared individually on host branches in the laboratory feed up to 12 times a day (T. D. Fitzgerald and C. R. Visscher, pers. observ.; Figs. 6.16, 6.17). This difference indicates that, rather than stimulating feeding activity, coloniality may actually constrain individual activity. Strong selection pressure for synchronous foraging appears to have given rise to group-mediated behavioral patterns that override the tendency of individuals to follow a foraging schedule dictated by individual hunger level (Casey et al. 1988). The proximate mechanism for this constraint is unknown, but it is likely to involve tactile or chemical cues associated with resting siblings.

7

Tent-Building Behavior and Thermoregulation

The modification of the labial glands to produce silk and the associated evolution of the spinneret provided caterpillars with the material and the means to alter their environment adaptively. Although other immature insects also use silk to modify their surroundings, the caterpillars of moths and butterflies are among the most prolific of the shelter-building insects. They roll, fold, or tie leaves with silk; envelop leaves and branches with silk to create loose "nests"; or create their shelters from silk alone. The tightly woven tents formed by tent caterpillars are among the largest and most conspicuous of these structures. They stand apart from a colony's food supply and serve as staging areas from which siblings launch forays in search of food. The tents promote aggregation and serve as the focal point of communal activities, prominent among which are daily episodes of sychronized resting, silk spinning, and thermoregulatory behavior.

Shelter-Building Habits

The tents built by colonies of the different species of *Malacosoma* vary in number and size. Of the American species, *M. constrictum* and *M. tigris* build a series of small tents each time they molt (Plate 1B); *M. americanum*, *M. californicum*, and *M. incurvum* build permanent or semipermanent tents that are expanded as the colonies grow (Fig. 7.1, Plate 1C, D). *M. disstria*, a highly nomadic species, is the only tent caterpillar known to build no tent at all. The caterpillars of this species aggregate

Figure 7.1. The tent of a colony of fifth-instar *Malacosoma americanum* caterpillars.

and molt on mats of silk spun on leaves and branches or on the trunk of the tree (Plate 1A).

The European lackey moth, *M. neustrium,* is reported to build a series of communal webs and to rest and molt on the outside of the structures (Tutt 1900, Balfour-Browne 1933). Colonies of *M. indicum* and *M. parallelum* occupy permanent webs that are expanded as the colonies grow (Joshi and Agarwal 1987, Li 1989, Bhandari and Singh 1991). The ground lackey, *M. castrensis,* makes periodic forays in search of food from tents constructed in low-lying vegetation (Tutt 1900, Goater 1991). Both *M. alpicolum* and *M. castrensis* molt within their tents but periodically abandon them and construct new tents elsewhere (Martin and Serrano 1984).

Caterpillars in some other genera of the Lasiocampidae also build tents. Larvae of the European eggar moth, *Eriogaster lanestris,* build large conspicuous tents similar to those of *M. americanum* (Balfour-Browne 1933, Carlberg 1980). The lasiocampids *Gloveria howardi* and *Eutachyptera psidii,* which occur in the American Southwest and Central America, respectively, also build large nests. *E. psidii* makes tents up to 80 cm long (Fig. 2.5) and the tent is reported to be one of the sources of the silk used by the indigenous Indians of Central America (Borah 1943, Peigler 1993).

Whether a tent serves a colony through its lifetime, is periodically abandoned, or, in the case of *M. disstria,* is not constructed at all is likely to be a consequence of features of the seasonal history and ecology of the species. The extent and intensity of solar radiation may be such that no single fixed base is optimal throughout the life cycle of the colony. An unpredictable food supply—a possible problem for generalist feeders such as the forest tent caterpillar, which feeds on trees with markedly different leafing phenologies—may favor nomadism and preclude the establishment of a permanent resting site. In similar fashion, a patchy food resource may favor the establishment of a series of staging areas, each of which is abandoned when the patch is exhausted.

The susceptibility of tent caterpillars to disease may disfavor the building of a permanent tent, because the structure may become a focal point for the spread of disease. Wellington (1974) reported that colonies of *M. californicum pluviale* that abandoned their tents frequently were less susceptible to disease than were colonies that built fewer tents. Bucher (1957) determined that the spores of the bacterium *Clostridium* sprayed directly onto the tent to simulate the regurgitations of diseased caterpillars readily infected a colony. Fungal (MacLeod and Tyrrell 1979) and viral agents capable of surviving on the damp interior surfaces of the tent may also act as selective agents influencing the evolution of tent-building behavior.

Functions of the Tent

The tents of lasiocampids are multifunctional. Perhaps the most immediately apparent utility of tents and resting mats is the secure purchase they afford caterpillars during ecdysis or during periods of prolonged cold when the insects may have limited muscular control. When they are firmly attached to their silk by their crochets, the caterpillars may withstand even the most driving rains that might otherwise wash them from the tree.

The surface area of a tent is adequate to hold the entire colony at once, allowing the caterpillars to rest, molt, and thermoregulate in synchrony. For those species that rest within the structure, the humid microclimate of the tent may also protect the molting caterpillars from desiccation (Sullivan and Wellington 1953). It has been reported that the larvae of *M. californicum* occurring in damp regions along the northern Pacific Coast commonly rest outside of their tents, whereas populations of the same species found in the dry Mojave Desert usually rest inside (Stehr and Cook 1968).

Tents also serve as staging areas from which colonies launch periodic en masse forays in search of food and as communication centers where successful foragers alert siblings to the discovery of new food finds, a function discussed in Chapter 6. The importance of the tent in thermoregulation is treated in detail later in this chapter.

Studies using exclusion netting indicate that tents may be only marginally effective in reducing the overall impact of predators and parasitoids (see Tents as Shelters, Chapter 8). Nonetheless, when they rest within the tent, caterpillars are secure from entomophages that cannot penetrate the silk. Moreover, caterpillars assembled on the outside of tents respond with a defensive body-flicking display when threatened (Ancona 1930, Myers and Smith 1978). Such displays, which set up vibrations in the tent, appear to alert nearby siblings, and the behavior radiates rapidly through the assemblage. Group displays are likely to amplify the deterrence value of the behavior and thereby to discourage predators.

Tent Site

In his studies of the honey bee, Lindauer (1967) determined the characteristics of potential hive sites that the insects evaluate when swarming. These include protection from the elements, particularly wind, the size of the nest cavity, and distance from the old hive. Moreover, consensus among scouts is required before one site is selected over another. Clearly, nothing that sophisticated occurs among tent caterpillars, but, in some species at least, an element of decision making appears critical to the establishment of the tent site.

The caterpillars of the eggar moth, *Eriogaster lanestris*, typically construct their tent over the egg mass so that the adult effectively selects the site. Adults of this species oviposit preferentially on the southern—sunny—side of the tree, but when an egg mass is deposited in a shaded area, the larvae migrate to a sunny spot to build their tent. The tents of this species typically are constructed on branches a meter or so from the trunk (Carlberg 1980). In contrast, the tent site of *M. americanum* typically is selected by the caterpillars. Tents are rarely constructed over the egg mass. In one survey, the caterpillars in 78 of 82 colonies of *M. americanum* migrated from their egg masses, which were deposited near branch tips, to construct tents at or near the trunk of the tree, an average of 157 ± 2.9 cm (SE) from the oviposition site (Fitzgerald and Willer 1983).

How do the caterpillars settle on a tent site? After abandoning their egg mass, the caterpillars may initially aggregate to bask on surfaces of the tree that are warmed by the sun. The normal silk-trailing behavior of the caterpillars may give rise to an incipient tent at nearly any site where the silk strands span small gaps between branches. Something like the process that underlies the initiation of column building by termites (E. O. Wilson 1971) may be involved in directing the collective spinning efforts of the caterpillars to sites where silk has begun to accumulate and lead to the eventual establishment of a substantial shelter.

Exposure to the sun appears to be a critical factor in the selection of aggregation sites among tent caterpillars, but it is not the only consideration. Species such as the eastern tent caterpillar, which makes large tents that are inhabited through maturity, require a site that provides for the repeated expansion of the tent. Indeed, caterpillars of this species tend to construct their tent at the point where a branch joins the trunk of the tree or at a place where sizable branches meet. Such sites nearly always allow for the growth of the tent. Sometimes, however, colonies initiate tent building at a site that later proves to be inadequate, and they are forced to move to a new site later on. Observations of laboratory colonies suggest that eastern tent caterpillars will abandon a partially constructed tent if it is not expansive enough to allow them to rest inside. Thus, tents constructed between parallel wooden dowels (Fig. 7.2) provide adequate internal space for small instars but cannot house the later instars. When reared on such frames, the growing caterpillars become

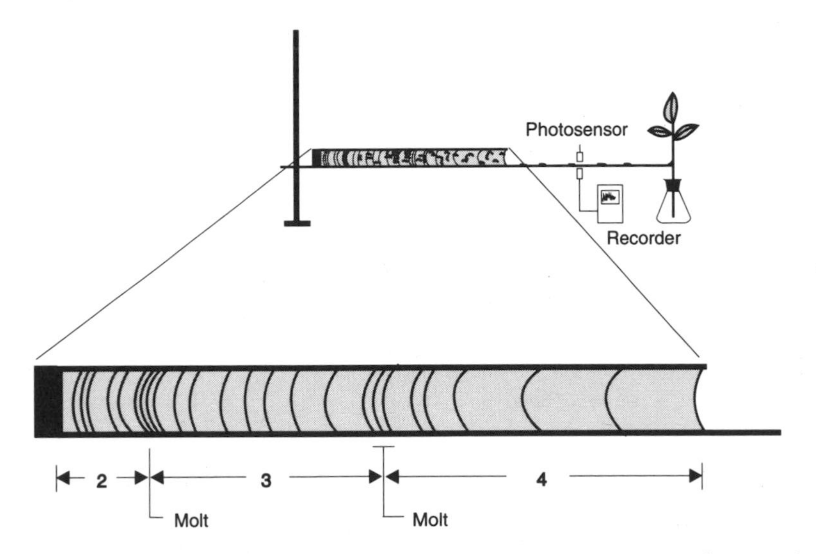

Figure 7.2. Experimental tent of a colony of eastern tent caterpillars, showing the relationship between activity periods and tent expansion. The colony was continuously monitored for 192 hours during which the caterpillars advanced from the second to fourth instar. New layers of silk (concave lines) were added to the tent during 25 of the 30 activity periods that occurred during the study. Caterpillars were held under a 14:10 light:dark regimen, and additions to the tent during the dark phase occurred under red light that provided approximately 3 lux at the tent. Numbers indicate instar. (T. D. Fitzgerald, unpubl. data.)

restless, refuse to settle after feeding, and eventually attempt to construct a new tent elsewhere.

Laboratory studies show that eastern tent caterpillars concentrate their spinning activity on the most intensely illuminated side of their tent, so that the most developed face of the structure lies perpendicular to the dominant light source (Fig. 7.3). Caterpillars that build permanent tents behave similarly under field conditions (Balfour-Browne 1933, Carlberg 1980, Fitzgerald and Willer 1983). Compass bearings taken perpendicular to the most developed face of 69 tents of the eastern tent caterpillar had an average facing direction of 122 ± 3.3°. The southeast orientation of these tents appears to reflect the direct influence of the position of the sun during the morning and afternoon on the spinning behavior of the photopositive caterpillars. Gravity plays little or no role in the tent-building behavior of the caterpillars. When tents are illuminated from below, the caterpillars build their tents upside down. Moreover, when so illuminated, the caterpillars build their trails on the underside of the branches that connect with the tent and walk upside down to and from the structure.

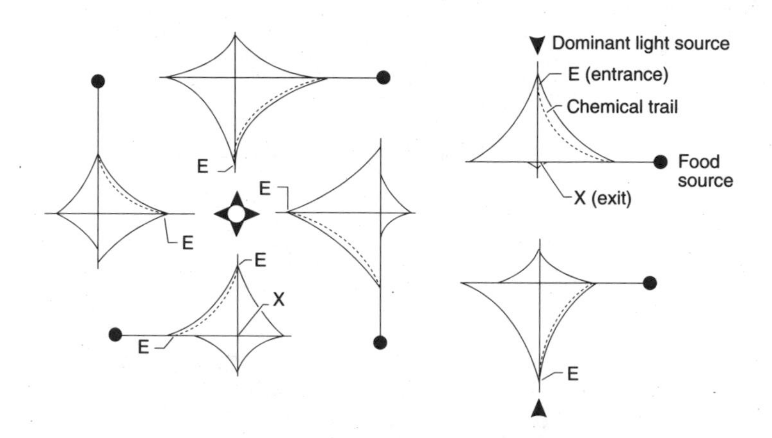

Figure 7.3. A laboratory experiment demonstrating the effect of light on the orientation of the tent of the eastern tent caterpillar. Although the tents are bathed in diffuse light from all sides, the larvae concentrate their tent-spinning activity on the side that faces a small lamp, the brightest object in their field of vision. The larvae also move toward the lamp when returning from feeding bouts, accounting for the position of the chemical recruitment trails and entrance locations. (Reprinted, by permission of the Kansas Entomological Society, from Fitzgerald and Willer 1983.)

Construction of the Tent

Most of our knowledge of the tent-building behavior of lasiocampids is based on studies of *E. lanestris* and *M. americanum* (Balfour-Browne 1933, Carlberg 1980, Fitzgerald and Willer 1983). Both species build tents that are enlarged daily and that typically are occupied until the caterpillars near maturity. The structures are built of multiple layers of silk that are separated by gaps within which the larvae rest (Fig. 7.4). Under laboratory conditions, the caterpillars of *M. americanum* typically add silk to the tent before each of the four bouts of foraging that occur daily (Fig. 7.2). The caterpillars spin en masse, and substantial additions to the tent are rarely made at other times. Whole-colony spinning bouts last up to an hour or more, but individuals typically spin for only 15 to 30 minutes.

The intermittency of whole-colony spinning bouts accounts for the layered structure of the tent. Although Balfour-Browne (1933) suggested that the gaps between layers of the tent of *Eriogaster* might be excavated by the caterpillars, studies of *M. americanum* indicate that they are a natural consequence of the physical and chemical properties of the silk. Caterpillars draw long, continuous strands of silk from their spinnerets by rapidly swinging their bodies from side to side as they move over the surface of the tent (Fig. 7.5). As it is drawn, the silk gels irreversibly

Figure 7.4. The tent of a mature colony of *Malacosoma americanum* cut open to expose its layered internal structure.

and loses water, and, because of its small diameter, it comes into almost instantaneous equilibrium with the ambient relative humidity (Work 1985). The drawn strand is elastic, and the caterpillars stretch it slightly beyond its equilibrium length before fastening it. If the fastened strand is cut in the middle, it snaps away from the cut, behaving much like a slightly stretched rubber band. Thus, each strand generates a contractile force along its axis as it attempts to return to its equilibrium length. It is the cumulative effect of the minuscule contractions of strands spun in all directions during a single bout of tent building that accounts for the taut, concave faces of the tent. During these episodes, enough silk is deposited at one time to form a large contiguous mat. Attached firmly at its periphery to the supporting framework of branches, the mat rises from the subjacent layer of silk as it contracts, forming a sizable gap as it separates (Fig. 7.4).

But more is involved. The equilibrium length of the silk strand of the tent caterpillar, like that of some spiders, is determined by the relative humidity (RH) of the atmosphere. When spider silk, spun at 50 to 60% RH, is wetted, it instantly contracts to almost 50% of its original length, a process known as supercontraction (Work 1985). The silk of the eastern

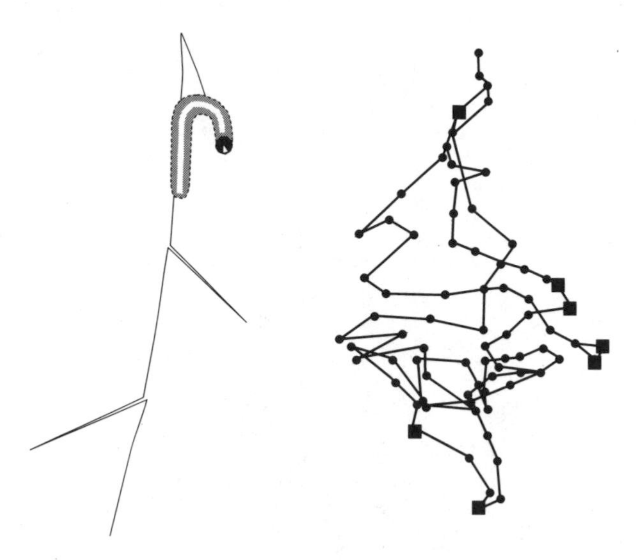

Figure 7.5. Pattern of spinning employed by larvae of the eastern tent caterpillar while adding silk to the tent (right) and movement pattern of a caterpillar while spinning on the tent surface. Dots = one-second intervals, squares = the edge of the tent. (After Fitzgerald and Willer 1983, by permission of the Kansas Entomological Society.)

tent caterpillar also contracts when wetted, but to a lesser extent. Strands spun at 65% RH contract to an average of 78% of their original length when wetted (Fitzgerald et al. 1991). The amount of force generated during this moisture-dependent contraction has not been measured, but studies with spider silk indicate that it is likely to be less significant than the forces due to axial stretching of the fiber when it is first spun. Nonetheless, when tents are exposed to high humidities or are wetted by rainfall or overnight condensation, the additional contractive forces that are generated can be expected to contribute to the tautness and stratification of the structures.

Some species of tent caterpillars such as *M. neustrium*, *M. constrictum*, and *M. tigris* build small tents. They molt on the outside of the structures and rarely, if ever, enter them (Balfour-Browne 1933, Stehr and Cook 1968). But the large-tent-building species move in and out of their tents, and the structures have discrete entrance and exit sites (Figs. 7.3, 7.6). Openings may occur wherever branches jut from the structure, but they are most common at the upper apex of the tent. The openings are made and maintained by caterpillars as they emerge from the tent. The openings in the successive layers of the tent are aligned because the caterpillars follow the same supporting branches to the tent surface. During

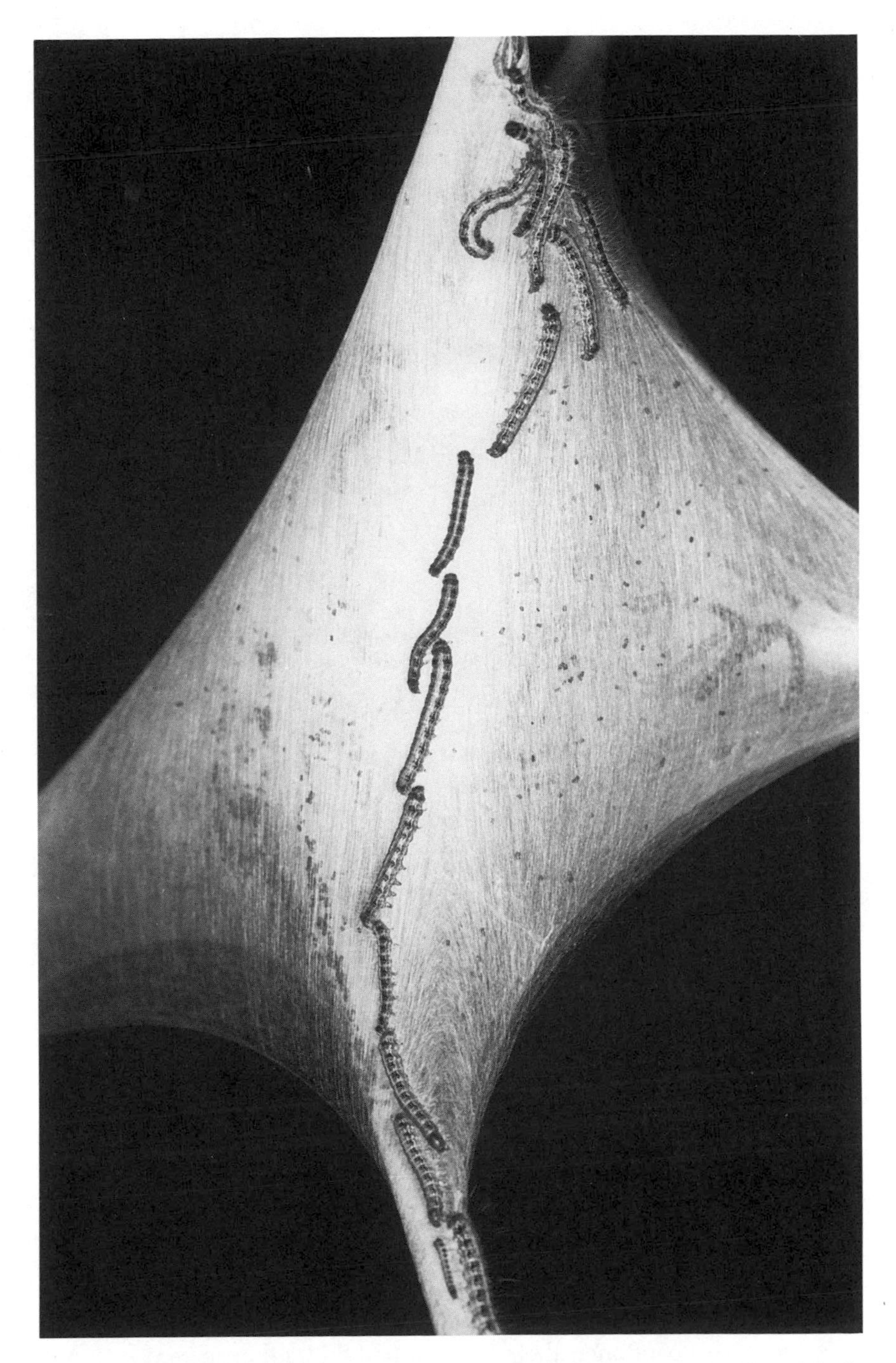

Figure 7.6. Fifth-instar eastern tent caterpillars following a recruitment trail laid down on the surface of their tent as they return from feeding. Caterpillars at the top are backed up at the tent entrance.

bouts of spinning, the old openings may be covered with silk, but late-emerging caterpillars chew new openings as they move to the surface.

When eastern tent caterpillars return to the tent after feeding they are strongly photopositive on the tent surface. As a consequence, the tent opening closest to the dominant light source is most often used as the entrance. Under field conditions this is the opening uppermost on the tent, because the caterpillars orient toward the sky or the sun. Laboratory experiments show that caterpillars can be made to enter the tent at almost any site by manipulating the position of a light source set up near the tent. The returning foragers also lay down a chemical recruitment trail as they orient over the tent surface, and they use this as an additional guide (Figs. 7.3, 7.6).

Tent Sharing among Families

There is presently no indication that tent caterpillar larvae recognize kin or in any way discriminate against members of other colonies. At low population densities, a single cohort of tent caterpillars is likely to inhabit a tent. But when local population density is higher, colonies from several egg masses may share a single tent. In one survey of 78 field colonies of eastern tent caterpillar occurring in an area of moderate population density, 49% of the tents housed a single colony, 31% had two colonies, 15% had three to four colonies, and 5% were shared by five or more colonies (Fitzgerald and Willer 1983). Contributing to the multiple use of a single tent is the failure of adults to discriminate against potential ovipositional sites that have an egg mass nearby. Some 24% of the egg masses found in this same survey were deposited within 3 cm of another egg mass on the same branch. In addition, when eclosion precedes leaf flush, the newly emerged caterpillars search widely for food, resulting in a coalescing of trail systems and tent sharing (Fig. 5.12). In general, as the season progresses and the caterpillars expand their foraging ranges, the likelihood increases that caterpillars will encounter the trail systems of other colonies that share the same tree and will follow them to their origins (Costa and Ross, 1993; Fig. 7.7), and the degree of relatedness of caterpillars sharing tents declines.

Tent sharing by different families of tent caterpillars has also been reported for *M. incurvum discoloratum* (Baker 1970) and *M. californicum* (Stelzer 1968). Tent sharing in both of these species was reported to occur after colonies abandoned their original tent and either moved into one already occupied or joined up with other colonies to build a new tent. Indeed, the willingness of tent caterpillars to form multifamily aggre-

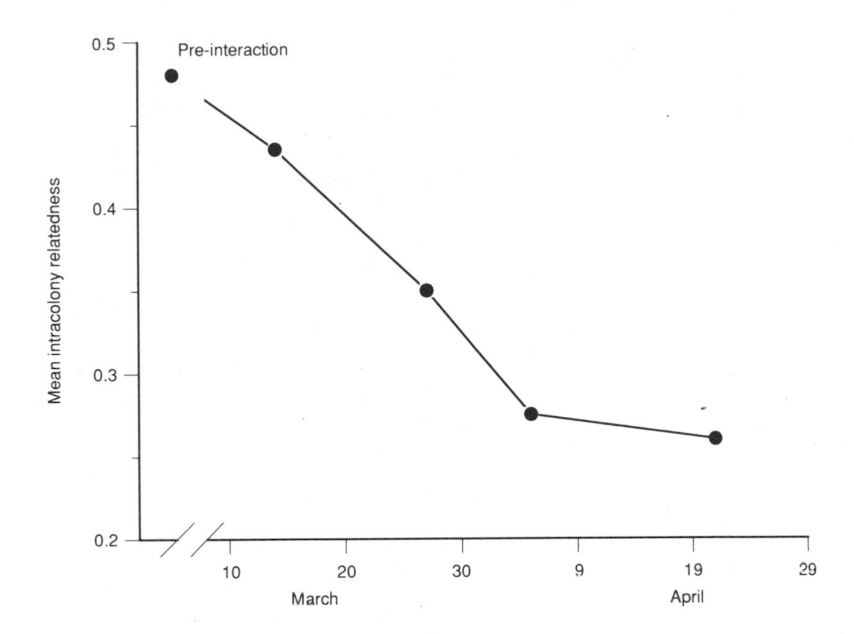

Figure 7.7. Seasonal decline in the degree of relatedness among eastern tent caterpillars that share the same tent. The decline is due to the merging of colonies that occur on the same tree. (Data from Costa and Ross 1993.)

gates may lead the caterpillars to aggregate with members of other species. Tutt (1900) reported an instance in which a colony of *M. neustrium* shared a tent with a colony of *Euproctis* (Lymantriidae) caterpillars and foraged with them. Colonies of forest tent caterpillars have also been reported to establish temporary aggregations with colonies of eastern tent caterpillars (Fitzgerald and Edgerly 1979a).

The Tent and Thermoregulation

Tent caterpillars are active at a time of year when the average daily ambient temperature (T_a) is often below the minimum temperature required for their growth and development (Hodson 1941, Knapp and Casey 1986). Nonetheless, the caterpillars are highly active foragers, and they grow rapidly, processing their food at an elevated body temperature (T_b) they achieve by basking in the sun (Fig. 7.8). Studies show not only that the dark-bodied caterpillars are effective behavioral thermoregulators but also that the tent may play a particularly important role in the thermal ecology of caterpillars.

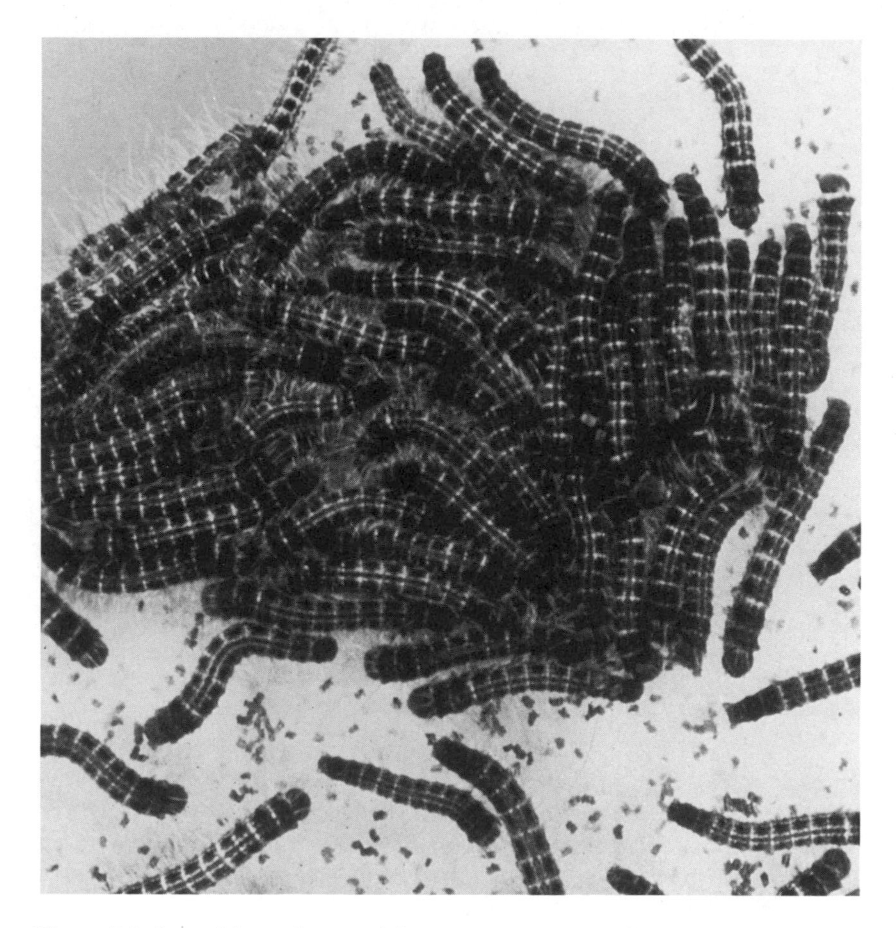

Figure 7.8. Second-instar larvae of the eastern tent caterpillar aggregated to bask in the sun.

Freeze-dried larvae with thermocouple probes inserted into their abdomens gain and lose heat much like living caterpillars and have been used to map the thermal contours of the tent of the eastern tent caterpillar (Joos et al. 1988). Although the temperature of living caterpillars can also be measured, models have the distinct advantage that they can be precisely oriented, singly or in groups, and can be held in one place for extended periods of time. Average body temperature excesses ($T_b - T_a$) reported for models of the eastern tent caterpillar show that a broad range of site-dependent temperatures can be achieved by the basking larvae (Fig. 7.9).

Models of solitary caterpillars placed on branches in the sun achieve

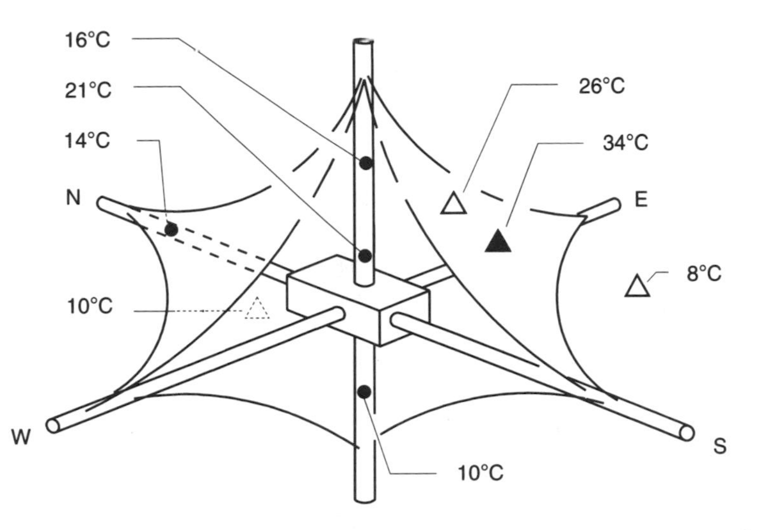

Figure 7.9. The experimental tent of a colony of eastern tent caterpillars established on a wooden frame to facilitate a study of the effect of solar radiation on thermoregulation. The thermocouples used to measure the temperature of the empty tent (filled circles) were mounted before tent construction began to assure that they would be buried within the layers of the structure. Thermocouples were mounted in solitary models prepared from freeze-dried caterpillars (open triangles) or in the model at the center of an aggregate of other models (filled triangle), to measure the full range of temperatures achievable by the caterpillars. Temperatures given in this diagram are excesses of the empty tent ($T_{tent} - T_a$) and the models on or near the tent ($T_b - T_a$) measured at noon on a clear, sunny day with solar radiation of approximately 900 W/m^2. The model in the northwest (dotted triangle) lies on the shaded undersurface of the tent, and the one in the southeast on a branch adjacent to the tent. Other models lie on the surface of the tent in full sunlight. Temperature excesses recorded in this study are for a tent in full sunlight and exceed those of naturally occurring tents, which typically are partially shaded by the host tree. (After Joos et al. 1988.)

mean temperature excesses of 0.4–11.2°C (mean = 4.8 ± 2.1°C [SD]) depending on their orientation. Studies of the gypsy moth show that when caterpillars position themselves so that the long axes of their bodies are perpendicular to a radiant heat source, they gain at least three times as much heat as caterpillars oriented parallel to the source (Casey and Hegel 1981). Resting on the tent further enhances the ability of the caterpillar to gain heat, and solitary models placed on the outside of the tent achieve temperature excesses of 3.5–27.5°C (mean = 16.2 ± 4.6°C [SD]). The gain in T_b over caterpillars that bask off the tent occurs because the caterpillar rests within the boundary layer of the tent, a layer of unstirred air that effectively reduces convective heat loss. In addition, the tent itself gains heat and provides a warmer substrate than the branches of the tree. Indeed, solitary models placed on the underside of the tent, completely shaded from direct sunlight, achieve temperature excesses of 2.5–15.7°C (mean = 8.0 ± 2.5°C) solely by conducting heat away from the tent.

Air currents can result in substantial loss of heat to basking caterpillars, particularly when a strong wind is present. But the long setae of caterpillars serve as effective barriers to air currents and can greatly reduce convective heat loss (Fig. 7.10; Casey and Hegel 1981). Studies with models show that loss of heat due to such convection can be further reduced by aggregation. A caterpillar at the center of an aggregate resting on the tent is buffered from the wind by the bodies of its siblings. Models, simulating aggregated caterpillars resting on the southeast face of the tent, achieve temperature excesses of 6.5–43.9°C (mean = 24.1 ± 7.2°C). It is estimated that approximately one-third of the temperature excess of aggregated eastern tent caterpillars is attributable to aggregation, one-third to direct exposure to sunlight, and from 14 to 47%, depending on orientation to incident sunlight, to the high ambient temperature and boundary layer of the tent (Joos et al. 1988).

The mapping of the thermal environment of the eastern tent caterpillar with models shows not only that caterpillars can attain physiological optimal T_b but also that they may be exposed to temperatures that exceed their upper physiological threshold. Unlike models, living caterpillars move from site to site to achieve or maintain T_b, and they actively avoid temperature extremes. Body temperature ranges of living caterpillars (measured with a thermocouple needle or by infrared thermography, a noninvasive procedure that allows T_b to be measured at a distance from the subject) are much narrower than those of models. While the range of T_b's of the four caterpillar models shown in Figure 7.9 was 25–55°C, living caterpillars had T_b's under similar conditions of 40–48°C. When T_b's exceed about 40°C the caterpillars cease to aggregate and rest in the shade, either inside the tent or on the underside of the structure. Under extreme conditions, the caterpillars hang by their abdominal prolegs from the shaded side of the tent, keeping their bodies clear of the structure, to maximize convective heat loss (Fig. 7.11, Plate 2C).

Tents function like greenhouses, trapping solar radiation while greatly reducing convective heat loss by blocking the wind. Temperature excesses of tents ($T_{tent} - T_a$) measured with thermocouples buried at various locations within the structure vary from as little as 4°C to as much as 23°C. Sites deep within the tent are warmer than peripheral sites, and the side of the tent facing the sun is warmer than the shaded side (Fig.

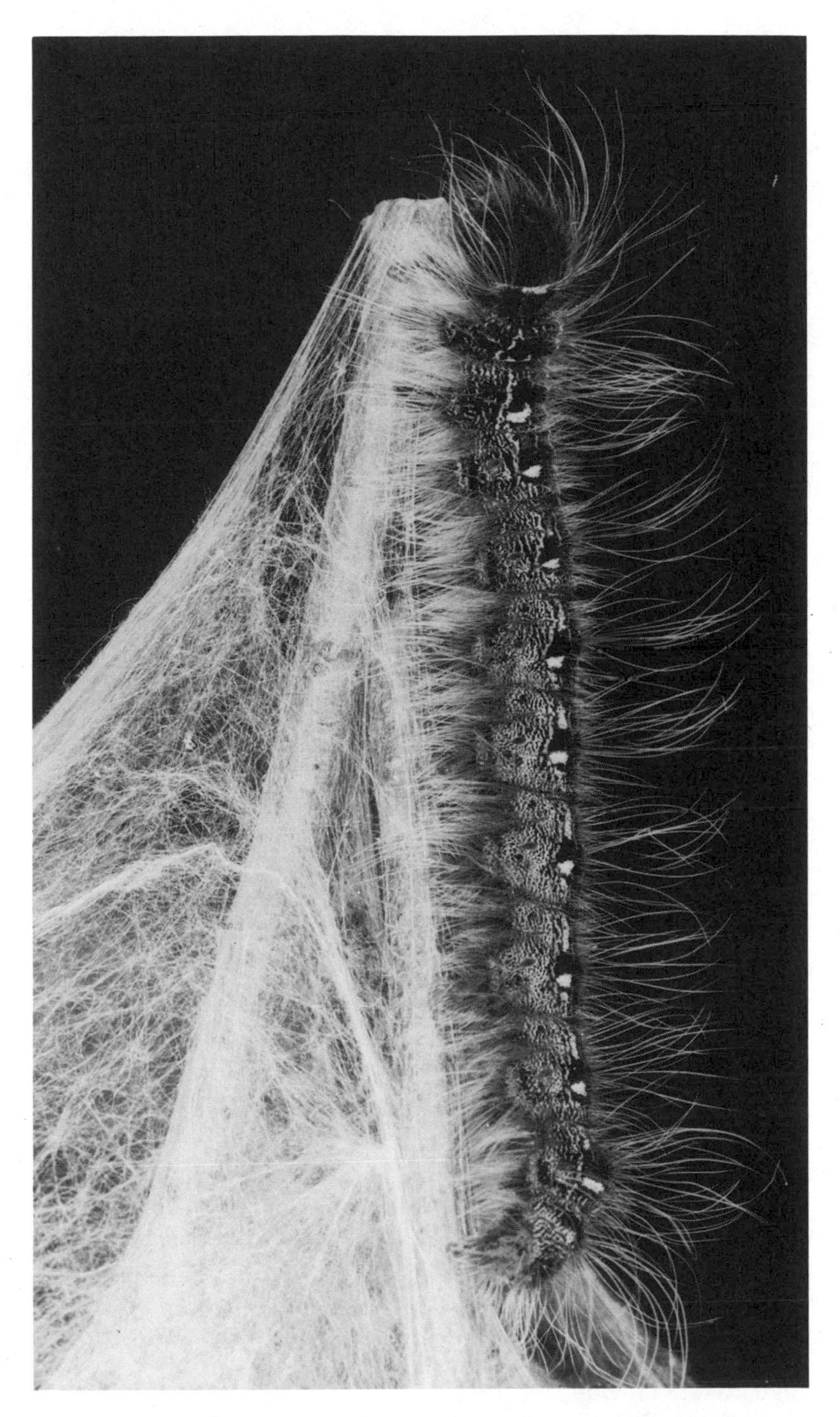

Figure 7.10. An eastern tent caterpillar resting on the surface of its tent. The long setae of the larvae disrupt air currents and reduce convective heat loss during thermoregulation, particularly when the caterpillars aggregate.

Figure 7.11. Larva of the eastern tent caterpillar hanging by its abdominal prolegs from the shaded side of the tent to facilitate convective heat loss and prevent overheating.

7.9). The mean temperature gradient found within 39 tents exposed to full sunlight was 7.3 ± 3.1°C, providing a thermally heterogeneous environment within which caterpillars could thermoregulate (Joos et al. 1988).

The earlier instars of the eastern tent caterpillar typically aggregate to bask just beneath the outer layer of the tent. During cool spring mornings, the small caterpillars warm rapidly, gaining the heat necessary to initiate activity and to process food collected during the morning foray (Fig. 7.12, Table 7.1). Although the layered structure of the tent might be expected to provide some insulation, helping caterpillars to retain T_b's during cloudy intervals or in the evening, there is little indication of

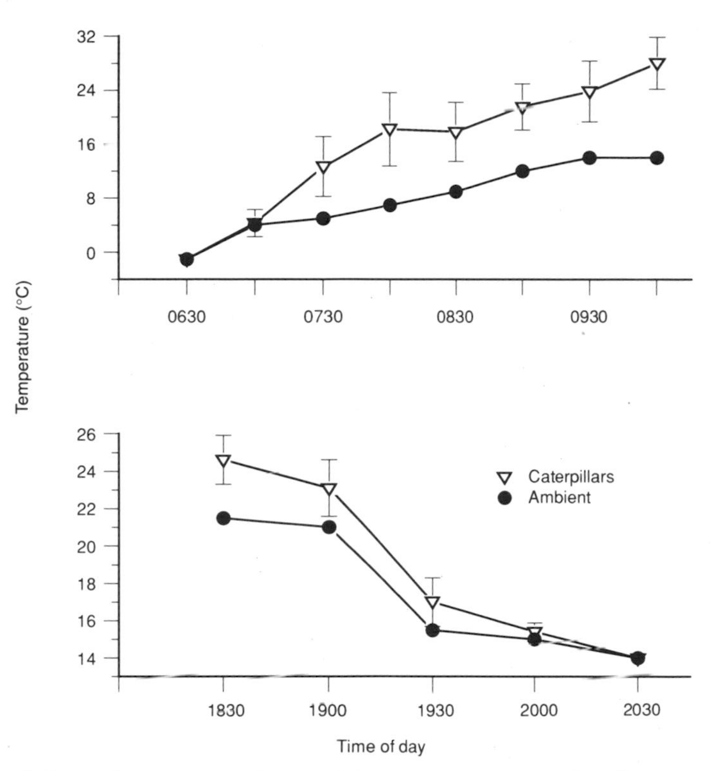

Figure 7.12. Ambient temperature and temperature measured at the center of an aggregate of third-instar eastern tent caterpillars lying just underneath the surface of their tent on a sunny day in mid-May. Values are means ± SD. The tent received solar radiation from approximately 0730 to 2000. (S. C. Peterson and T. D. Fitzgerald, unpubl. data.)

Table 7.1. Temperature of early instars of the eastern tent caterpillar aggregated just under the surface of their tent in sunlight

Date	No. of individuals	Body temperature (T_b) Mean ± SD (°C)	Body temperature (T_b) Range (°C)	Ambient temperature (T_a) (°C)
4 May	8	28.4 ± 1.7	24.8–30.8	18.5
7 May	11	37.0 ± 3.7	28.6–44.2	19.0
10 May	7	32.4 ± 1.7	30.0–35.6	17.0
12 May	7	34.2 ± 3.8	31.0–40.2	17.0

Source: From "Thermal Ecology, Behavior, and Growth of Gypsy Moth and Eastern Tent Caterpillars," by R. Knapp and T. M. Casey, *Ecology*, 1986, *67*, 598–608. Copyright © 1986 by Ecological Society of America. Reprinted by permission.

Table 7.2. Dry-matter budgets and growth of sixth-instar eastern tent caterpillars fed black cherry leaves and maintained from the second instar at three different temperatures

	Rearing temperature (°C)		
	27	22	17
Leaf ingested (dry mass, mg)	1000	940	540
Percent of ingested leaf assimilated	51	42	40
Percent of ingested leaf converted to caterpillar mass	19	16	16
Days to complete stadium	7	10	18
Final dry mass of caterpillar (mg)	218	188	103

Source: After Schroeder and Lawson 1992.
Note: All values are means for 10 caterpillars.

such an effect. Tents, and the caterpillars in them, cool to the ambient temperature shortly after the sun sets. Carlberg (1980) reported similar rapid cooling of the tent of *E. lanestris* in the evening. Moreover, temperature measurements taken at the center of dense aggregates of caterpillars lying within the structure after sunset indicate that there is no endothermic component to temperature regulation (Knapp and Casey 1986).

Thermoregulation and Growth

Tent caterpillars appear to have no special metabolic adaptations for feeding at low temperatures. Schroeder and Lawson (1992) reared caterpillars at three differerent temperatures and found that ingestion rate and assimilation efficiency decreased with decreasing temperature (Table 7.2). As a result, caterpillars held at low temperatures took markedly longer to develop and achieved much smaller final body masses than caterpillars held at higher temperatures.

Under field conditions, tent caterpillars must rely on behavioral thermoregulation to achieve body temperatures conducive to rapid growth, and, when allowed access to the sun's radiant heat, they grow markedly faster than caterpillars kept in the shade. In one study, two groups of genetically similar eastern tent caterpillars were maintained under two different temperature regimes in the laboratory but otherwise were treated similiarly (Casey et al. 1988). Both groups were held in a room with an ambient temperature of 20°C, but a radiant heat source was placed near the tent of one of the experimental groups for 12 hours each

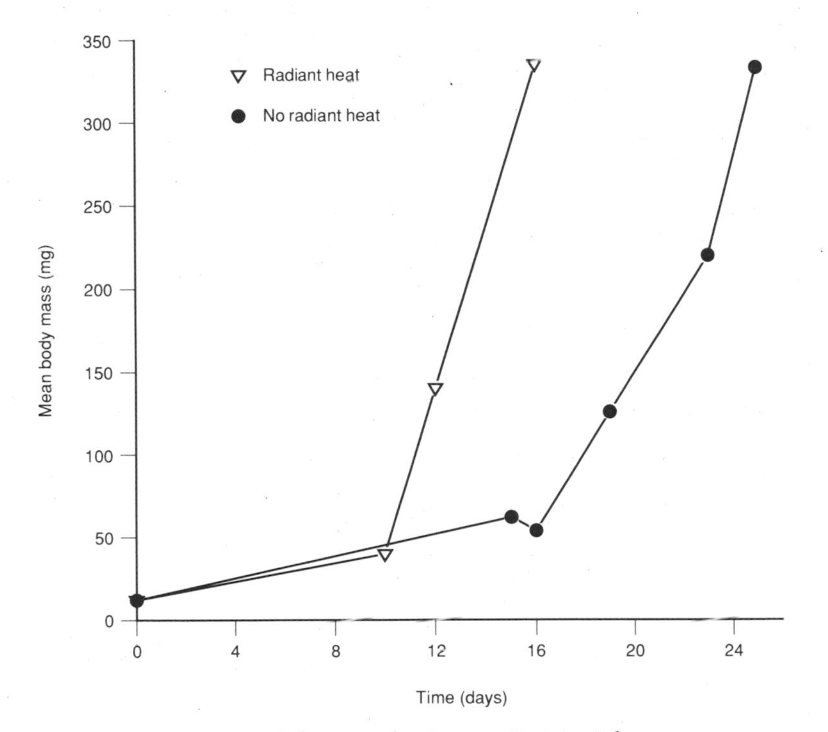

Figure 7.13. Comparison of the growth of two colonies of the eastern tent caterpillar held under different temperature regimes. Beginning on the 10th day of the study, one colony was provided with a radiant heat source for half of each day. (Reprinted from T. M. Casey et al., *Physiological Zoology* 61[1988]:372–377, © 1988 by the University of Chicago, by permission of the University of Chicago Press.)

day. The caterpillars in that tent were able to raise their T_bs to 35°C when they rested under the heat source, and they grew about 50% faster than the group held at a constant 20°C (Fig. 7.13). They grew faster not because they foraged more frequently (both groups ate four times each day) but because they processed the food stored in their gut more rapidly and, as studies by Schroeder and Lawson (1992) suggest, somewhat more efficiently. Caterpillars allowed to thermoregulate had guts that were nearly empty at the beginning of each activity period, and they were packed with new food during each foray. In contrast, caterpillars held at a constant 20°C began each bout of foraging with a gut still partially filled with food from the previous foray. They, therefore, had less room for new food and had lower overall ingestion rates. Indeed, when caterpillars were held for six hours at 15°C after a bout of foraging their foregut and midgut were still completely packed with leaf fragments, and they showed no indication of having processed food. Caterpillars held at 32°C for the same length of time had both the foregut and midgut completely empty (Casey et al. 1988). The exact temperature threshold for digestion is likely to vary at different locations in the caterpillar's range. Thus, Mansingh (1974) found that fifth-instar eastern tent caterpillars obtained from egg masses collected in Ontario continued to grow, albeit slowly, when held at a constant 15°C and were able to complete their development.

Not all species of spring-feeding caterpillars are thermoregulators. The gypsy moth, for example, is a thermal conformer, whose body temperature closely approximates T_a (Knapp and Casey 1986). Differences in the life histories of caterpillars appear to account for the evolution of one strategy over the other. The advantage to the eastern tent caterpillar of rapid larval development appears to be twofold. As discussed in Chapter 5, tent caterpillars are highly sensitive to food quality. They forage at a time when leaves are maturing rapidly and, in the process, becoming increasingly less palatable to the caterpillars. It is to the caterpillars' advantage to complete their development while their food supply is still of high quality, a circumstance that is likely to favor rapid development. In addition, tent caterpillars are attacked by many parasitoids and predators and are subject to various bacterial and viral diseases. Rapid development reduces the total amount of time that the caterpillars are exposed to depredation and disease agents. Moreover, the greater agility of caterpillars warmed by the sun may enable them to evade predators more readily than caterpillars with cooler body temperatures.

8

Predation and Antipredation

Tent caterpillars are large, conspicuous insects, and they are attacked in all stages of development by predators, parasitoids and parasites. They are eaten by birds and, on occasion, mammals and amphibians, and are infected by microsporidians and nematodes. But by far the largest toll is taken by entomophagous insects (Table 8.1, Fig. 8.1). Little is known of the biology and ecology of many of the organisms that attack tent caterpillars, and, indeed, most are encountered only rarely during field studies. Some, however, are common associates of tent caterpillars, and a few have obligatory relationships and life cycles that are tightly sychronized with those of their tent caterpillar hosts. The life histories of the more important and better known of these and the antipredatory responses of tent caterpillars to attacks by entomophages are reviewed here.

Predators and Parasitoids

All entomophagous insects that attack tent caterpillars function as either predators or parasitoids. Predatory entomophages typically attack and kill several to many caterpillars during their lifetime. The need to search actively for successive prey items favors mobility, and the predators of tent caterpillars are mobile larval, nymphal, and adult forms. Parasitoids require only a single host caterpillar to complete their development and, for the most part, are relatively immobile larval forms. Parasitoids may be either specialists, confining their attack only to tent

Table 8.1. Taxonomic distribution of parasites / parasitoids (PA) and predators (PD) of North American species of tent caterpillars

Taxon	Egg		Larva & pupa	
	PA	PD	PA	PD
Microsporidia			3[a]	
Nematoda			1	
Acari		3		
Insecta				
Dermaptera				1
Hemiptera				
Miridae				1
Reduviidae				4[b]
Pentatomidae				15
Coleoptera				
Carabidae				5
Staphylinidae				1
Elateridae				1
Coccinellidae				2
Diptera				
Muscidae			5	
Calliphoridae			1	
Sarcophagidae			5	
Tachinidae			41	
Hymenoptera				
Braconidae			11[c]	
Ichneumonidae			44[d]	
Trichogrammatidae	2			
Eulophidae	4		3	
Encyrtidae	2			
Eupelmidae	2			
Torymidae	1		2	
Pteromalidae			3	
Chalcididae			1	
Scelionidae	3			
Formicidae				7
Vespidae				3[e]
Sphecidae				1
Amphibia				2
Aves		3		63[f]
Mammalia		1		2

Source: Compiled from Witter and Kulman 1972 except as otherwise noted.

[a]*Nosema disstriae, Pleistophora schubergi,* and *Vairimorpha necatrix* from *Malacosoma disstria* (H. M. Thomson 1959, G. G. Wilson 1984)

[b]*Banasus banasus* from *M. californicum pluviale* (Adams 1989); *Zelus exsanguis* from *M. disstria* (Knight et al. 1991)

[c]Mason (1979) described *Rogas malacosomatos* as a parasitoid of *Malacosoma* spp. and stated that previous references to *R. stigmator* as a parasitoid of tent caterpillars are incorrect.

[d]*Enicospilus cushmani* from *M. disstria* and *M. americanum* (Gauld 1988); *Coccygomimus disparis* from *M. americanum* (Schaefer et al. 1989); *C. sanguinipes erythropus* from *M. disstria* (Knight et al. 1991).

[e]*Polistes fuscatus* from *M. americanum* (Fitzgerald et al. 1988); *Vespula vespula* from *M. californicum pluviale* (Adams 1989) previously identified only to genus.

[f]*Larus pipixcan* (Shepherd 1979); *Dendroica castanea* (Sealy 1979); *Larus delawarensis* (Bloome 1991).

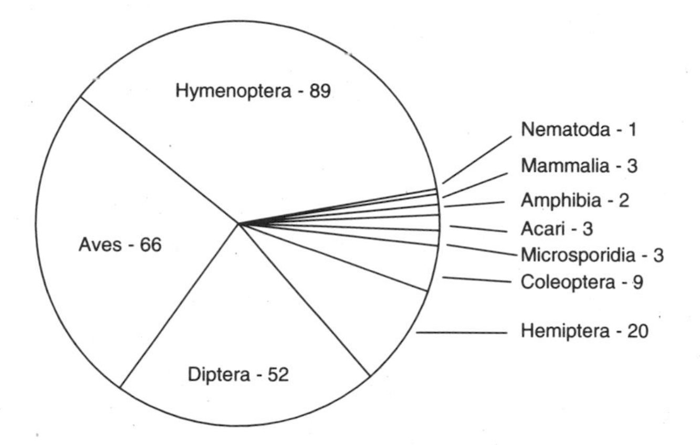

Figure 8.1. Taxonomic distribution of known predators, parasitoids, and parasites of North American tent caterpillars. Data from Table 8.1.

caterpillars, or opportunistic generalists, attacking a broader array of insects.

Adult parasitoids deposit eggs in, on, or near the host, and their offspring consume the victim. Parasitoid larvae feed from within the caterpillar (endoparasitoids) or consume the host from the outside (ectoparasitoids). They typically confine their attack to a particular point in the life cycle of the host. Thus, whereas the smallest wasp parasitoids of tent caterpillars obtain all the food required for their development from the contents of a single egg, large parasitoid flies attack only the mature caterpillars and pupae. The parasitic habit is well developed in the Insecta, and the parasitoids of tent caterpillars may themselves become victims. Secondary parasitism is common among the entomophages of tent caterpillars and involves 33 known species of hyperparasitoids (Witter and Kulman 1972).

The entomophagous insects associated with tent caterpillars occur in the orders Coleoptera, Hemiptera, Diptera and Hymenoptera. All the Coleoptera and Hemiptera function as predators. In addition, more than 60 species of birds prey on tent caterpillars. All the known Diptera associated with tent caterpillars function as ecto- or endoparasitoids. With the exception of the ants and the vespid and sphecid wasps, all the hymenopterans are parasitoids. Most of the entomophagous insects that attack tent caterpillars occur in the families Pentatomidae, Tachinidae, and Ichneumonidae (Table 8.1). Although North American tent caterpillars are attacked by 171 known species of entomophagous insects, most are relatively rare and typically only a few have a significant impact on the populations of any given species of tent caterpillar. For ex-

Table 8.2. Relative abundance of the most common parasitoids of the forest tent caterpillar during three population outbreaks

Relative abundance	Species	Habits
	Sarcophagidae	
1	*Sarcophaga aldrichi*	Specialist on tent caterpillars. Parasitizes pupa; overwinters in soil.
	Tachinidae	
2	*Patelloa pachypyga*	Specialist on tent caterpillars. Microtype eggs ingested by late instars. Parasite larva emerges from host pupa and overwinters in soil.
3	*Lespesia frenchii*	Multiple hosts. Macrotype eggs oviposted on later instars. Parsite larva emerges before or just after host pupation; forms puparium in soil.
4	*Leschenaultia exul*	Specialist on tent caterpillars. Deposits microtype eggs on foliage which are ingested by larvae; overwinters in soil.
	Ichneumonidae	
5	*Itoplectis conquisitor*	Multiple hosts. Attacks pupal stage.
6	*Trichonotus analis*	Multiple hosts. Attacks late instars; pupates in eviscerated host.
	Braconidae	
7	*Rogas* spp.	Specialists on tent caterpillars. Attack small caterpillars; emerge from eviscerated 4th-instar host.

Source: Data from Sippell 1957, Witter and Kulman 1979, Eggen 1987.

ample, Witter and Kulman (1979) recovered 41 entomophagous insects from the forest tent caterpillar, but fewer than one-quarter of them are abundant during outbreaks (Table 8.2).

Predators

Coleoptera

Several species of Coleoptera have been reported to be associated with colonies of tent caterpillars (Witter and Kulman 1972), but most appear to be inquilines or scavengers. Beetles in the genus *Calosoma*, however, are well-known predators of caterpillars, and some species attack tent caterpillars. The best known of these is *C. sycophanta*, which was studied extensively by Burgess and Collins (1915). In the early 1900s approximately 4000 of the beetles were captured in Europe, packed singly in match boxes and shipped to the United States, where they or their progeny were eventually released in New England, New Brunswick, California, and New Mexico for the control of various species of caterpillars, principally the gypsy moth. The beetle adapted well and has become firmly established in North America. Although the seasonal life history of *C. sycophanta* is not well synchronized with that of tent caterpillars, late-stage larvae and pupae of *Malacosoma americanum* and *M. disstria* are still present when the beetles emerge, and they are attacked by the predator (Witter and Kulman 1972).

The adult *C. sycophanta* is a colorful metallic green beetle, 25–30 mm long. It has an unusually long life span of two to four years, but except for approximately seven weeks each year, the beetle spends its time diapausing in the soil. Beetles emerge in early June from an overwintering cell formed in the soil the previous summer. They comb the branches of trees in search of caterpillar prey and may disperse by flight if food is scarce. Caterpillar prey are usually grasped at their middle and cut in half with sharp mandibles; the beetles often injure more caterpillars than they actually eat. They have voracious appetites. Burgess and Collins (1915) determined that a single beetle killed an average of 239–328 caterpillars from the time it emerged in June until it returned to the ground in mid-July, depending on its age. The beetles also search for and kill pupae.

The beetles mate in early June, and the female oviposits in the soil soon thereafter. The eggs hatch in 3–10 days. Larvae search the branches of trees for caterpillar prey. The larva is highly active and mobile. One individual, followed in a specially designed apparatus for 72 continuous hours, traveled a total of nearly 2.9 km (Burgess and Collins 1915). Like the adult, the larva is a bold predator and even the first instar is able to successfully attack and kill full-grown caterpillars. In one experiment, the larval predator killed an average of 41 full-grown gypsy moth caterpillars during the 14 days between emergence and the time it stopped eating to prepare for pupation. The larvae also tear open cocoons of tent caterpillars and attack the pupae.

By mid-July the shiny black larvae are about 25 mm long. They cease feeding and burrow into the soil to form pupal cells in which they overwinter. Surviving adults also move to the soil and form overwintering cells at this time. In addition to *C. sycophanta,* several other species of *Calosoma* are known predators of tent caterpillars, but little is known of their impact on populations of their hosts.

Hemiptera

The most important hemipteran predators of tent caterpillars are the pentatomids, or stink bugs. Fifteen species of Pentatomidae are known to prey on tent caterpillars. These bugs are all generalist predators,

Figure 8.2. The pentatomid bug *Podisus* feeding on an eastern tent caterpillar.

though some specialize on the caterpillars of beetles, sawflies, and lepidopterans. They are timid predators that dispatch their prey by injecting them with saliva that contains paralytic substances and then feed by drawing the liquefied body contents of the prey through their beaks. The adults are often seen feeding or walking about on the branches of caterpillar-infested trees with a flaccid larval prey dangling from their beaks (Fig. 8.2).

Downes (1920:28) described the habits of *Apateticus crocatus*, a generalist predator of caterpillars, including *M. americanum*:

> It was interesting to observe the extreme timidity and caution which marked the attitude of the bugs when attacking their prey. Not being possessed of any weapon to aid them, such as the powerful grasping forelegs of those species that are solely predatory, they are forced to await a propitious time for attack. On scenting game the beak is immediately extended and the bug advances toward its quarry. Usually weak and sickly caterpillars are selected or one that is in such a position as to be unable to escape. When within half an inch or so the rate of advance is cautiously slackened and the progress of the extended beak towards the caterpillar becomes so slow as to be scarcely perceptible. Should the caterpillar make the slightest movement, the bug immediately retreats, advancing again and again until at last from sheer weariness the caterpillar permits the beak to be inserted. Once this occurs there is no escape. The barbed tips of the maxillae give the bug a hold that is not readily shaken off, and despite wriggling and squirming the beak turns in the wound without withdrawing. Not until the caterpillar is sucked dry does the bug withdraw its beak, and it may retain its hold for as long as twelve hours. It is only when pressed by hunger that they show any boldness.

Four sympatric species of the stink bug genus *Podisus* prey on the eastern tent caterpillar in the northeastern United States (E. W. Evans 1983). They overwinter as adults and attack the caterpillars in the early spring. Three of these bugs may co-occur on the same infested trees that grow in open fields, but they appear to hunt in only partially overlapping feeding zones. *P. placidius* preys on caterpillars inside the tent. *P. maculiventris* primarily attacks caterpillars that are outside the tent or on branches nearby. *P. modestus* feeds at distances exceeding 30 cm from the tent. All three limit their predation to caterpillars 20 mm or less in length, though they are occasionally able to kill larger prey. Nymphs appear later in the season, but the tent caterpillars are usually too big to be attacked by then, and only occasionally have stink bugs been observed to feed on them. There have been no attempts to evaluate the effectiveness of these insects as predators of tent caterpillars, but, when abundant, they may inflict significant damage on colonies. *P. maculiventris*, one of the most common of tent caterpillar predators, was credited in one study with consuming 122 army worm larvae during a nine-week period (C. P. Clausen 1972).

Hymenoptera

Several species of ants in the genera *Myrmica*, *Camponotus*, and *Formica* are predators of tent caterpillars. Because their nests are perennial and well stocked in the early spring, ants have the potential of devastating incipient colonies of tent caterpillars. The ant *Formica obscuripes* appears to be one of the more significant ant predators of eastern tent caterpillars (Ayre and Hitchon 1968, Tilman 1978). As discussed in Chapter 5, ants of this species visit cherry trees in the early spring, apparently to obtain sugars secreted from leaf-margin nectaries. When they encounter tent caterpillars, the ants kill them and take them back to their nest. Predation is limited to caterpillars in the early instars. The ants, which average 7.5 mm in length, are nearly three times the size of the small first-instar caterpillars and have no difficulty subduing them, but when the caterpillars reach a length of 15–18 mm they are too formidable a prey, and the ants no longer bother with them. Thus, only if a caterpillar colony is discovered early enough can the ants inflict serious damage.

Observations by Ayre and Hitchon (1968) suggest that *F. obscuripes* ants that find a tent may recruit nestmates to their discovery and systematically remove the caterpillars. They attributed the disappearance of all the larvae in 13 of 16 experimental colonies they established near ants' nests to depredation by this species. Proximity of the ants to the tent caterpillars appears to bear strongly on the likelihood of predation by this ant. Tilman (1978) found that none of the tents in trees that were within 10 m of ant nests produced late-instar larvae, and those at greater distances had proportionally greater numbers of mature caterpillars. Significant loss of whole colonies of early instars has also been attributed to the ants *F. fusca* and *F. lasioides*. *Myrmica americana* has been observed to kill small eastern tent caterpillars, but it apparently is not as effective a predator as the *Formica* ants. The carpenter ant *Camponotus herculeanus* preys on the forest tent caterpillar until the larvae reach the fifth instar and has been reported to be capable of destroying whole colonies (Green and Sullivan 1950).

In addition to ants, the large wasps in the family Vespidae are predatory on a broad range of insect prey. Although tent caterpillars might well be a favored prey of these wasps because of their social habits and large size, the wasps are seasonal nesters and most have only incipient colonies at the time tent caterpillars are active or small enough to subdue. Thus, although vespid wasps have been reported to devastate colonies of caterpillars that occur later in the year, such as web-builder *Hyphantria cunea* (R. F. Morris 1972), there have been few observations of predatory wasps attacking tent caterpillars in North America (Witter and Kulman 1972).

That tent caterpillars are attractive to vespid wasps was demonstrated inadvertently during a field study of the eastern tent caterpillar that required that colonies be established later in the season than they naturally occur (Casey et al. 1988, Fitzgerald et al. 1988). Although no depredation by predatory wasps was observed during the normal field season, it was necessary to build wire exclusion cages around tents established in July to prevent *Polistes fuscatus* wasps from killing and carrying off fourth- and fifth-instar caterpillars. Wasps that discovered experimental colonies returned repeatedly, and at least one unprotected colony was completely destroyed. The wasps killed individuals as they rested on the outside of the tent or as they moved to and from feeding sites, but they did not extract caterpillars from within the tent (Plate 2A). Adams (1989) also found that colonies of the western tent caterpillar that he attempted to rear late in the season, after naturally occurring colonies had already completed their life cycle, were heavily attacked by *Vespula wasps*. In Japan, *Polistes* wasps have been reported to be significant predators of late-instar larvae of *M. neustrium testacum*. Shiga (1979) observed that once wasps had located a colony they returned continually and removed the caterpillars until few remained.

H. E. Evans (1987) observed the solitary sphecid wasp *Podalonia occidentalis* preying on the caterpillars of *M. californicum* in Colorado. The wasps sting the caterpillars along the midventral line then drag them to nest sites constructed in the ground. Nests are provisioned with only a single caterpillar, and the wasps seem to prefer full-grown larvae. The life cycle of the wasp is synchronized with the life cycle of the prey, and the predator appears to provision its nest only with tent caterpillars. Wasps, which appeared about the time the *M. californicum* caterpillars were in the last instar and dispersing, were no longer seen flying after the caterpillars had pupated.

Birds

More than 60 species of birds have been reported to feed on tent caterpillars in North America (Witter and Kulman 1972). The actual impact of bird predation on the population dynamics of tent caterpillars is largely unknown, but the numbers of caterpillars that occur during even the incipient stages of outbreaks is likely to exceed the capacity of insectivorous birds to contain them. Most of the significant bird predators are territorial and are likely to be thinly distributed relative to the density of the caterpillar. Moreover, birds have relatively low reproductive rates and, unlike entomophagous insects, are not capable of the magnitude of numerical response needed to regulate caterpillar populations. Field studies also show that birds exhibit little or no apparent functional response to increasing tent caterpillar density during the incipient phases of caterpillar outbreaks. Most species of birds find any but the smallest tent caterpillars unpalatable, and even when they are surrounded by huge numbers of potential prey, tent caterpillars are likely to constitute only a fraction of the food intake of the resident birds. The impact of songbirds on populations is also lessened by the habit of tent caterpillars to have mostly completed their larval development before the eggs of many songbirds hatch, producing young that create a manifold increase in the demand for food. Nonetheless, studies show that many birds consume tent caterpillars, and small populations of the insects may on occasion suffer significant mortality and setback due to bird predation.

The eggs of tent caterpillars are susceptible to predation longer than any of the other life stages, but eggs constitute only a trivial portion of the diet of birds. The eggs have tough chorions and are cemented into tight aggregates. Those of most New World species are also covered with spumaline and appear to be a largely unappealing food item to any but the hungriest of birds. During difficult times, winter foragers are likely to take whatever food is available, and some winter residents of the northeastern United States include some tent caterpillar eggs in their diets. Stomach analyses performed on black-capped chickadees (*Parus atricapillus*) by Weed (1898) showed that during the winter months approximately half the food taken consisted of insects. The stomachs of 3 of the 41 chickadees studied contained eggs of the forest or eastern tent caterpillar, but the eggs constituted no more than 6% of any one bird's diet. In contrast, the soft eggs of aphids were found in the stomachs of 31 birds and made up as much as 35% of the diet of a given individual. McAtee (1927) reported that titmice (*Parus* sp.), also ate the eggs of tent caterpillars but provided no quantitative data. He also noted that the stomachs of two blue jays (*Cyanocitta cristata*) contained 845 and 1047 tent caterpillar eggs, relatively significant numbers, but there have been no subsequent reports of predation on tent caterpillar eggs by this species.

In contrast to the few reports of predation on tent caterpillar eggs, there are numerous accounts of bird predation on tent caterpillar larvae. Many of the early field observers were enthusiastic champions of birds, and they filed anecdotal reports of birds decimating populations of tent caterpillars. McAtee (1927) quoted the contents of a letter sent to him by an amateur ornithologist, dated April 7, 1915:

> Some years ago I worked as Entomologist through three seasons of serious infestations by forest tent-caterpillars in a small Vermont town, and my daily, even hourly, watching of birds which ate, or fed their young, any stage of that pest, enabled me to add a few facts in favor of the Baltimore oriole, which not only devoured larvae in all stages, and moths, but wholly cleared many trees of the pupae, ripping cocoons so fast that the young were given hundreds of pupae in a day. The redwing blackbirds were a good second and cleared low trees and shrubs. Chebecs, wood pewees, and phoebes snapped up the larvae which were falling by threads, and ate the moths which fly before dark. Cuckoos and downy woodpeckers did their share and nuthatches ate larvae by scores.

Rigorous investigations of the feeding behavior of individual bird species provide equivocal support for the contention that birds, with the exception of the cuckoos and a few other species as noted below, are the voracious predators of tent caterpillar larvae that the early and undocumented literature suggests. For the most part, birds that attack larval tent caterpillars are opportunistic feeders and appear not to be predisposed to deal with any but the smallest larvae.

Fashingbauer et al. (1957) studied the response of woodland bird populations to outbreaks of the forest tent caterpillar. Fifteen species were observed to capture tent caterpillars, and larvae were found in the stomachs of about half of the birds dissected. Most of the birds had eaten only the smaller caterpillars, even though full-grown larvae were abundant. Red-eyed vireos (*Vireo olivaceus*), veerys (*Catharus fuscescens*), and black-billed cuckoos (*Coccyzus erythropthalmus)* were the most significant predators. In addition to those species, Witter and Kulman (1979) also found larger forest tent caterpillars in the stomach of the robin (*Turdus migratorius),* and the brown-headed cowbird (*Molothrus ater*), but most of the other birds they observed fed only on small caterpillars.

Despite the huge number of tent caterpillars available during the course of outbreaks, Fashingbauer et al. (1957) showed that the birds did not gorge on them, and they constituted only a fraction of the birds' diets. Moreover, the investigators noted that woodland birds are likely to have an insignificant impact on outbreak populations of tent caterpillars because the number of breeding pairs per acre (3–4) is too few to exert significant predation pressure on the population. In addition, the birds did not begin to establish nests until the caterpillars were far along in their development, and by the time the nestlings of most of the bird species created a demand for food the caterpillars had already pupated or emerged. Sealy (1979) made similar observations on the lack of temporal synchrony between the demand for food by nesting bay-breasted warblers (*Dendroica castanea*) and availability of suitably sized forest tent caterpillars (Fig. 8.3).

Studies by Root (1966) showed that most of the woodland birds he observed included few if any tent caterpillars in their diets. He documented the predation of caterpillars by birds in California in an area experiencing an outbreak of *M. constrictum.* Although the diet of blue-gray gnatcatchers (*Polioptila caerulea*), warbling vireos (*Vireo gilvus*), Hutton vireos (*Vireo huttoni*), orange-crowned warblers (*Vermivora celata*), and plain titmice (*Parus inornatus*) included many different species of caterpillars, direct observation and gut analysis indicated that these birds did not prey on the larvae of *M. constrictum,* even though the tent caterpillar was very abundant, conspicuous, and well within the size range of the caterpillars of other species eaten by the birds.

Songbirds that attack colonies of the eastern tent caterpillar often do so primarily at the outset of the breeding season when caterpillars are still small, and they may inflict considerable damage on the tents and their occupants (Fig. 8.4). During the spring of 1917 Saunders (1920) observed that many of the new tents of the eastern tent caterpillar she found had been torn open by newly arrived songbirds. She observed a warbler (*Parula americana*) and a yellow-breasted chat (*Icteria virens*) feeding on the small caterpillars and noted that many of the tents were completely empty, apparently because of bird predation. It is not uncommon, however, to find the small tents of the eastern tent caterpillar torn open by birds early in the season but left unmolested by territorial

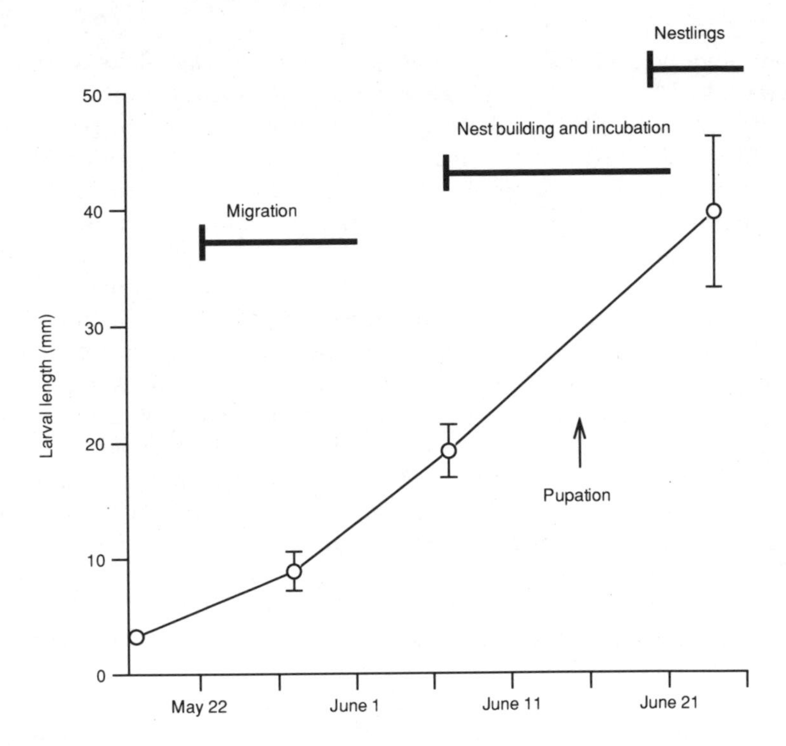

Figure 8.3. The relationship between the seasonal chronology of nesting bay-breasted warblers (*Dendroica castenea*) and the size of forest tent caterpillar larvae. Although the birds feed on the smaller caterpillars during May and early June, the caterpillars have grown too large to eat and have nearly completed their development by the time the nestlings create a demand for food. (Data from Sealy 1979.)

pairs thereafter, presumably because the caterpillars prove to be less palatable than other available prey items.

Most authors have attributed the avoidance of tent caterpillars by birds to their hairiness. Caterpillar hairs accumulate in the guts of birds, block the stomach lining, and eventually interfere with digestion. To avoid setae, some birds may skin caterpillars before ingesting them. Root (1966) observed Bullock orioles (*Icterus galbula*) eating the late-instar larvae of tent caterpillars on two occasions. In each instance, the birds removed the integument of the larva and ingested only the insides. Similar behavior was reported for the pine siskin (*Carduelis pinus*) feeding on full-grown forest tent caterpillars in British Columbia (Grant 1959). The

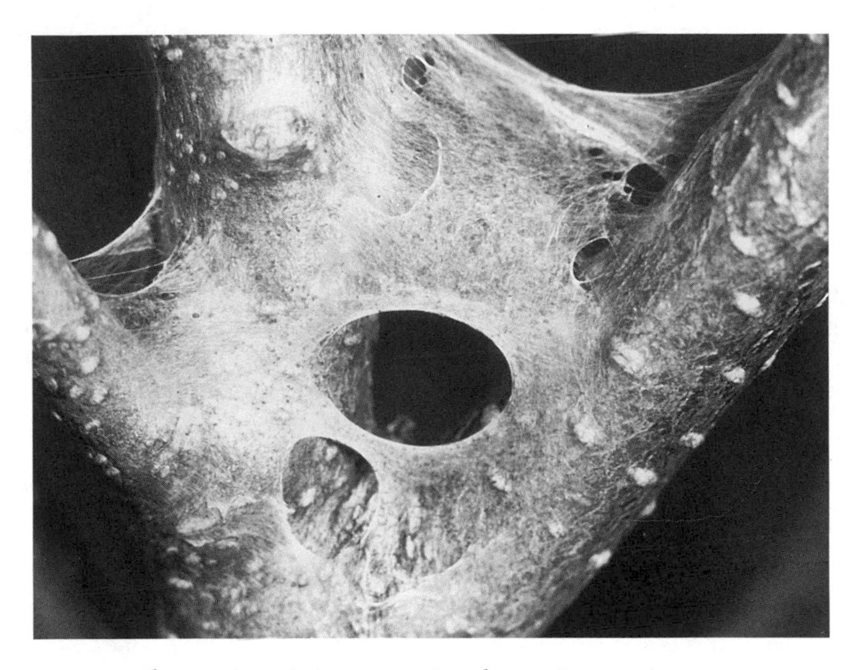

Figure 8.4. Characteristic damage caused to the small tents of eastern tent caterpillars by songbirds.

siskin was reported to stand on the body of the caterpillar, grasping the head in its beak, and pulling until the gut was drawn out of the integument. The bird fed selectively on only a portion of the exposed viscera. It is likely that some birds reject hairy species solely on the basis of tactile or visual assessment and, thus, avoid the caterpillars entirely.

Birds that ingest whole caterpillars are also feeding on plant material that is packed into the victim's gut. In certain cases, the material may constitute an additional deterrent to bird predation. Eastern tent caterpillars, for example, fill their gut with the highly cyanogenic leaves of black cherry (see Chapter 5). Recently fed caterpillars are likely to be bitter as well as toxic, but it has not yet been demonstrated that these properties cause birds to reject them. If the gut is bitter and toxic and the integument hairy, there would seem to be little left for the bird to enjoy.

The most important bird predators of tent caterpillar larvae in North America are the cuckoos. The yellow-billed cuckoo (*Coccyzus americanus americanus*) and its western subspecies (*C. americanus occidentalis*) occur over most of the Austral zone of North America. The black-billed cuckoo

(*C. erythropthalmus*) occurs east of the Rocky Mountains in the northern half of the United States and southern Canada. Both species overwinter in South America.

Cuckoos arrive in the north early in the spring while tent caterpillars are still in the larval stage, and the caterpillars make up an appreciable portion of their diet. The birds tear open the tents of the caterpillars and can eliminate entire colonies. Indeed, investigators have reported that the bird is capable of decimating whole populations of the caterpillars. Cuckoos are well adapted to feeding on hairy species of caterpillars and include even in their diets species such as the *Io* moth, which has urticating setae. Although the hairs of caterpillars accumulate in their stomachs to the point where they might interfere with the process of digestion, cuckoos circumvent this problem by periodically shedding the lining of their stomach and growing a new one Bent (1940).

The voracious feeding behavior of the cuckoo was documented by E. H. Forbush, as quoted by Bent (1940:59):

> The cuckoos are of the greatest service to the farmer, by reason of their well-known fondness for caterpillars, particularly the hairy species. No caterpillar is safe from the Cuckoo. It does not matter how well they may be protected by webs. Often the stomach of the Cuckoo will be found lined with a felted mass of caterpillar hairs, and some times its intestines are pierced by the spines of the noxious caterpillars that it has swallowed. Wherever caterpillar outbreaks occur we hear the calls of the Cuckoos. There they stay; there they bring their newly fledged young; and the number of caterpillars they eat is incredible.

Forbush, cited in Bent (1940:78), went on to say:

> In seasons when caterpillars of any species are abundant, cuckoos usually become common in the infested localities. They follow the caterpillars, and where such food is plentiful, the size of their broods seems to increase. During an invasion of forest tent caterpillars in Stoneham, Massachusetts, in May,1898, Mr. Frank H. Moser watched one of these birds that caught and ate 36 of these insects inside of five minutes. He saw another in Malden eat 29, rest a few minutes and then eat 14 more. . . . The late Professor Walter B. Barrows, of Michigan, an extremely conservative ornithologist, is responsible for the statement that in several instances remains of over 100 tent caterpillars have been taken from a single cuckoo's stomach.

The pupae of tent caterpillars appear to be the most palatable of the life stages and may be actively sought by some species of birds that are opportunistic foragers. Hodson (1941) observed crows (*Corvus brachyrhynchos*) feeding on the pupae of the forest tent caterpillar during an outbreak in Minnesota. The birds removed cocoons from their attachment sites and tore them open by holding them with their feet. Estimates made one day after the crows were observed indicated that they had taken about 20% of the local population of pupae. Shiga (1979) studied predation by grey starlings (*Sturnus*) on the pupae of *M. neustrium testacum.* Over an eight-year period, starlings took from 2 to 95% (mean = 32%) of the pupal population of his study area each year. The absolute number of pupae taken by the birds increased with prey density. The percentage of the prey population cropped also increased each year over the last five years of the study. An individual's past season's experience and observational learning by new birds may have accounted for the season-to-season increase in foraging efficiency of the starlings.

Sealy (1980) documented an apparent numerical response of breeding pairs of northern orioles (*Icterus galbula galbula*) during an outbreak of forest tent caterpillars. In the first year of the outbreak, he found 52 nests in his study area. In the second year the number increased to 101 nests, but the outbreak of the caterpillars collapsed and relatively few tent caterpillars were available to feed the nestlings. Tent caterpillars were virtually absent on the study site in the third year, and the number of oriole nests dropped to 51. During the first year of the infestation orioles were observed feeding both larvae and pupae of tent caterpillars to their nestlings and were observed capturing flying tent caterpillar moths. Approximately 63% of the fledgling diet was estimated to consist of larval and pupal forest tent caterpillars. The caterpillars had for the most part pupated by the time the eggs of the orioles had hatched, so pupae constituted approximately 78% of this component of the diet. The parents extracted the pupae from the cocoons and fed them whole to the nestlings.

A curious incident of opportunistic predation on cocoons of *M. disstria* was reported by Bloome (1991). Flying ring-billed gulls, *Larus delawarensis,* were observed to swoop in on the upper branches of a tree standing near the shore of a lake in which tent caterpillars had spun numerous cocoons among the leaves and twigs. A strong headwind allowed the gulls to hover while they plucked cocoons from the tree. The gulls were observed to swallow the whole cocoons and the enclosed pupae after they flew away with them.

Relatively little is known of the impact of birds on the tent caterpillar moths. The moths of tent caterpillars are inconspicuous when they rest motionless during the daylight hours and are unlikely to be discovered except by the most diligent of predators. Moths that are active only at

Figure 8.5. An undetermined species of tachinid parasitoid about to parasitize an eastern tent caterpillar. The caterpillar has been previously parasitized and carries the small white egg of a tachinid on its thorax.

night are also protected from birds. Thus, only those species that fly during the day, like the forest tent caterpillar, are likely to suffer appreciably from bird predation. Indeed, Bieman (1980) found that the slow-flying moths of this species are easy targets for birds and may be taken in large numbers. In another study, crop analysis of 11 Franklin's gulls (*Larus pipixcan*) captured in late morning while flying above a stand infested with forest tent caterpillars showed that the gulls had eaten 1490 tent caterpillar moths (Shepherd 1979). Ninety percent of the moths were males, the more active of the sexes. Krivda (1980) observed house sparrows (*Passer domesticus*) capturing adult eastern tent caterpillar moths that had alighted in large numbers on a brick wall after swarming the previous evening. Predation was observed throughout the afternoon hours, and some birds took one moth after another, discarding the wings as they fed.

Parasitoids

Diptera

Tachinidae. Most of the dipteran parasitoids of tent caterpillars belong to the family Tachinidae. The flies are obligatory parasitoids, and many are univoltine species that have evolved life cycles tightly synchronized with those of the host species. Adult flies deposit eggs or larvae on the caterpillar host (Figs. 8.5, 8.6) or on the substrate in areas frequented by caterpillars. The latter type has extremely high fecundity.

One species that deposits eggs on leaves is the tachnid *Leschenaultia*

Figure 8.6. Heavily parasitized eastern tent caterpillar. The 30 eggs attached to the caterpillar were deposited by an undetermined species of tachinid parasitoid. Most of the eggs are attached on the anterior portion of the body, where they cannot be bitten off by the caterpillar.

exul, an obligatory parasitoid of *M. americanum* and *M. disstria* in the eastern United States. The insect overwinters in a puparium constructed in the soil and emerges in late April and early May in synchrony with the seasonal emergence of the tent caterpillar host. During their lifetime, the adults produce up to 5000 microtype eggs, which they fasten singly to the undersurface of the leaves of trees colonized by tent caterpillars. Oviposition is apparently nonrandom: in laboratory studies, Bess (1939) found that the presence of caterpillars or clipped leaves stimulated flies to deposit eggs. The eggs remain viable for as long as 26 days, and some are eventually ingested by tent caterpillars as they feed on the contaminated leaves.

The eggs of the parasitoid hatch within a few hours of ingestion, and the young maggots move from the gut to the salivary gland. The first stadium lasts 8–10 days and appears to be completed in the gland. Second- and third-instar maggots develop in an integumental funnel, usually found in the posterior abdomen, that allows them to obtain oxygen from outside the host. The second and third stadia together last 5 to 6 days and then the maggot emerges from the eviscerated host. The larva moves to the soil, forms its puparium, and remains there until the following spring. Only a single fly usually develops in a caterpillar, but as many as four flies can sometimes obtain enough nourishment from a single host to complete their larval development (Bess 1939). Field-collected pupae of *M. americanum* and *M. disstria* indicate parasitism rates by this species from less than 1% to 23% (Bess 1939, Sippell 1957, Eggen 1987).

Another tachinid, *Compsilura concinnata,* is a generalist parasitoid of caterpillars that was introduced into the United States from Europe in the early 1900s for the control of the gypsy moth. In Europe, the fly attacks *M. neustrium* and has been reported from *M. americanum, M. disstria,* and *M. californicum* in the United States. This is a larviparous species. The eggs hatch while they are still within the elongated uterus of the parent and can be stored there until they are deposited. The fly produces about 100 larvae during its lifetime and it injects them into the host after penetrating its integument with a sharp, piercing organ (Fig. 8.7). This species was reported to be rare or absent in collections of parasitized larvae of *M. disstria* by Sippell (1957) and Witter and Kulman (1979) in the Great Lakes states but was the fifth most common species collected by Eggen (1987) from populations of *M. disstria* in New York State. Eggen attributed this difference, in part, to the co-occurrence of both the tent caterpillar and the gypsy moth in his study plots.

Sarcophagidae. Several sarcophagid flies are also prominent parasitoids of tent caterpillars. Those associated with tent caterpillars lack the finely tuned host-parasitoid relationship found among the tachinids and are only facultative parasitoids.

Hodson (1939b) conducted a detailed study of *Sarcophaga aldrichi* after the insect appeared in huge numbers during an outbreak of the forest tent caterpillar in Minnesota and established that it attacks and kills otherwise healthy pupae. The fly has a single generation each year and overwinters in a puparium in the soil. It emerges in late May, and the females are ready to larviposit by late June, just as the tent caterpillars are constructing their cocoons. The adult larviposits on the cocoon of the caterpillar, and after working its way through the silk the maggot burrows into the pupa. Shortly thereafter, the pupa is reduced to a liquefied mass within which the maggot feeds, much as it would feed on decomposing carrion. The maggot appears to complete its development in 10–12 days under favorable conditions, but in dry years it may remain in the cocoon for more than a month before burrowing into the soil. Usually only one maggot develops in each pupa, but Hodson (1939b) occasionally found that as many as three developed in a single pupa. Studies of this species by Sippell (1957), Witter and Kulman (1979), and Eggen (1987) all found that this parasitoid was a superior competitor,

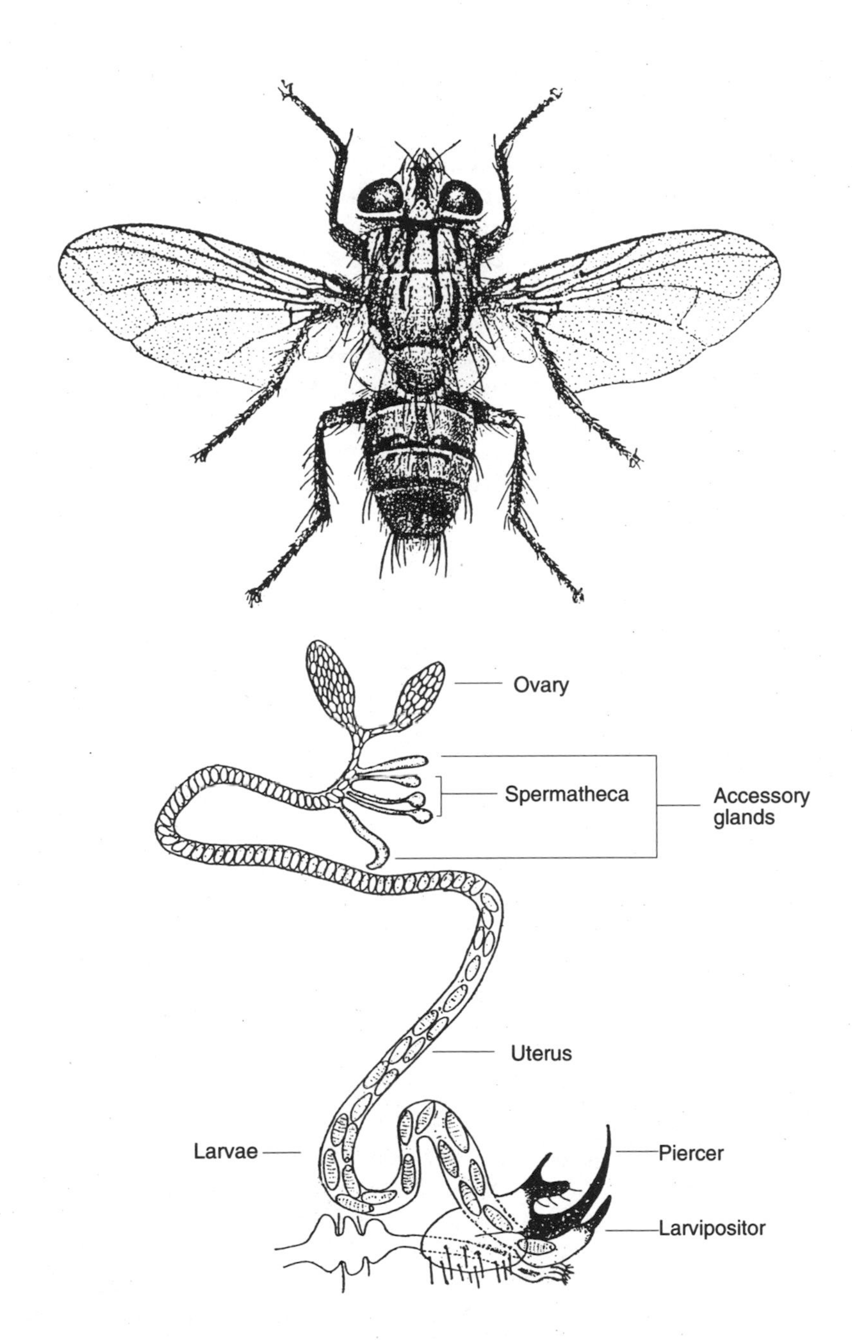

Figure 8.7. The generalist tachinid parasitoid *Compsilura concinnata* and the mature reproductive tract of the larviparus female. (From Culver 1919.)

destroying other species of parasitoids that occurred along with it in the same pupa.

During the infestation of the forest tent caterpillar in Minnesota in the late 1930s, *S. aldrichi* was the dominant parasitoid and in some samples 100% of the tent caterpillar pupae collected had been killed by the insect. The fly was also the most abundant parasitoid found in other studies of the forest tent caterpillar by Sippell (1957) and Witter and Kulman (1979). The fly increases in frequency during successive years of outbreaks and is often extremely abundant in the year preceding the collapse (Fig. 8.8). The newly emerged flies are strongly attracted to humans and are themselves a nuisance. Hodson (1939b) reported capturing as many as 2100 flies in a bottle trap baited with rotting meat during one two-hour period. The impact of the fly on populations of the forest tent caterpillar is discussed in Chapter 9.

Hymenoptera

Egg parasitoids. Fourteen species of wasps occurring in six families of the Hymenoptera oviposit in the eggs of North American tent caterpillars (Table 8.1). These are among the smallest of insects and are able to complete their entire development, from egg to adult, within a single egg of the host. The most common egg parasitoids of tent caterpillars are *Testrastichus malacosomae, Telenomus clisiocampae, Ooencyrtus clisiocampae,* and *Ablerus clisiocampae.*

Egg parasitoids may be specialists on tent caterpillars or generalists that have alternative hosts. The specialist species have life cycles that are closely synchronized with that of the host. Thus, the generalist egg parasitoid *Trichogramma minutum* emerges in advance of unparasitized *M. californicum* caterpillars (Langston 1957), but specialist parasitoids typically stay in the host egg for at least several weeks after the nonparasitized host larvae eclose. This delay allows time for the tent caterpillars to mature and to produce new egg masses that the parasitoids attack. Hodson (1939a) found that *O. clisiocampae, Telenomus clisiocampae,* and *Testrastichus sylvaticus* emerge from mid-June to mid-July. The parasitoids are able to feed, and even early emergers can survive until the first few weeks in July when the tent caterpillar moths oviposit. The adults of *Testriastichus malacosomae* also wait in the egg until well after unparasitized host larvae emerge. Stelzer (1968) observed them flying in large numbers two weeks after peak egg production of their host *M. californicum.* Obligatory egg parasitoids of *M. americanum* emerge 85–118 days after the nonparasitized caterpillar larvae, in near synchrony with the period of host oviposition (Liu 1926).

Egg parasitoids that oviposit in the newly deposited eggs of *Malaco-*

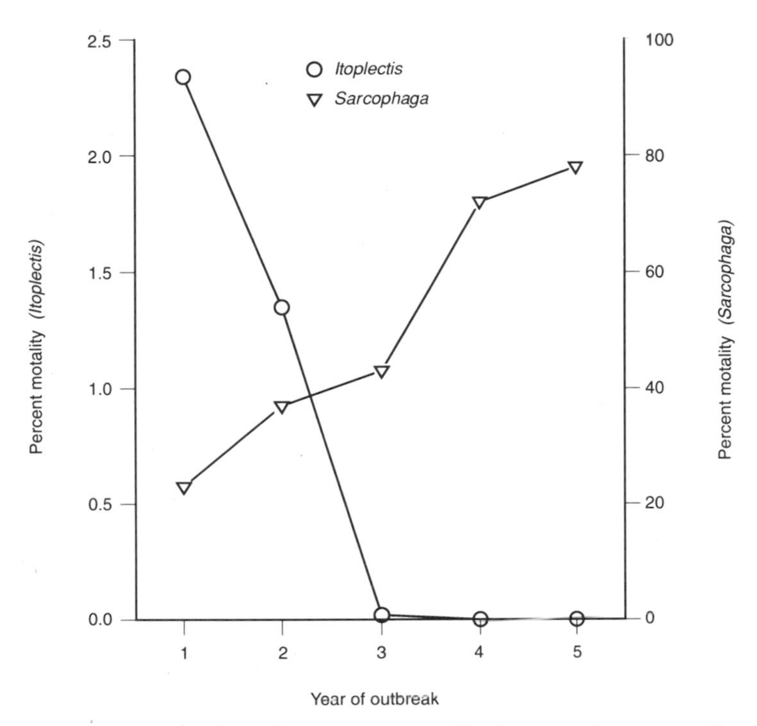

Figure 8.8. Percent mortality of forest tent caterpillar larvae and pupae attributable to *Itoplectis conquisitor* (Hymenoptera) and *Sarcophaga aldrichi* (Diptera) during successive years of an outbreak. (Data from Sippell 1957.)

soma may subsist entirely on the yolk of the egg. But because tent caterpillar embryogenesis proceeds rapidly, some egg parasitoids may be able to feed on the embryo or newly formed caterpillar. Although virtually nothing is known of the developmental biology of these species, some species of *Ooencyrtus* have been shown to oviposit in fully developed embryos of other species, and the parasitoid larva can kill and consume the pharate host caterpillar (Maple 1937)

The impact of egg parasitoids on tent caterpillars has been assessed by several investigators (Table 8.3). Typically, 50% or more of the egg masses in a local population of tent caterpillars are attacked. On average, fewer than 13% of the eggs in an egg mass are affected, but individual egg masses occasionally suffer much higher rates of parasitism. Smith and Goyer (1985) found that 61% of the eggs in a sample of five egg masses of *M. disstria* were parasitized by *A. clisiocampae* and *O. clisiocampae.*

Egg masses of *M. californicum* analyzed by Stelzer (1968) had up to

Table 8.3. Rates of parasitism by egg parasitoids of tent caterpillars

Host	Parasitoids[a]	Parasitism of egg masses: Rate[b]	Parasitism of egg masses: (no. of egg masses)	Parasitism of eggs: Rate	Parasitism of eggs: (no. of eggs)	Reference and notes
Malacosoma americanum	1,4,6,8			.03[b]	(>1000)	Liu 1926; parasitism rates for six different locations
				.06[b]	(?)	
				.007	(?)	
				.08	(104)	
				.02	(100)	
				.04	(12)	
M. americanum	4,5,13	—	—	.04	(12)	Leius 1967; parasitism rates for three different locations
				.02	(12)	
				.01	(12)	
M. americanum	1,4	—	—	.02	(50)	Stacey et al. 1975
M. americanum	1,4,6,8	.89	(90)	.07	(10)	Darling & Johnson 1982
M. californicum	12,*1,10,5,7,4*	—	—	<.01	(124)	Langston 1957
M. californicum	1,4	.86	(338)	.07	(140)	Stelzer 1968
M. disstria	6,4,2,*8,11*	.39	(>700)	.02	(>700)	Hodson 1939a; parasitism rates for two different years
		.73	(>700)	.07	(>700)	
M. disstria	4,6,2	.72	(2025)	.07[c]	(2025)	Witter & Kulman 1979
M. disstria	8,6	—	—	.13[b]	(292)	Smith & Goyer 1985
M. disstria	6,4,8,*3,5*	—	—	.004	(66)	Knight et al. 1991
M. neustrium	7,5,*13,10,9*	—	—	.17[d]	(82)	Shiga 1979

Note: — = data not available.

[a]Species are listed in order of abundance, italics designate very low frequency. 1 = *Tetrastichus malacosomae,* 2 = *Testrastichus silvaticus,* 3 = *Testrastichus* sp., 4 = *Telenomus clisiocampae,* 5 = *Telenomus* sp., 6 = *Ooencyrtus clisiocampae,* 7 = *Ooencyrtus* sp., 8 = *Ablerus clisiocampae,* 9 = *Ablerus* sp., 10 = *Anastatus* sp., 11 = *Trichogramma evanescens,* 12 = *Trichogramma minutum,* 13 = *Trichogramma* sp.

[b]Averaged over 2 years.

[c]Averaged over 6 years.

[d]Averaged over 8 years.

48% of their eggs parasitized by *Testrastichus malacosomae* and *Telenomus clisiocampae,* though the average proportion of eggs per egg mass attacked for the population as a whole was only about 7%. Rates of egg parasitism reported for *M. neustrium testacum* in Japan ranged from 9.5

to 32.4% over an eight-year study period and averaged approximately 17%, a higher average rate of parasitism than that recorded for North American species. Leius (1967) found that the rate of egg parasitism on *M. americanum* was greater in apple orchards with an abundance of wildflowers than in those with fewer flowering plants. He suggested that the nectar of the plants provided a food base for the adult parasitoids, prolonging their life and enhancing their fecundity.

Some investigators have observed that egg parasitoids of tent caterpillars prefer to attack eggs that lie at the margin of the egg mass. Hodson (1939a) studied *O. clisiocampae, Telenomus clisiocampae* and *Testrastichus sylvaticus,* common parasitoids of *M. disstria,* and found that approximately 50% of the wasps of these species emerged from marginal eggs, and another 33% emerged from the eggs in the second row of the egg mass. Darling and Johnson (1982) reported that 73% of the wasps of *T. malacosomae* they recovered from the eggs of *M. americanum* emerged from marginal eggs. In addition, Darling and Johnson (1982) determined that 43% of *O. clisiocampae* and 52% of *T. clisiocampae* emerged from marginal eggs of the egg masses of *M. americanum.* On the basis of their calculation that the average egg mass in their sample had 56 marginal eggs and 190 interior eggs, one would expect over three times as many parasitoids in interior eggs than in marginal eggs if the parasitoids were distributed randomly. Analysis of their data indicates that significantly more parasitoids occur in marginal eggs than expected, corroborating Hodson's findings for *M. disstria.* Although differential survival of parasitoids in marginal and interior eggs could account for this pattern, it has been assumed that the pattern results from selective ovipositional behavior.

It is generally thought that the spumaline that covers the eggs of many species of tent caterpillars protects them from attack by egg parasitoids. Indeed, Darling and Johnson (1982) found that eggs of the eastern tent caterpillar that lacked a spumaline coating were parasitized at a disproportionately high rate. They found, however, that marginal eggs were usually covered as well as interior eggs, so that the high rate of parasitism experienced by these eggs could not be solely due to exposure. Thus, if we assume that the emergence patterns of egg parasitoids are not a matter of differential survival but reflect actual egg-laying patterns, it remains to be determined why the wasps prefer to oviposit in marginal eggs. One uninvestigated possibility is that the wasps find it easier to oviposit through the side of the egg. The eggs of tent caterpillars are packed tightly, and it is only on the margins that the sides are not protected by other eggs. Previous studies that compared the rates of parasitism of exposed and spumaline-covered eggs have suffered from the fact that eggs found exposed in the spring, when the studies were conducted, may not have been exposed when the parasitoids were flying the previous summer. Indeed, it is not uncommon for egg masses to lose much of their spumaline cover after spending nine months exposed to the elements. Future investigators might evaluate the importance of spumaline as a barrier to parasitoids by removing the covering from egg masses at the time the parasitoids are flying, comparing the rate of parasitism of eggs so treated with that of a control group.

Despite the lack of strong supportive evidence, it seems likely that spumaline forms a mechanical barrier that interferes with oviposition. Does the cellular structure of the material cause the ovipositor to be deflected? Is the ovipositor of some species too short to reach eggs covered with a thick layer of spumaline? Does the spumaline obscure olfactory or gustatorial cues associated with fertile eggs? Lacking answers to these questions, we might obtain some insight into the role of spumaline from a comparative study of the rates of egg parasitism of species that do and do not produce a spumaline covering. The few hard data bearing on this matter at the present time fail to show any consistent pattern. For instance, although Shiga (1979) found that in Japan the eggs of *M. neustrium testacum,* a species that does not cover its eggs with spumaline, had a high rate of parasitism, a seven-year study of the same species in Sardinia by Delrio and colleagues (1983) showed that at peak populations, only 6.2% of the eggs were parasitized, a value well within the range reported for species that cover their eggs with spumaline. Moreover, Stehr and Cook (1968) observed that *M. tigris,* the only New World *Malacosoma* that does not cover its eggs with spumaline, has a relatively low rate of egg parasitism, but hard data bearing on this point are not available.

Larval Parasitoids. At least 11 species of braconid wasps occurring in seven genera attack North American *Malacosoma.* The most common is the wasp *Rogas,* which has been reared from tent caterpillars throughout North America (Fig. 8.9). It is presently unclear how many species of *Rogas* wasps attack tent caterpillars. Most publications refer the parasitoid only to the genus, though some investigators have identified the species as *stigmator. R. stigmator,* however, superparasitizes caterpillars, with as many as 20 or more parasitoids supported by a single caterpillar, and does not appear to include *Malacosoma* among its hosts (Mason 1979). The only fully identified species of *Rogas* known to attack tent caterpillars is *R. malacosomatos,* which deposits eggs singly. It has been reared from *M. americanum, M. disstria, M. californicum pluviale,* and *M. californicum lutescens* (Mason 1979).

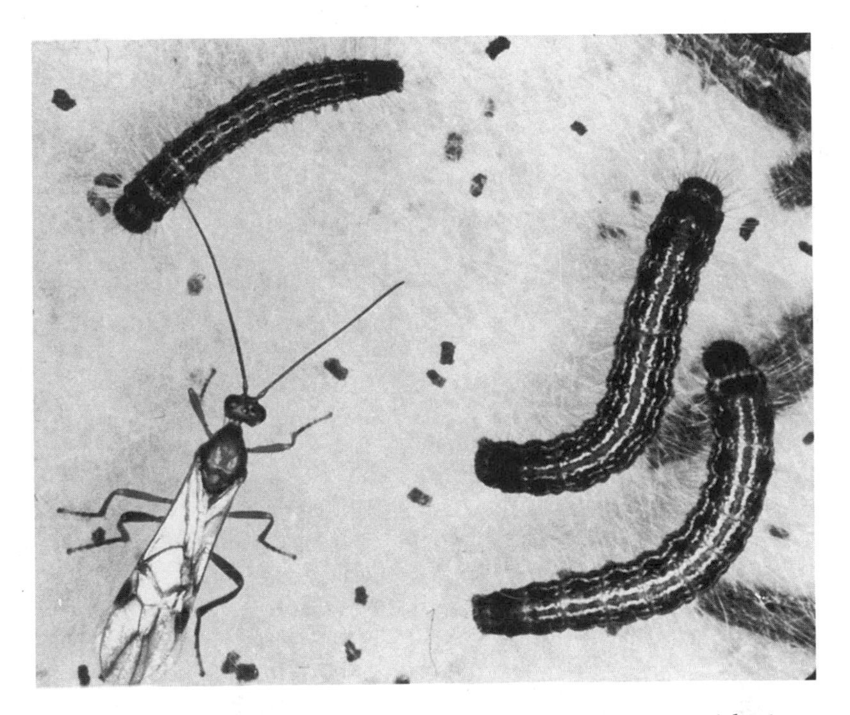

Figure 8.9. *Rogas malacosomatos* (Braconidae) on the tent of a colony of third-instar eastern tent caterpillars.

The adults of *Rogas* are active early in the season, parasitizing larvae principally in the second or third instar (Fiske 1903, Langston 1957). The endoparasite feeds on the internal organs of the living host, retarding the caterpillar's growth and eventually killing it before it has advanced beyond the fourth instar. When fully fed, the parasitoid larva secretes silk through a hole cut on the ventral surface of the eviscerated victim, securely fastening the cadaver to the substrate. By this time, the body of the caterpillar has shrunken markedly in length, taking on a squat, cigar-shaped appearance (Fig. 8.10). The parasitoid spins a cocoon within the mummified exoskeleton of the host and pupates. Eight to fourteen days later, the newly formed adult escapes through a hole it cuts in the dorsum near the tip of the abdomen (Langston 1957).

Most of the hymenopterous parasitoids of tent caterpillars occur in the family Ichneumonidae. One of the more common is *Itoplectis conquisitor*, which has been reared from *M. americanum*, *M. disstria*, and *M. califor-*

Figure 8.10. The shrunken cadavers of eastern tent caterpillar larvae parasitized by *Rogas malacosomatos.*

nicum (Witter and Kulman 1972). The adult female, which is black and about 15 mm long, is a generalist parasitoid of lepidopterous pupae. Its survival is dependent on a seasonal succession of caterpillars. After overwintering in the cocoon of one of its alternate hosts, the wasp emerges early in the spring and completes another generation in the pupa of a second species of caterpillar before seeking tent caterpillars. Wasps that parasitize tent caterpillars emerge from their hosts in early July, after a developmental period of about 18 days. The wasp must then pass through two additional generations in other species of caterpillars before parasitizing its overwintering host (Fiske 1903). Thus, the success of the insect depends on the year-to-year abundances of caterpillar species required to support its five annual generations.

Itoplectis oviposits through the cocoon of the tent caterpillar, frequently searching abandoned tents for caterpillars that have pupated within. The parasitoid larva develops singly in the pupa of the caterpillar and completely fills the cuticle of the victim when fully grown. At maturity, the parasitoid shrinks to about one-half its former size then spins a thin cocoon and pupates. The adult wasp chews its way to the outside. Fiske (1903) noted that about 20% of the *Itoplectis* larvae he observed were victims of the hyperparasitoid *Theronia atalantae fulvescens*, itself an ichneumonid, which feeds externally on the *Itoplectis* larva.

Itoplectis, as well as other polyphagous parasitoids, do much better during periods between outbreaks of the tent caterpillar. Studies of larval and pupal parasitoids of *M. disstria* by Sippell (1957), Witter and Kulman (1979), and Eggen (1987) all show that the incidence of parasitism by *Itoplectis* is greatest at the outset of infestations then declines in subsequent years until the wasp virtually disappears from the parasitoid complex (Fig. 8.8). Sippell (1957) attributed this trend to the disappearance of alternate hosts of the wasps, a consequence of the severe defoliation caused by the forest tent caterpillars during later stages of the outbreak.

Parasites

Nematodes

While conducting a study of the eastern tent caterpillar as a doctoral student at Cornell University, Liu (1926) was surprised to find remarkably long nematodes emerging from the full-grown caterpillars. The worms emerged by boring a hole through the victim's cuticle, and the host caterpillar died a few days later. Liu determined that mature females of the parasite lie coiled within the hemocoel of the victim to accommodate their thread-thin, but elongate body, which averages 33 cm in length. The nematode was subsequently determined to be a new species and was named *Hexamermis microamphidis* (Mermithidae). During one year of his study, Liu found that the nematode was the most significant control agent affecting the caterpillar, with parasitism rates as high as 25%. But the following year the parasite had virtually disappeared from his study population. Parasitism by this or other undetermined species of nematodes has been mentioned only occasionally in the literature since then, indicating that nematodes are infrequent parasites of tent caterpillars. The rarity of nematode parasitism is not surprising because the infective stage of the worm occurs in the soil and desiccates rapidly if exposed to the air. The relatively small incidence reported for tent caterpillars likely reflects the fact that the caterpillars have little or no ground contact before dispersal. It is possible, but untested, that when the bark of the tree is wetted with rain, nematodes may make their way to the resting sites of the caterpillars, particularly if such sites are near the base of the tree.

Microsporidians

Thomson (1959) described the microsporidian parasite *Nosema disstriae* from infected *M. disstria* larvae. The parasite infects many of the larval tissues including the silk glands, epidermis, midgut epithelium, tracheae, and Malphigian tubules. Laboratory studies show that microsproridian infections decrease host vigor and cause increased larval and pupal mortality (G. G. Wilson 1977). When applied to leaf discs in laboratory experiments, the LD_{50} for third-instar *M. disstria* larvae was 230,000 spores per larva (G. G. Wilson 1984). The forest tent caterpillar is also susceptible to two less frequently encountered microsporidians, *Pleistophora schubergi* and *Vairimorpha necatrix*. The LD_{50} for tent caterpillar larvae fed *P. schubergi* is an order of magnitude lower than that of *N. disstriae*. *V. necatrix*, the most virulent of the three species, has an LD_{50} of approximately 1400 spores per larva (G. G. Wilson 1984).

The impact of microsporidians on natural populations of the forest tent caterpillar is largely unknown, but the incidence of infection has been stated to increase during successive years of population outbreaks. In the only field test of the potential of microspridians to control populations of tent caterpillars, a mixture of both *N. disstriae* and *P. schubergi* was applied at the rate of 1.8×10^{11} spores per tree to aspen infested with natural populations of *M. disstria* (G. G. Wilson and Kaupp 1977). Follow-up studies showed that larvae became infected with *P. schubergi*, but there was no increase in the incidence over background levels of *N. disstriae*. No mortality of caterpillars attributable to either species was reported.

N. disstriae also infects field populations of *M. americanum*. Nordin (1975) found that the parasite is transmitted transovarially from one generation to the next. The parasite infected an average of more than 70% of the egg masses inspected during a three-year study of field populations. The infected egg masses had a significantly lower hatching rate and a lower larval survival than uninfected egg masses (Nordin 1976). *Pleistophora* sp. also infects field populations of *M. americanum*, but compared with *Nosema* the parasite is relatively rare and appears to cause little mortality. *M. neustrium* has been shown to be vulnerable to infection by *P. schubergi* (Simchuk 1980).

Antipredator Defense

Agonistic Behavior

Tent caterpillars typically respond to the presence of predators and parasitoids by thrashing the anterior portion of their body from side to side, a common defensive behavior of caterpillars in general. Such displays may involve a single, isolated caterpillar or may escalate into group displays involving, in some cases, most of the caterpillars in the colony. Group displays may occur while caterpillars are assembled at the resting site or while at feeding sites. The behavior is readily elicited by loud sounds such as those produced by clapping or coughing near a colony. In contrast, substrate vibrations generated by tapping a branch on which the caterpillars are feeding usually cause the caterpillars to become immobile.

Myers and Smith (1978) found that the caterpillars of *M. californicum pluviale* were responsive to airborne synthetic sounds having frequencies of 300–2100 Hz. They suggested that the flight sounds of predators, such as the common tachinid *Tachinomyia similis*, might trigger the response. The frequencies of flight sounds produced by this parasitoid range from about 375 to 775 Hz and are, thus, within the range of frequencies to which the caterpillar is sensitive. Thrashing responses, which persist for some time after the immediate cause of the disturbance, may interfere with the ability of entomophages to accurately place their eggs on the prey or, in the instance of group displays, serve to distract or disorient predators or parasitoids. Stamp (1982, 1984) reported that the thrashing of Baltimore checkerspot (*Euphydryas phaeton*) larvae sometimes knocked predatory wasps away. Similarly, E. W. Evans (1982) reported that vigorous thrashing by the larger instars of the eastern tent caterpillar was adequate to drive away predator bugs. Sullivan and Green (1950) observed one encounter between eastern tent caterpillars and a *Podisus* bug in which the disturbed caterpillars managed to entangle the predator with silk to the point where it was webbed into the tent. Most predators and parasitoids are likely to move too quickly to be victimized by caterpillars in that way.

Thrashing on the tent may originate at a single site then rapidly radiate through the assemblage until most of the caterpillars on the tent are simultaneously engaged in the behavior. Caterpillars that are not directly affected by the disturbing element may thrash after physically contacting a thrashing sibling, and the behavior can pass in this way from caterpillar to caterpillar. It is unclear how caterpillars at sites distant from a disturbance are recruited, but one untested possibility is that the activity sets up distinctive vibrations in the taut silk surface of the tent that can be perceived at a distance. Whatever the stimulus, this recruitment effect probably accounts for the observation of Myers and Smith (1978) that caterpillars on the tent are much more likely to thrash in response to airborne sounds than are solitary caterpillars away from the tent. They found that 65% of western tent caterpillars on tents responded to loud sounds having a frequency range of 600–900 Hz, whereas only about 15% of solitary caterpillars off the tent responded to the same sounds. It may be significant that caterpillars that aggregate to bask on the tent are able to elevate their temperature to a greater extent than are solitary caterpillars off the tent (see Chapter 7). Although the primary function of basking in tent caterpillars is probably the facilitation of digestion, the elevated temperatures that basking caterpillars achieve is likely to shorten their response time to entomophages and affect the intensity and duration of their responses.

Many species of caterpillar respond to entomophages by regurgitating toxic or sticky materials that repel the attacker. Little is known about the use of such substances by tent caterpillars or of their potential effectiveness. Tent caterpillars lack specialized diverticula of the gut, such as those found in sawflies that store plant resins; nonetheless, they readily regurgitate the contents of their foregut when disturbed. Peterson (1986a) found that regurgitations of the eastern tent caterpillar contain the cyanogenic glycoside prunasin, a component of the leaf of the host plant *Prunus serotina*. An enzyme present in the leaf is activated when the leaf is eaten, and the material is converted to benzaldehyde and hydrogen cyanide (see Chapter 5). It has been demonstrated under laboratory conditions that ants are repelled by these secretions and, if forced into contact with them, can be irreversibly poisoned, but the significance of defensive regurgitation by tent caterpillars under field conditions remains to be determined. Because the foregut of caterpillars is often fully packed with fragmented leaf material that is subjected to these enzymatic reactions, eastern tent caterpillars would appear to be a bitter component of the diets of predators that consume the whole insect, such as birds, but nothing is known of the potential of such a meal to act as a deterrent to further predation.

Crowding in Space

Aggregation in itself can serve as a form of antipredator defense because the probability that any one individual will be attacked is inversely proportional to the number of individuals in the aggregate. Moreover, individuals may gain protection from enemies by attempting to surround themselves with siblings (Hamilton 1971). Tostowaryk (1971) found that sawfly larvae that lie on the periphery of the aggregate are approximately twice as likely to be killed by the pentatomid predator *Podisus modestus* than are those toward the center.

Aggregation serves as a defense only if predators or parasitoids are not attracted to grouped prey in disproportionate numbers and the rate of attack by a given predator is independent of colony size. The extent to which these criteria are met is likely to vary among tent caterpillar species, depending on the relative conspicuousness of their assemblage and the habits of their predators and parasitoids. Thus, the ability of ants to recruit nestmates with chemical trails to an aggregate of tent caterpillars greatly increases the rate of depredation that the caterpillars suffer over that which would occur were the caterpillars scattered.

Tent-building social caterpillars are clearly more conspicuous than non-tent-building species, and among tent builders those that occupy a single tent for the duration of their larval life are liable to be easier to find than those that build and abandon a series of tents. Indeed, interspecific variation in tent-building habits may derive in no small part from selective pressure from parasitoids and predators. Predators with good memories for the locations of productive feeding sites, such as vespid wasps and birds, are, like ants, likely to profit from the aggregative tendency of tent caterpillars. Indeed, the ubiquitousness of these day-active, visual predators suggests that aggregation in tent caterpillars is not primarily an antipredator adaptation. Much less is known of the cues that solitary entomophages use to locate their prey, and it remains to be determined how aggregation of their prey influences their effectiveness. Parasitoids and predators, however, are likely to do particularly well if they are able to stay in contact with an assemblage once they locate it.

Nocturnal Foraging and Crowding in Time

Although assembly may serve to attract at least some of their enemies, tent caterpillars are far from passive victims. One antipredator behavior that tent caterpillars may adopt is to forage at times that minimize the likelihood that they will encounter enemies. Studies of the daily foraging pattern of the eastern tent caterpillar have shown that colony foraging bouts occur most often under the cover of darkness before and after sunset (Fitzgerald et al. 1988). As discussed in Chapter 5, this foraging schedule does not maximize the rate of energy intake. Rather, the schedule minimizes the time caterpillars spend away from the tent during the daylight hours, when hunters that depend wholly or in part on vision, such as tachinid flies, pentatomid bugs, and birds, are active.

It is unclear, however, whether the foraging schedules of tent caterpillars represent specific adaptations to minimize predation or whether the minimization of predation is simply a fortuitous side effect of a foraging schedule selected for other reasons. There are no hard data to show that individuals away from the tent are more vulnerable than those that rest in the open on the tent during the daylight hours. But, when disturbed, the caterpillars can move rapidly to the relative security of the interior of the tent, where few predators or parasitoids can reach them (but see Tents as Shelters, below). Predators may be less inclined to approach massed prey, particularly if they are simultaneously active, than individuals that are separated from the group. Moreover, Heinrich (1979) showed that predators, especially birds, may associate damaged leaves with prey and conduct particularly diligent searches for caterpillars at feeding sites, providing caterpillars with additional incentive to minimize the time they spend at such sites.

Tent caterpillars typically move from the relative safety of the tent to distant feeding sites en masse during discrete bouts of foraging (Chapter 6). Because predators or parasitoids that attack caterpillars when they are away from the tent can attack only so many prey per unit of time, the overall rate of colony depredation (and thus the odds that any one individual will suffer an attack) can be minimized if caterpillars crowd their enemies in time. Studies (Casey et al. 1988; T. D. Fitzgerald and C. R. Visscher, unpubl.) indicate that selection pressure for synchronous foraging has been strong enough to deter caterpillars from following foraging schedules dictated by their individual hunger levels. The intermittency of en masse foraging, and, thus, of prey availability, may also discourage predators and parasitoids from taking up residency in trees inhabited by tent caterpillars.

The ability of tent caterpillars to forage and grow at low temperatures in the spring may enable them to escape enemies that are not as well adapted to cold weather. E. W. Evans (1982) found that *Podisus* bugs that had been preying on tent caterpillars became largely inactive during a 10-day cold spell that occurred in central New York during the spring of 1977. During the period, temperatures averaged 9°C, and it snowed on one day. These conditions had a lesser effect on the tent caterpillars, and they were observed to forage and grow during much of the period. (Indeed, Joos [1992] observed eastern tent caterpillars foraging at body temperatures as low as 7°C.) The differential effect of

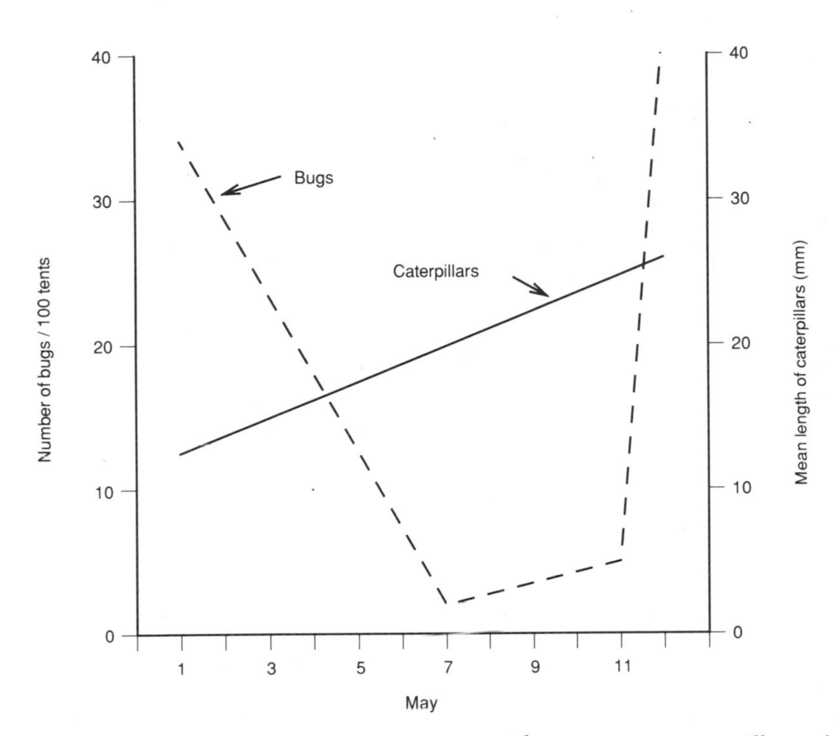

Figure 8.11. Number of *Podisus* bugs on tents of the eastern tent caterpillar and the growth in length of the caterpillars over a 12-day period in May. The average temperature during the period was 9°C, causing the bugs to become inactive while the caterpillars continued to grow. Snow fell on May 9. (Data from E. W. Evans 1982.)

cold on the prey and predator allowed the caterpillar to feed unmolested during the period. Furthermore, when warm weather returned, many of the caterpillars had grown too large to be subdued by the pentatomids (Fig. 8.11). Ayre and Hitchon (1968) reported a similar effect of low temperature on predation of caterpillars by ants. Although one species of ant was reported to forage in trees at temperatures as low as 5°C, three other species did not forage above ground when temperatures fell below 10°C. Moreover, the ability of predators to subdue prey when chilled to temperatures near their activity threshold may be markedly diminished. Although no specific reports exist, it is likely that during cold springs delays in nest establishment and breeding of insectivorous birds may reduce the demand of these predators for caterpillar prey.

Tents as Shelters

Little is known of the extent to which the large tents constructed by some species of *Malacosoma* provide protection from predators and parasitoids. Like the non-tent-building *M. disstria* and like those species that build only small temporary tents, caterpillars that construct tents large enough to accommodate the entire colony commonly spend considerable time basking in the sun while lying exposed on the outside of the structure. When so exposed, large-tent–building species are likely to gain no more protection from their tent than species that build no tent or less substantial structures. But caterpillars that build large structures commonly retreat into their tents when disturbed and can enjoy temporary refuge from enemies that cannot tear open the tent, follow them inside, or oviposit through the walls of the structure.

Some insight into the role that the tent plays in protecting colonies from predators and parasitoids can be derived from the results of studies that involved enveloping tents with exclusion netting. If (1) the tent constitutes a significant barrier to entomophages and (2) the caterpillars make use of its protective potential by moving inside rapidly when threatened, then exclusion netting would be expected to have little influence on colony survival. One such study involved *M. incurvum aztecum*, a species that builds a substantial tent. Filip and Dirzo (1985) covered the tents of some colonies of this species with fine netting and compared survival of the colonies with that of colonies that were not so enveloped. Life tables constructed for the colonies in both treatment groups showed that colonies protected with netting suffered significantly less loss to predators and parasitoids than did unprotected colonies (Fig. 8.12, Table 8.4). Thus, for this species at least, the tent does not appear to provide much protection. Shiga (1979) conducted a similar study with *M. neustrium testacum* in Japan. Although this species does not build as substantial a tent as *M. incurvum aztecum*, the tent, nonetheless, might be expected to provide some protection from enemies. To the contrary, his study showed that colonies protected with fine netting suffered significantly less mortality that those that were unprotected (Fig. 8.12, Table 8.4), indicating that the tent of this species provides only marginal protection from entomophages.

Aposematism

Gregariousness and bright coloration (aposematism) are common features of caterpillars that are venomous, distasteful, or too hairy for their predators to eat (Stamp 1980, Nothnagle and Schultz 1987). Gregarious

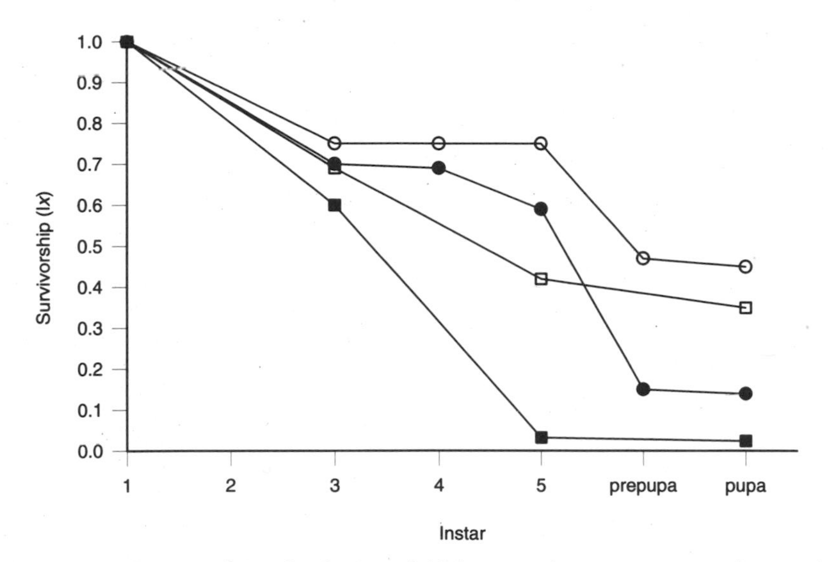

Figure 8.12. Survivorship of colonies of *Malacosoma incurvum aztecum* (squares) and *M. neustrium testacum* (circles) attributable to parasitoids and predators when their tents were either covered (open symbols) with exclusion netting left uncovered (filled symbols). (Data from Shiga 1979, Filip and Dirzo 1985.)

and conspicuously colored caterpillars are typically avoided by birds (see Wiklund and Järvi 1982). Many studies have shown that these predators associate conspicuous features of potential prey with their undesirable properties and, after as little as a single encounter, learn to avoid them. All species of tent caterpillars are gregarious, and the later instars are typically brightly colored. It is plausible that the bright colors of the caterpillars function as warning colors directed at avian predators that find the large caterpillars too hairy for their taste. Thus, Heinrich (1979) observed that the dispersed late-instar larvae of the forest tent caterpillar were able to feed in the open during the day, unmolested by birds that scoured the trees for other more cryptic (and palatable) species. As noted above, other investigators have recovered the remains of mature tent caterpillars from the stomachs of bird predators, so it would seem that such protection is not absolute.

The conspicuousness of some species of tent caterpillars is greatly increased when they rest gregariously at silk shelters. Even small tents can be seen at distances of tens or even hundreds of meters, particularly when the sun strikes them at certain angles. Such blatant gregariousness may serve to further enhance the conspicuousness of undesirable prey and increase the probability that, once educated, predators will leave them alone. Gregariousness may also reduce colony losses if predators

Table 8.4. Mortality in colonies of two species of tent caterpillars protected or not protected with exclusion netting

Species	Instar	Percent mortality during instar			Major sources of differential mortality
		No exclusion net (m_{np})	Exclusion net (m_p)	Differential mortality ($m_{np} - m_p$)	
Malacosoma neustrium[a]	1 + 2	29.8	22.8	7.0	?
	3	0.4	0.4	0	?
	4	15.5	0	15.5	*Polistes*, birds
	5	74.4	38.0	36.4	*Polistes*, birds
	Prepupa	6.3	3.9	2.4	Parasites
	Pupa	45.0	1.0	44.0	Parasites
M. incurvum[b]	1–3	39.7	31.4	8.3	Predation
	3–5	94.7	38.9	55.8	Predation and parasitism
	Pupa	24.1	15.6	8.5	Parasitism

[a]Shiga 1979.
[b]Filip and Dirzo 1985.

use proximity as a cue, avoiding caterpillars near the sampled and rejected victim. Furthermore, when there is a delayed negative effect, aggregation increases the probability that subsequent prey eaten will be of the same type as the one that caused the sickness (see review by Guilford 1990). Although generalist invertebrate predators may be capable of associative learning, the potential value of aposematic coloration is much less clear in their case. Young (1983), for example, found that distasteful and aposematically colored Neotropical caterpillars are little protected from predatory wasps.

Fisher (1958) considered the relationship between gregariousness, distastefulness, and aposematic coloration of caterpillars. While recognizing that solitary caterpillars may be distasteful and warningly colored and that differential survival of individuals sampled by predators could lead to the evolution of distastefulness, he favored a kin selection model to explain the origins of distastefulness. He suggested that even though a conspicuously colored and distasteful caterpillar may be fatally wounded by a bird, the sickened bird would thenceforth avoid the victim's nearby siblings and thus favor the evolution of the trait. He argued that distastefulness preceded the evolution of warning coloration and that gregariousness was a prerequisite for the evolution of distastefulness. More recent studies have given additional support to the possibility that individual caterpillars can survive attack by predators (Boyden

1976, Järvi et al. 1981, Wiklund and Järvi 1982) and that individual selection can give rise to aposematic coloration (Sillen-Tullenberg and Bryant 1983). The current consensus appears to be that both individual selection and kin selection play a role in the evolution of aposematism, but the relative importance of each remains to be determined (Guilford 1990).

9

Population Dynamics and Economic Impact

Tent caterpillars have irruptive population dynamics and are widely recognized because of the attention they draw during periods of population expansion. Regional records dating back to the 1600s show that tent caterpillars have population cycles of irregular interval and duration characterized by episodes of short-term explosive growth followed by the abrupt decline and the apparent extinction of local populations. During outbreaks, populations often defoliate host trees over extensive areas, and the overflow populations may decimate other plants that lie within the infestation zone but are not normally attacked. Although their full ecological and economic impact remains to be determined, studies show that repeated defoliation by tent caterpillars causes significant reduction in the growth of affected trees and, on occasion, the death of particularly susceptible individuals.

History of Outbreaks

Although all species of North American *Malacosoma* appear to be capable of achieving outbreak status (Stehr and Cook 1968), *M. disstria, M. americanum, M. californicum,* and *M. incurvum* are the most economically important outbreak species. Records of forest insect infestations show that outbreaks of these species occur nearly every year in some region of the United States, and some forests are almost continually defoliated (Fig. 9.1).

Outbreaks of tent caterpillars were first documented in the mid-1600s (W. E. Britton 1935), but until the early 1900s records of outbreaks are

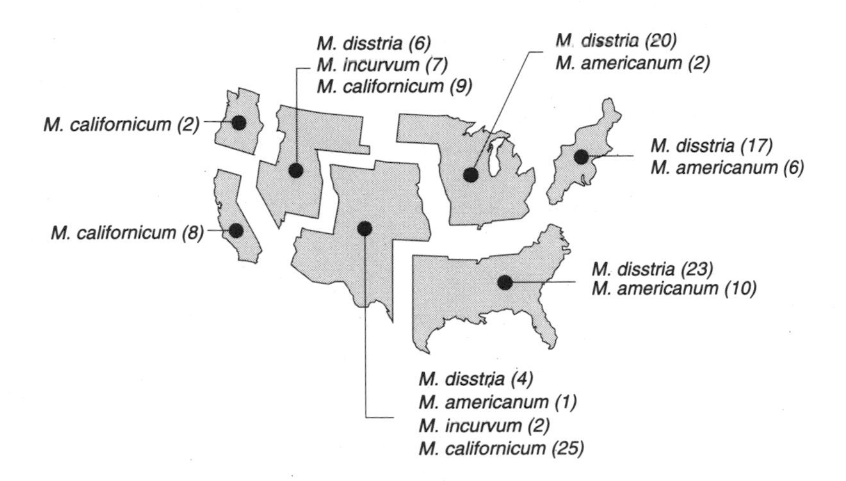

Figure 9.1. Number of years in which populations of tent caterpillars achieved outbreak status in seven regions of the United States from 1953 to 1983. (Data from 1953–1983, U.S. Department of Agriculture, "Forest Insect and Disease Conditions in the United States.")

typically sketchy, providing few reliable data regarding year-to-year changes in the density of the caterpillar populations, the size of outbreak zones, the dates of onset and collapse of the outbreak, and the severity of the damage. Despite these deficiencies, the historical record shows that, particularly in the colder regions of North America, tent caterpillars exhibit a recurring pattern of irruptive population dynamics (Fig. 9.2). Outbreak populations appear suddenly, persist for a few years then virtually disappear.

Although it is often stated that tent caterpillars in these regions have population cycles with a period of about 10 years, the records indicate considerable variability in the interval between outbreaks. Provincewide infestations of the forest tent caterpillar occurred in Ontario every 9–16 years during the period from 1867 to 1987 (Anonymous 1960–1990, Sippell 1962). From 1923 to1957 outbreaks of forest tent caterpillars at six locations in Saskatchewan and Manitoba occurred at intervals of 6–16 years. In four other locations in those provinces, however, outbreaks occurred only once during the 34-year period (Hildahl and Reeks 1960).

Records of populations of the forest tent caterpillar in northern Minnesota compiled by Hodson (1941) indicate that regionwide outbreaks persist from 1 to 8 years (mean = 3.1 years). Records of outbreaks of the same insect in Ontario from 1867 to 1987 indicate a duration of 3–9 years (mean = 6.4 years). Disparity of the mean values derived from these two sets of records is more likely attributable to the imprecise and incomplete

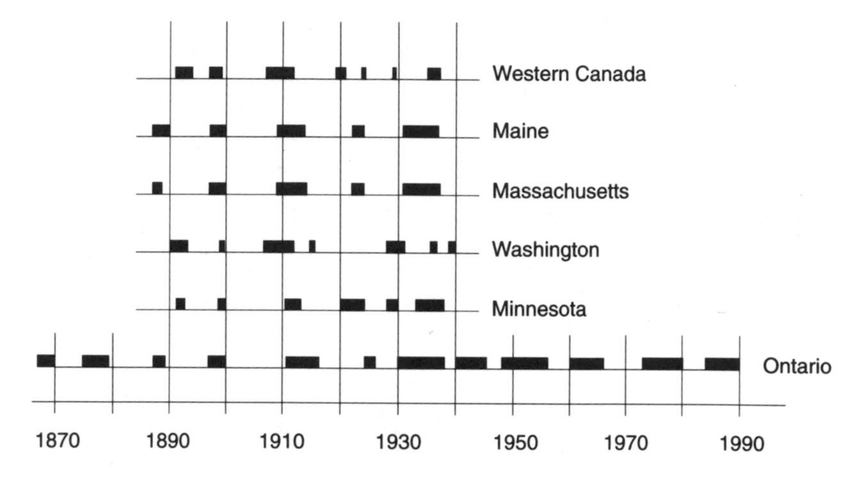

Figure 9.2. Periods of regionwide outbreaks of the forest tent caterpillar in Canada and the northern United states. The exact duration of many of the earlier outbreaks is uncertain. (Data from Hodson 1941, Anonymous 1960–1990, Sippell 1962.)

nature of the historical record rather than to any real differences among the populations. After analyzing forest service data on forest tent caterpillar outbreaks in Ontario, Roland (1993) concluded that the duration of outbreaks was positively correlated with the extent of forest fragmentation. He speculated that disease and entomophagous insects might be less active in fragmented areas than in intact areas, but it seems equally probable that increased opportunities for basking might be involved. Populations of *M. neustrium* in oak forests of Sardinia have episodic population surges that last 7–10 years (Delrio et al. 1983).

Regionwide outbreaks of tent caterpillars typically begin at one or a few sites, and the zone of infestation expands as moths disperse to new areas (Fig. 9.3). Thus, populations may be declining in one area while they are expanding, amoebae-like, into another. Sippel (1962) reported that local populations of the forest tent caterpillar characteristically increase in size during a two- to three-year period of incipience. This initial phase is followed by a one- to two-year period of excess, when caterpillar densities are greater than necessary to defoliate trees, and a one-year period of decline, leading to collapse of the local population. Local populations of the western tent caterpillar in British Columbia, studied by Myers (1990), exhibited a similar pattern of incipience and decline (Fig. 9.4). Other investigators have reported that localized infestations of northern populations of the forest tent caterpillar last as few as two years to as many as nine or more years (Duncan and Hodson 1958, Hildahl and Reeks 1960, Shepherd and Brown 1971; Fig. 9.5).

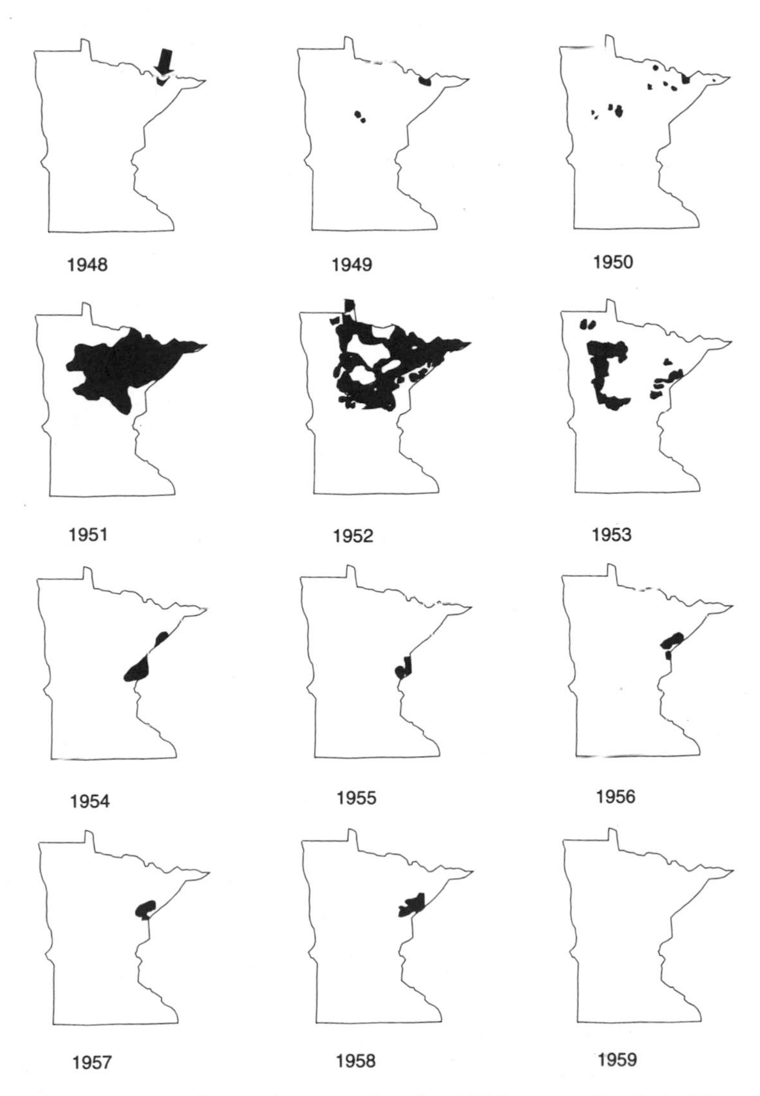

Figure 9.3. Spread of the 1948–1958 outbreak of *Malacosoma disstria* in Minnesota as indicated by defoliation of host trees. (Redrawn from Hodson 1977, courtesy of the University of Minnesota Agricultural Experiment Station.)

Although forest tent caterpillars attract considerable attention during outbreaks, relatively little is known of the status of local populations between outbreaks. For the most part, caterpillar damage is apparently nonexistent or so light in the first several years following outbreaks that populations seem to disappear entirely. Indeed, many investigators have

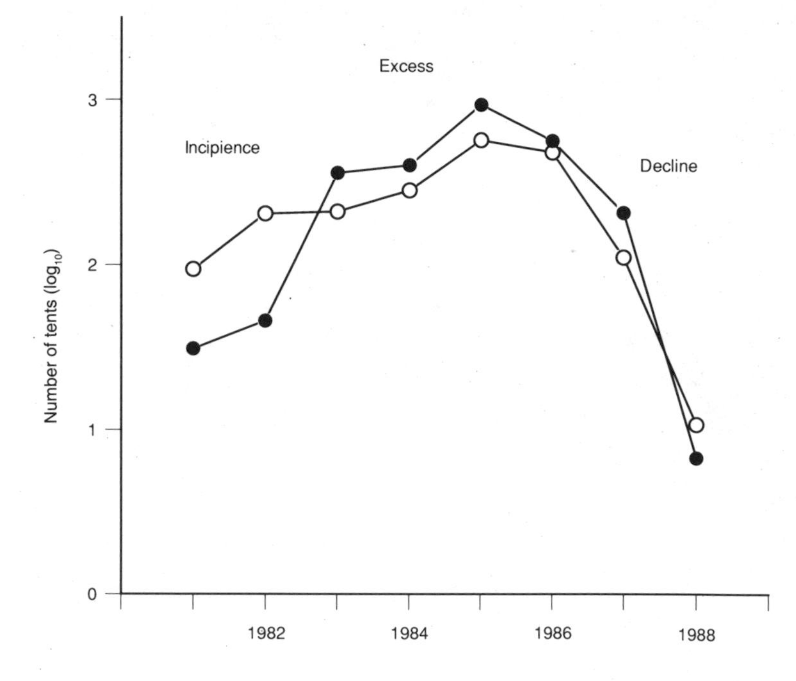

Figure 9.4. Characteristic time course of the growth and decline of tent caterpillar outbreaks. Data are for two local populations of the western tent caterpillar in British Columbia. (After Myers 1990.)

been forced to suspend studies on the caterpillars during such periods because they cannot locate field populations. Studies by Hodson (1977) indicate that local populations may not become completely extinct after the collapse of an outbreak. In the years following the 1948–1958 infestation in Minnesota, Hodson conducted annual light-trap surveys in areas defoliated during the outbreak. Traps were set at 19 locations, and collections were made during July. With a few exceptions, moths were captured at all sites each year, indicating that small, residual populations of the insect persist in scattered pockets throughout former outbreak zones (Fig. 9.6).

In contrast to the episodic outbreaks of the forest tent caterpillar in northern forests, decades-old, chronic infestations of the insect have occurred in some forests in the southern United States. In swamp forests in southern Louisiana and southwestern Alabama, stands dominated by water tupelo are known to have been defoliated annually for 20 to 30 successive years (Batzer and Morris 1978, Smith and Goyer 1986). Some stands are defoliated every year; others have two- to four-year cycles of

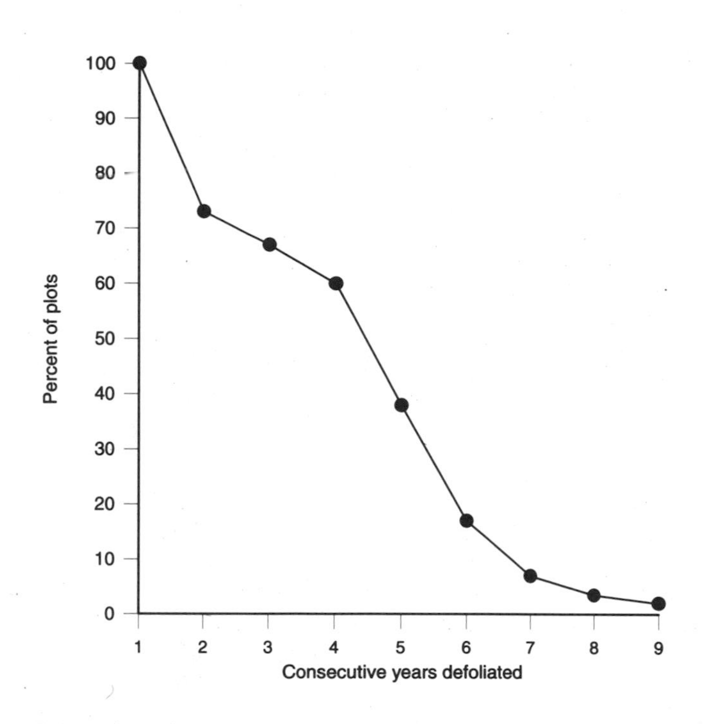

Figure 9.5. Number of consecutive years of defoliation of aspen trees in 61 plots by *Malacosoma disstria*. Plots were established in aspen forests of central and northern Alberta during a region-wide outbreak of the caterpillar from 1957 to 1965. (After Shepherd and Brown 1971.)

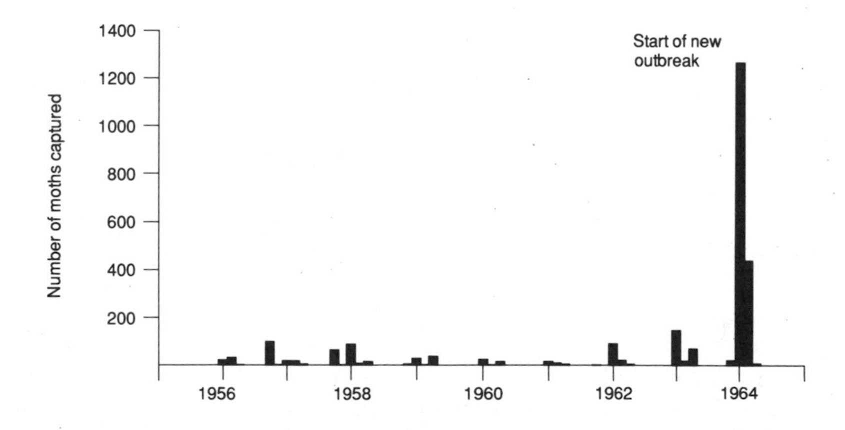

Figure 9.6. Number of moths taken at each of five light traps during a period between outbreaks of the forest tent caterpillar in Minnesota. Outbreak populations at these five sites collapsed by 1954, and the outbreak collapsed everywhere by 1959 (see Fig. 9.3). A new outbreak, indicated by a marked rise in moths collected, began in 1964. (After Hodson 1977.)

abundance that are repeated continually. As noted below, these cycles appear to be driven by larval starvation (Harper and Hyland 1981). A relatively mild climate and annual flooding of the stands are thought to account for the marked difference in the dynamics of tent caterpillars in these forests and more northern populations.

Causes of Population Surges and Declines

Explosive population dynamics are often characteristic of pest insects introduced from other countries or are attributable to the disruption of natural ecosystems by human activities. All North American species of tent caterpillar, however, are native insects, and the historical record shows that outbreaks of forest and eastern tent caterpillars occurred long before humans had significant impact on the forest ecosystem (Baird 1917, W. E. Britton 1935). In more recent times, outbreaks of tent caterpillars have been recorded in remote mountainous regions largely uninfluenced by humans (Stehr and Cook 1968). But tent caterpillars have clearly profited from human activity too. The principal host of the eastern tent caterpillar, black cherry, is shade-intolerant and originally occurred as a scattered forest species. The tree is now common in abandoned fields and along roadside clearings, and the caterpillar flourishes in these areas. The introduction of the apple tree from Eurasia has also broadened the food range of this species. Widespread replacement of conifers by aspen after logging or forest fires has markedly increased the likelihood of outbreaks of the forest tent caterpillar (Hodson 1941).

What factors might predispose tent caterpillars to have boom-and-bust population dynamics? Species of *Malacosoma* and other caterpillars that make up the estimated 2% of North American tree-feeding Macrolepidoptera that are considered outbreak species have similar life histories, most notably univoltinism, a tendency to overwinter as an egg and to feed on early spring leaves, and larval gregariousness (Nothnagle and Schultz 1987). Nearly 90% of the North American forest pest species with irruptive population dynamics have a single generation each year. Approximately half overwinter as eggs and feed on young host tissue in the spring. Although only 7% of the tree-feeding Lepidoptera are estimated to lay their eggs in batches of 10 or more (Herbert 1983), approximately 70% of the forest species that periodically achieve outbreak status are batch layers (Nothnagle and Schultz 1987). Of these, over half are gregarious as larvae. Sudden surges in population growth under favorable conditions may be facilitated by larval gregariousness, which may minimize losses to predators though group defense tactics and speed development by enhancing the ability of the larvae to thermoregulate and to forage efficiently.

Phytophagous insects having a single generation each year are likely to be physiologically adapted to feed within a relatively narrow phenological window. They probably are particularly sensitive to variation in food quality, doing well when their development is synchronized with that of the host, but suffering when they are out of phase. Early spring feeders are apt to be out of phase with host plant development because of the vagaries of the spring weather regime. Thus, it might be expected that the year-to-year population dynamics of early spring feeders would show considerable variation, with populations doing exceptionally well in some years and declining or even disappearing in others.

The average fecundity of forest Lepidoptera that periodically achieve outbreak status ranges from 49 to 880 eggs, indicating that high fecundity, in itself, is not a major force driving population expansion (Nothnagle and Schultz 1987). Even species with relatively low fecundities have the capacity to achieve astronomical numbers in remarkably short periods under appropriate conditions. In the Great Lakes states, forest tent caterpillars produce egg masses containing an average of 175 eggs (Witter et al. 1975). In the southern United States, the same species may average as many as 400 eggs per egg mass (Smith and Goyer 1986). If we assume a 50:50 sex ratio, a single pair of moths from these regions could in theory give rise to populations containing 10^8 to 10^9 individuals in only four years (Fig. 9.7).

Hodson (1941) estimated that at average fecundity, populations of the forest tent caterpillar in northern regions must have 98.7% mortality for stability; greater survival rates lead to population expansion. Smith and Goyer (1986) found that generational survival rates equal to or greater than 0.7% result in population increase in the southeastern United States. Generational survival of *M. neustrium testacum* in Japan measured over eight successive generations was also less than 1% (Shiga 1979; Fig. 9.8). Thus, the survival of only a small fraction of a population is adequate to sustain population growth.

The inherent capacity for growth of populations is opposed by density-dependent and density-independent mortality factors that suppress the growth of tent caterpillar populations and may ultimately drive them to local extinction. Factors that affect the population dynamics of tent caterpillars include food quality and quantity, the direct and indirect effects of weather (particularly temperature), parasites, predators, and pathogens. The most frequently cited causes of the collapse of outbreaks are low spring temperatures, disease, pupal parasitism, and starvation (Table 9.1).

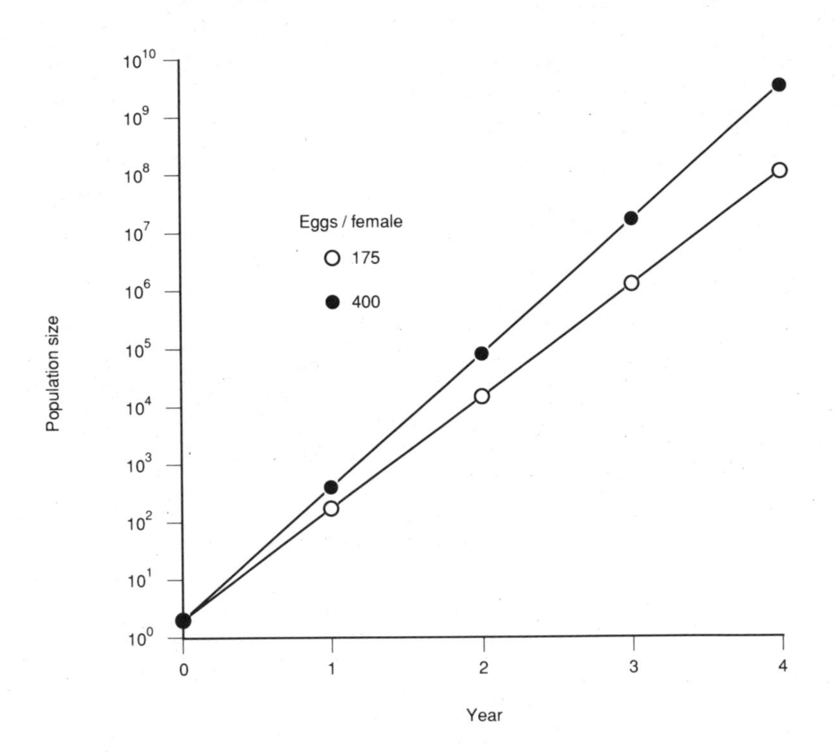

Figure 9.7. Theoretical reproductive potential of a single pair of forest tent caterpillar moths at fecundities characteristic of northern (open circles) and southern (filled circles) populations.

Weather

Nothnagle and Schultz (1987) argued that the mortality risks from weather that larvae subject themselves to in order to hit an early phenological window drive the population dynamics of tent caterpillars and other spring-feeding Lepidoptera. A review of the records of forest insect outbreaks in Canada and the northern regions of the United States corroborates this view. Temperature-related mortality is the most frequently cited cause of the collapse of outbreaks of tent caterpillars in northern regions (Table 9.1).

The overwintering pharate larvae of northern populations of tent caterpillars are extremely cold-hardy, but during unusually cold winters larvae may die. Complete mortality of the pharate larvae of the forest tent caterpillar was reported along the Salmon River basin in New Brunswick when temperatures during the winter of 1922 went as low as −43°C. Normal hatching of eggs in other areas of southern New Bruns-

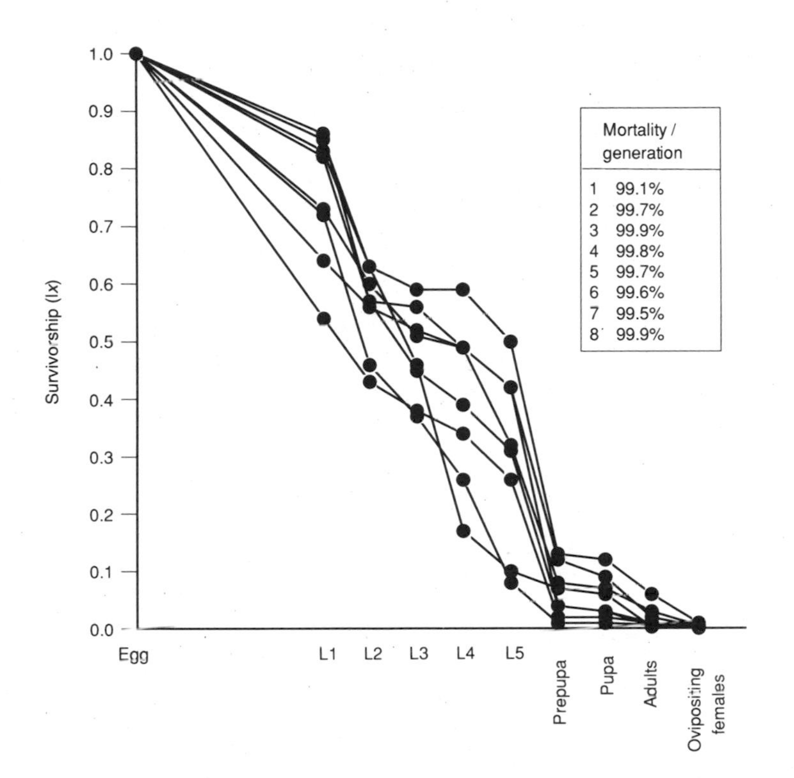

Figure 9.8. Survivorship curves for eight successive generations of *Malacosoma neustrium testacum*. Box shows mortality from egg to ovipositing female. (After Shita 1979.)

wick, where winter temperatures were more moderate, indicated that the basin population had frozen to death (Gorham 1923). Witter et al. (1975) reported high levels of mortality among the pharate larvae of forest tent caterpillars in northern Minnesota during the winters of 1966, 1968, and 1972, when temperatures dropped below −42°C. The collapse of the nine-year infestation in 1972 was attributed to the loss—most of which appeared to be weather-related—of more than 75% of the pharate larval population.

Much of the tent caterpillar mortality attributable to weather occurs in the spring and is not as direct as freezing. After investigating the relationship between temperature and 60 outbreaks of the forest tent caterpillar in the prairie provinces of Canada, Ives (1973) concluded that cold winters and warm springs were most often correlated with the onset of outbreaks. Cold winters reduced demand for yolk by the postdiapause caterpillars, and warm springs allowed the caterpillars to synchronize their development with that of the host leaves. In contrast,

Table 9.1. Factors reported as causing or contributing to the collapse of *Malacosoma* outbreaks

Source of collapse	Species	Reference
Direct or indirect effects of weather	*M. americanum*	Blackman 1918
	M. disstria	Tothill 1918
	M. disstria	Gorham 1923
	M. americanum	Sweetman 1940
	M. disstria	Sweetman 1940
	M. disstria	Hodson 1941, 1977
	M. disstria	Prentice 1954
	M. disstria	Blais et al. 1955
	M. californicum pluviale	Wellington 1962
	M. disstria	Sippell et al. 1963, 1966, 1972, 1974, 1976[a]
	M. disstria	Gautreau 1964
	M. disstria	Witter et al. 1972
	M. disstria	Ives 1973
	M. disstria	Anonymous 1980, 1983, 1984, 1986[b]
	M. disstria	Smith & Goyer 1986
Disease	*M. americanum*	Davis 1903
	M. californicum fragile	Clark 1955
	M. californicum	Stelzer 1968
	M. disstria	Anonymous 1986[b]
	M. californicum	Anonymous 1987[b]
Pupal parasitoids	*M. disstria*	Hodson 1941
	M. disstria	Sippell 1957
	M. disstria	Witter et al. 1972
	M. disstria	Witter & Kulman 1979
	M. disstria	Smith & Goyer 1986
	M. disstria	Anonymous 1986, 1987[b]
Starvation of larva	*M. disstria*	Hodson 1941, 1977
	M. californicum	Stelzer 1968
	M. disstria	Sippel et al. 1976[a]
	M. disstria	Harper & Hyland 1981
	M. disstria	Stark & Harper 1982
	M. disstria	Anonymous 1984, 1986[b]
	M. disstria	Smith & Goyer 1986

[a] "Forest Insect and Disease Conditions in Ontario." In *Annual Report of the Forest Insect and Disease Survey*. Ottawa: Canadian Forest Service.
[b] "Forest Insect and Disease Conditions in Canada." Ottawa: Canadian Forest Service.

population declines were often associated with cool springs, which acted to desynchronize caterpillar and host tree development.

Blackman (1918) documented the weather-driven collapse of a three-year infestation of the eastern tent caterpillar in Syracuse, New York, during the spring of 1917. From April 18 to April 22 the temperature reached daily highs of 19–22°C and led to widespread eclosion of caterpillars. During the next 26 days it rained incessantly, and the maximum daily air temperatures rarely exceeded 13°C. Observations made on May 26, 38 days after eclosion, indicated that over 99% of the caterpillars had died. Blackman (1918) attributed the collapse of the outbreak to an indirect effect of cold weather: the forestalling of the development of the leaves of the trees and subsequent starvation due to the prolonged absence of food. Although no more details are available, it is likely that the lack of sunshine also prevented the caterpillars from basking to achieve the body temperatures necessary for locomotion and digestion. The force of sustained driving rains may have contributed to mortality by destroying small tents and washing caterpillars to the ground.

Hodson (1977) compiled temperature records for the period before, during, and after two major outbreaks of the forest tent caterpillar in Minnesota from 1933 to 1938 and from 1948 to 1959. His analysis showed an apparent relationship between the rise and fall of populations and an index based on the number of degree-days during which the maximum temperature was less than 15°C during the three-week period following egg hatch. Both outbreaks began after two to three years of mild weather during the posthatch period and terminated abruptly after prolonged cool temperatures during the posthatch period.

One of the most thoroughly documented weather-related population collapses of an outbreak of the forest tent caterpillar occurred in central Canada during the spring of 1953 (Blais et al. 1955). The outbreak began in 1948. From 1950 to 1952, the infested area increased by several thousand square miles each year, and egg counts made during the spring of 1953 indicated a potential for massive defoliation. Trees sampled at 31 areas within the infestation zone had an average of 47 egg masses attached to their branches; one tree had 389 egg masses. Unusually high temperatures in early May caused most of the eggs in the outbreak zone to hatch just as the buds on the aspen trees began to open. Several days later, the weather changed abruptly. Temperatures fell below freezing, and snow and hail pelted the newly eclosed larvae. Near the end of the three-day storm, observers found as many as 20 larvae per square foot lying on the snow-covered ground. It was estimated that 95% of the population had perished. Much of the young foliage was also killed, and new leaves produced from adventitious buds did not appear until nearly a month later. Consequently, larvae that were not killed by the storm eventually starved. In addition, the parasitoid *Sarcophaga aldrichi,* which had increased to large numbers over the course of the outbreak, was not affected by the storm and contributed to the mortality. Follow-up sur-

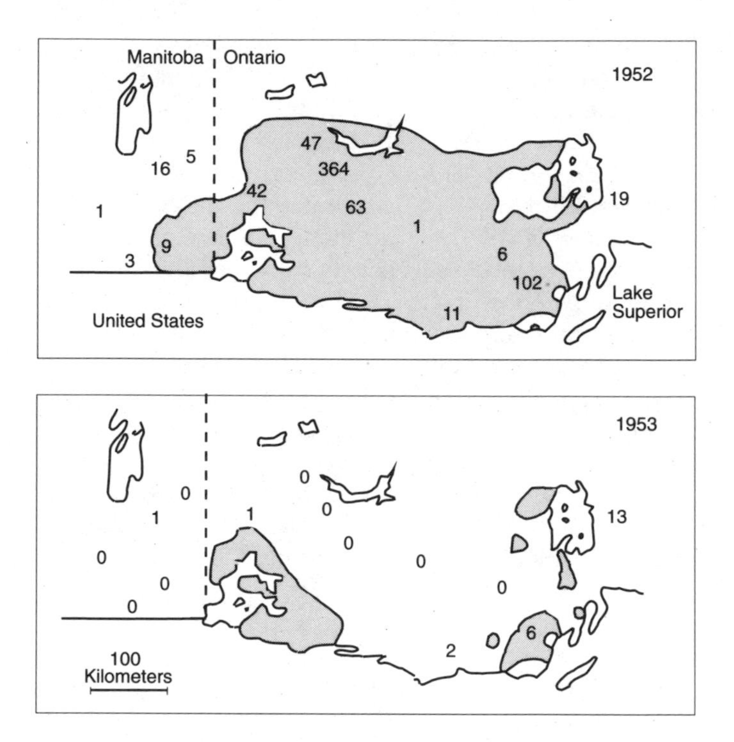

Figure 9.9. Areas of central Canada having medium to heavy infestations of the forest tent caterpillar during 1952 and 1953. The collapse of the population in 1953 was attributed largely to inclement weather. Average number of egg masses per tree are given for several observation points following the 1952 and 1953 infestations. (After Blais et al. 1955, by permission of the Entomological Society of Canada.)

veys at 39 locations spread throughout the 11 million hectares affected by the outbreak during the summer of 1953 showed that the infestation had collapsed to only 16% of its former size (Fig. 9.9). Counts made in the fall indicated that egg production was less than 5% of that of the previous year.

Unusually high temperatures during the period of flight and oviposition were blamed for the widespread failure of eggs to develop during an outbreak in Minnesota in 1936 (Hodson 1941). During the first two weeks of July, temperatures in forested regions went as high as 42°C, and some areas had as many as nine days during which the temperature exceeded 38°C. These high temperatures coincided with the period of flight and oviposition of the forest tent caterpillar. Surveys to assess egg fertility in areas that experienced these high temperatures showed that an average of 33% of the eggs failed to develop. At some sites, up to 93% of the eggs in individual egg masses were inviable. The following year, when temperatures were normal, an average of only 9% of the eggs were so affected. It is unclear whether temperature influenced the ability of the ovipositing females to produce viable eggs or whether there was a failure in embryogenesis. Unfortunately, there have been no laboratory studies to confirm that temperatures within the observed range can influence egg viability, and the evidence for such an effect remains largely circumstantial.

Weather-related pharate larval mortality is much rarer in the southern range of the forest tent caterpillar, though it does occasionally occur. Smith and Goyer (1986) attributed the loss of populations of first- and second-instar forest tent caterpillars to unusually cold temperatures in Louisiana in the spring of 1983. In general, however, spring weather in the forests of the southeastern United States is moderate and predictable, and populations of the forest tent caterpillar, as noted above, are less subject to the pronounced swings in density characteristic of northern populations. Populations of *M. neustrium testacum* in the warmer regions of central Japan also tend to be relatively stable from year to year (Shiga 1979).

An unusual potentiality for weather-related mortality was reported for *M. indicum* populations that attack *Betula utilis*, a tree found at elevations of 3500–4800 m in the Himalayas. The tree grows to the limits of the treeline, and Bhandari and Singh (1991) reported that the adults of *M. indicum*, which do not emerge until late August or early September, may be killed before they oviposit when winters set in early. Mortality of the larvae in the spring was also attributed to prolonged winters that adversely affected the newly eclosed caterpillars.

Biological Mortality Agents

Life tables constructed for outbreak populations of tent caterpillars show that they are reduced by predators, parasitoids, and epizootics of disease agents. In his analysis of the population dynamics of *M. neustrium testacum* in central Japan over eight successive generations, Shiga (1979) found generational mortality (egg to ovipositing adult) in excess of 99% in each generation (Fig. 9.8). When I eliminated mortality due to known cases of predation and parasitism from three of these generations (generations in which disease was not a complicating factor) by the simple expedient of removing the entries for these factors from the life tables, generational survivorship to the adult stage ranged from 49 to 69%, indicating that mortality from predation and parasitism accounted for the loss of 31–51% of the populations (Fig. 9.10). The significance of these biotic agents in population regulation, however, is difficult to assess by life table analysis alone. As Price (1987) cautioned, correlation is easily confused with causation, and experimention is needed to assess the true impact of parasitoids and predators. The most signifcant problem involves compensatory mortality factors. A population not reduced by the activity of parasitoid or predator, for example, may be more susceptible to another agent, such as disease or starvation, later on.

Studies in which Shiga (1979) and Filip and Dirzo (1985) enveloped colonies in netting to exclude predators and parasitoids provide less ambiguous evidence for the impact of natural enemies on tent caterpillars. As shown in Table 8.4 and Figure 8.13, colonies protected from these mortality agents suffered significantly less mortality than those that were not protected. These and other studies discussed below show that the impact of predators and parasitoids is not evenly distributed throughout the life cycle of a colony and that some developmental stages are more vulnerable than others.

Tent caterpillars are vulnerable to biological mortality agents in the egg, larval, and pupal stages. The rate of parasitism of the eggs of tent caterpillars rarely exceeds 10%, and loss of pharate larvae to egg parasitoids cannot be considered an effective population-regulating mechanism (Stelzer 1968, Witter and Kulman 1979). Similarly, although braconid wasps and tachinid flies may destroy large numbers of tent caterpillar larvae during the later stages of outbreaks, there is little evidence that larval parasitoids regulate population growth.

Pupal parasitoids appear to have more impact on the population dynamics of tent caterpillars than egg or larval parasitoids. Hodson (1941) established an apparent density-dependent relationship between host abundance and the rate of pupal parasitism. He found that losses to populations of caterpillar pupae increased through successive years of an outbreak in Minnesota that extended from 1933 to 1938. During the final three years, pupal mortality averaged 67.3%, 81.0%, and 99.6%, respectively. Up to 97% of the total parasitism was attributed to *Sarcophaga aldrichi*. A similar pattern of increasing parasitism with duration of outbreak was recorded during the subsequent outbreak that extended from 1948 to 1958 (Hodson 1977; Fig. 9.11). Witter et al. (1972) subsequently constructed life tables for two generations of the forest tent caterpillar in the same region and also identified pupal parasitism as one of the most important factors influencing the population dynamics of the caterpillar (Table 9.2). Mortality of the populations was 99.4% in the first year and 97.8% in the second year. Life table analysis indicated that, although loss of pharate larvae to unknown causes and death of first-

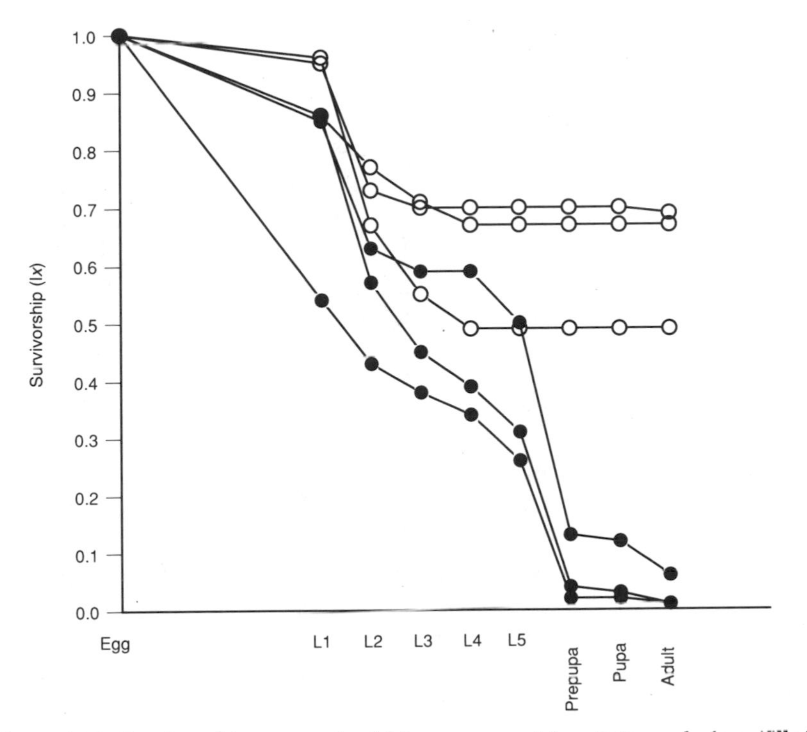

Figure 9.10. Survivorship curves for *Malacosoma neustrium testacum* before (filled circles) and after (open circles) removal of life table entries for predation and parasitism. (Data from Shiga 1979.)

instar larvae due to spring frosts played important roles, 50–74% of the pupal population was lost to parasitism by *S. aldrichi*.

Pupal parasitism also appears to be an important mortality agent in southern populations of *M. disstria* found in drier areas, but it plays less of a role in swamp forests because of the apparent inability of the dominant parasitoid, *S. houghi*, to find dry litter in which to overwinter (Stark and Harper 1982, Smith and Goyer 1986). Indeed, annual flooding of these swamp forests may not only inhibit the development of parasitoids and predators that pupate in the soil but may also interfere with the year-to-year survival of pathogens, making these populations relatively immune to mortality agents that plague more northern populations (Harper and Hyland 1981).

Tent caterpillar larvae are particularly susceptible to infection by nuclear polyhedrosis viruses. Stairs (1972) showed that a diseased first-instar larva may contain enough viruses to infect as many as 1 million other first instars. The viral particles can be transmitted from larva to

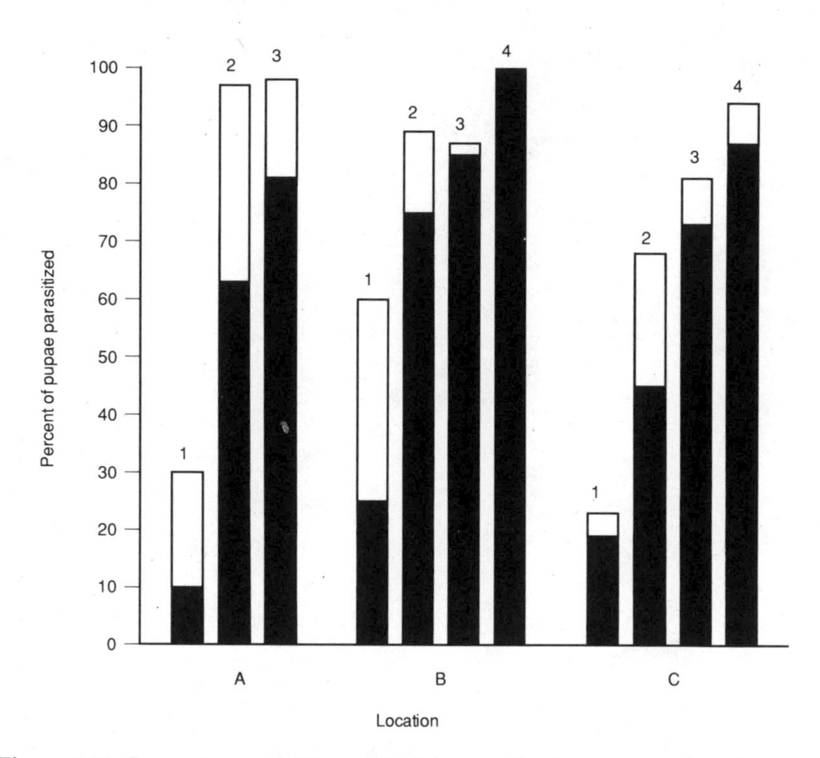

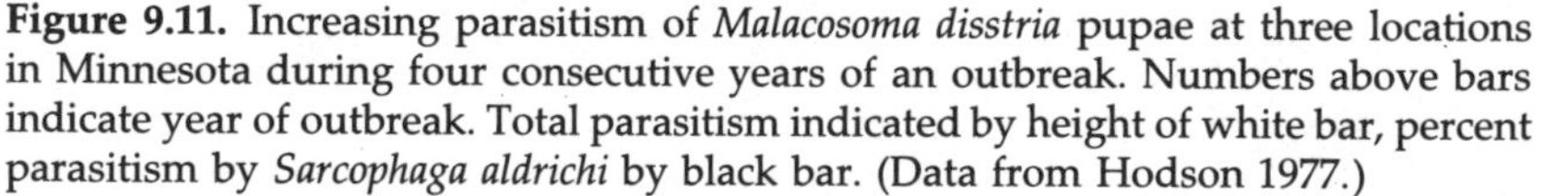
Figure 9.11. Increasing parasitism of *Malacosoma disstria* pupae at three locations in Minnesota during four consecutive years of an outbreak. Numbers above bars indicate year of outbreak. Total parasitism indicated by height of white bar, percent parasitism by *Sarcophaga aldrichi* by black bar. (Data from Hodson 1977.)

larva during the course of an outbreak and can be passed from one generation to the next on the egg mass. Thus, it is not surprising that massive infection sometimes accompanies the collapse of populations of tent caterpillars.

Clark (1955) documented one such viral epizootic during an outbreak of *M. californicum fragile*. In the fourth year of the infestation, the population covered several thousand acres, and most of the host trees were defoliated. The virus increased rapidly within the population, and by the end of the season most of the population had been infected. The following year, the population of tent caterpillars was estimated to be only 10% of its former size. Clark (1955, 1956a) conducted experiments that showed that overwintered egg masses and larval cadavers stuck to the trees contained viable viruses that reinitiated the infection in the residual population of caterpillars. Subsequent within-generation transmission among caterpillars led to the virtual collapse of the infestation by the end of the season.

Table 9.2. Life table for the forest tent caterpillar, *Malacosoma disstria*

Life stage (*x*)	Number alive (*lx*)	Cause of mortality (*DxF*)		Number dying (*Dx*)
Egg (avg. no./tree)	3105	Parasitoids		230
		Failure to eclose		1287
			TOTAL	1517
Larva	1588	Parasitoids		46
		Weather, starvation, disease, predators		1245
			TOTAL	1291
Pupa	297	Parasitoids		219
		Weather, starvation, disease, predators, migration		60
			TOTAL	279
Adult	18			

Generational mortality (egg–adult) = 1 − (18 ÷ 3105) = 99.4%

Source: After Witter et al. 1972, by permission of the Entomological Society of America.

Despite the potential for viruses to cause massive infection, there have been few other cases in which investigators have credited nuclear polyhedrosis viruses with bringing infestations of tent caterpillars to a halt. Studies of outbreak populations of the forest tent caterpillar by Hodson (1941) and Witter et al. (1972) indicated that viral infection, though present, had no significant bearing on the collapse of the infestations. Moreover, attempts to infect populations by introducing a pathogen have been only marginally effective (see Chapter 10). Disease has also been shown to persist within populations for years without appreciable impact. An infestation of the forest tent caterpillar on Prince Edward Island began in 1973 and collapsed in 1985. Disease was reported to be at high levels from 1976 to 1982, and the population was expected to crash each year. Instead, populations grew steadily and exploded in 1983, when the infestation expanded to 67,000 hectares, 3.5 times as large as it had been in the previous year (Anonymous 1960–1990). Myers (1988, 1993) proposed that sublethal infection with viruses or microparasites may drive the population dynamics of tent caterpillars and other forest Lepidoptera, but there are as yet no hard data bearing on this possibility.

Starvation

Although tent caterpillars commonly defoliate their natal trees during outbreaks, the dispersing caterpillars usually locate enough food to allow them to complete their larval development. But, in areas undergoing massive defoliation, dispersing caterpillars may find it difficult to locate isolated patches of palatable leaves and fail to achieve their full potential for growth. In fields and more open forest, the movement of caterpillars as they search for food may be greatly impeded by low-growing vegetation. Small plants and shrubs create a complex three-dimensional labyrinth that the caterpillars must negotiate, and it is not uncommon to find them stranded on nonpalatable species or on plants that offer only suboptimal nutrition.

Even though lack of adequate food may cause caterpillars in outbreak populations to pupate before they are fully grown and to produce small, less fecund adults, death of caterpillars due to starvation has been cited infrequently as a cause of population decline. One starvation die-off of the forest tent caterpillar occurred in Minnesota during the spring of 1936. Hundreds of thousands of caterpillars failed to find enough food to complete the fifth stadium and starved to death. In some places, their bodies were piled several inches deep in roadside ditches (Hodson 1941). Surviving caterpillars produced small adults that deposited egg masses having only 56% of the normal allotment of eggs, contributing to further decline of the population the following year. Starvation appears to be the major factor driving the two- to three-year cycles of abundance of the forest tent caterpillar in some swamp forests of the southeastern United States (Harper and Hyland 1981). At the peak of each cycle, the caterpillars run out of food and many perish. Those that survive produce few eggs, which give rise to a relatively small population of caterpillars the following year. Caterpillars in this remnant population experience little or no competition for food and there is high survival. Sufficient offspring are produced by this generation, or by the following generation, to once again strain the food resource, leading to population decline and the completion of a cycle.

Cultural and Economic Impacts of Outbreaks

Cultural Impact

Damage to economically important forest trees and forage plants is a major cause for concern and the subject of much of the remainder of this chapter, but the huge numbers of caterpillars and moths produced during outbreaks have often been a serious problem in their own right. During peaks of population expansion, as many as 20,000 tent caterpillars may inhabit a single tree (Stairs 1972), and the hordes of hungry caterpillars that disperse after completely stripping trees of leaves can be a major nuisance to the residents of communities bordering infested forests. Wandering caterpillars can suddenly appear in huge numbers on walkways, fences, and the walls of buildings, and mature larvae may enter houses in search of pupation sites. Dispersing caterpillars have been known to delay traffic and to short power lines (Sippell 1962). Spring campers and anglers have been driven from the woods by the swarming insects (Ghent 1958). During one outbreak of the forest tent caterpillar in Ontario in 1964, in which 7.7 million hectares were infested, so many moths descended on one small town that drivers were partially blinded by the moths at night. Fire hoses were used to wash down buildings, and the bodies of dead and dying moths were scooped up in snow shovels and carried away in a truck (Sippell et al. 1964). Swarms of parasitic flies often accompany the invading caterpillars and can be the cause of as many complaints from distraught residents as the caterpillars themselves.

An invasion of migrating forest tent caterpillars was documented during an outbreak in Saskatchewan (Casson 1979). The writer, the owner of a farm, had first observed defoliation in the surrounding aspen forest the previous year. In the spring of the current year, she noted that the infestation had spread closer to her farm and that numerous trees had been defoliated. Soon thereafter hungry caterpillars moved out of the trees and into her yard. Notwithstanding concerted attempts to kill them with an insecticide, the caterpillars kept coming and within a week had eaten all the leaves and flowers from her rose bushes, fruit trees, and shade trees except for those of the Manitoba maple (*Acer negundo*), which the caterpillars avoided. Twenty-seven wheelbarrow loads were collected, carted away, and burned. Despite aerial spraying, the caterpillars continued to come, and Casson (1979) described the situation some 10 days after the caterpillars first invaded her yard: "There were webs all around and caterpillars were dropping everywhere. As we looked through the denuded forest at sunset, the trees appeared to be draped in a beautiful silk veil. Their branches were black with clustered caterpillars and looked like huge black snakes writhing up into the air."

The enormous numbers of swarming caterpillars produced during outbreaks can interfere with the flow of traffic through infested areas and have on occasion brought locomotives to a halt. Riley and Howard (1890–1891:59) republished the following newspaper account of an incident involving a Canadian Pacific freight train that passed through a section of northern Maine during an outbreak of the forest tent caterpillar.

> When the train had proceeded a few miles, and when it was on a short grade, it was brought to a standstill by an army of small, gray caterpil-

> lars, greasing the track and driving-wheels to such an extent as to almost entirely suspend friction between the rails and the driving-wheels. In some places they were half an inch thick, and the army stretched out 11 miles. . . . [The train] began a series of small charges at that grade, which now had been liberally sprinkled with sand, but the animal life was so thick that various attempts were unsuccessful, and it was not until late night and the sun had gone down that the creeping things desisted in their march. . . . Nothing like it was ever known hereabouts before, but then sunlight was never before let into the wilds of Maine as the Canadian road has let it in, and there may be unknown difficulties to come consequent upon it.

In another account, Riley and Howard (1890–1891:477) related an incident originally published in the Washington, D.C., newspaper the *Evening Star* in 1891, involving swarming forest tent caterpillars in North Carolina.

> The rails on the Carolina Central Railroad were recently covered inches deep with caterpillars and, . . . for three days in succession trains were brought to a dead standstill, the driving wheels of the engines slipping round as though the rails had been thoroughly oiled. The engineers were obliged to exhaust the contents of their sand boxes before crossing the strip of swamp from which the caterpillars seemed to come. The rails and cross-ties were said to be obscured from sight, and the ground and swamps on each side of the track were covered with millions of crushed caterpillars, and from the mass an unendurable stench arose.

Economic Impact of Defoliation

Forest Tent Caterpillar

The forest tent caterpillar is the most widespread and economically important of the North American species of tent caterpillars, and its impact on host trees has been studied more extensively than that of any other tent caterpillar. During outbreaks, the insect commonly defoliates trees occurring over millions of hectares of contiguous land (Fig. 9.12). Although the insect is among the most polyphagous of the tent caterpillars, its importance is largely a consequence of the damage it causes to a few economically important trees species. The caterpillar causes severe defoliation of quaking aspen (*Populus tremuloides*) in Canada and the north-central United States, sugar maple (*Acer saccharum*) in the eastern United States and Canada, and black gum (*Nyssa sylvatica*) and swamp tupelo (*N. aquatica*) in the southeastern United States. Numerous

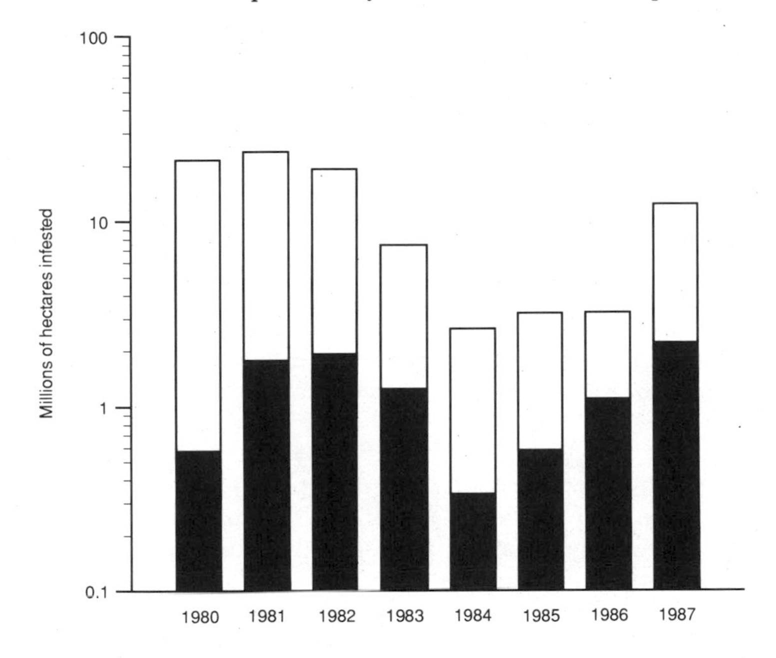

Figure 9.12. Total land area of Canada infested with the forest tent caterpillar during the 1980's. Aspen and sugar maple trees occurring within these areas were moderately to severely defoliated. White bars = prairie provinces, black bars = remaining provinces. (Data from 1980 to 1987, Canadian Forest Service, "Forest Insect and Disease Conditions in Canada," Ottawa.)

regional studies have documented the economic impact of the caterpillar on these host tree species.

Aspen. Quaking aspen, probably the most widely distributed tree in North America, is severely affected by the forest tent caterpillar. Fashingbauer et al. (1957:133–135) documented the seasonal development of an outbreak population of the forest tent caterpillar infesting aspen that occurred in Minnesota during the spring of 1951:

> [May 3:] Coincident with the first hatching of the tent caterpillar larvae, the buds of aspen growing on favorable sites had unfolded.
>
> [May 14:] Entire branches of many aspen were completely stripped.
>
> [May 18:] The terminal portion of most trees was defoliated. Scattered individual trees were almost completely bare.
>
> [May 24:] Extensive stands of aspen were completely denuded and hazel was stripped. The larvae were beginning to move up into the foliage of the birch trees . . . by the 30th, at which time almost all the larvae had attained the fifth instar, scarcely any aspen foliage remained.

[May 30:] The hazel and birch defoliation was almost as complete as that of the aspen and the larvae were swarming over the forest floor as the first signs of starvation were noted.

[June 4:] Herbaceous plants constituted the only remaining green vegetation in the area.... The forest floor and the roads into the infested regions were literally covered with caterpillars.

[June 14:] Cocoons were very common throughout the defoliated area and on the 21st the first completely developed pupae were found. For the past few days a greenish cast has been noted throughout the defoliated tree canopy as the development of new leaves progressed.

[June 26:] The defoliated region began to take on the appearance of early spring, with almost all of the new leaf buds of aspen now unfolded.

Although trees that are defoliated early in the season produce new leaves shortly after the caterpillars disappear, refoliated trees typically have smaller and fewer leaves. Defoliated trees produce less wood, store less food, and may have loss of photosynthetic area due to branch dieback (Gregory and Wargo 1986 and references therein). In extreme cases, trees may die following one or more episodes of defoliation.

One of a number of extensive outbreaks of forest tent caterpillars in the north-central United States erupted in Minnesota in 1948 and continued until 1958. During the infestation, Duncan and Hodson (1958) measured the impact of the caterpillar on aspen and associated plant species. After eliminating dead trees that showed evidence of disease, damage by other insects, suppression by overstory trees, or mechanical damage, they found no immediate evidence that aspen tree mortality occurred as a direct consequence of defoliation by the forest tent caterpillar. Six years after completing their study, Duncan and Hodson joined forces with two other investigators to look for long-term adverse affects on aspen that had been defoliated during the epizootic (Churchill et al. 1964). They recorded the incidence of aspen mortality that occurred from 1955 to 1961 in the previously infested stands and related this result to the extent of prior defoliation. Their study showed a relatively high incidence of unexplained mortality among dominant and codominant aspen in stands that had experienced three successive years of heavy defoliation during the outbreak. In addition, they found that the incidence of fungal disease and wood borer infestation was greatest in the stands that had experienced the most defoliation. Thus, their studies indicated that, although there was little apparent immediate increase in tree mortality after defoliation, some trees may be irreversibly affected and eventually succumb.

Other investigators report a negligible effect of the insect on aspen mortality. Hildahl and Reeks (1960) assessed the impact of the insect on aspen during a brief outbreak in Manitoba and Saskatchewan that occurred in the early 1950s. After comparing mortality of trees in stands experiencing no defoliation, light defoliation, and heavy defoliation, they found no evidence linking defoliation with aspen tree mortality. Ghent (1958) performed another retrospective study of the relationship between aspen mortality and tent caterpillar defoliation to determine if defoliation hastened the death of mature overstory aspen. He ascertained the year that trees had died by inspecting annual growth rings in samples of wood taken from trees determined to be dead for as long as 35 years. He then compared the rates of tree mortality in the years before, during, and after a serious outbreak of the forest tent caterpillar that occurred in western Ontario in the 1930s. Even though the overmature trees in his study plots were completely defoliated, he found no evidence that infestation by the tent caterpillar hastened their death.

The possibility that defoliation favors other mortality agents, and thus indirectly contributes to the death of aspen, could not be completely precluded in any of these studies. Although Duncan and Hodson (1958) found that during the course of the tent caterpillar infestation there was no annual increase in the mortality of aspen trees due to *Hypoxylon* canker, other investigators found that defoliation increased the likelihood that a tree would eventually develop a canker. Anderson and Martin (1981) kept records of the incidence of Hypoxylon infections in previously defoliated and control trees during the eight-year period following the collapse of the infestation in northern Minnesota. Their data indicate that the incidence of cankers varied directly with the number of years that trees had been defoliated (Fig. 9.13). Moreover, they found that about 96% of trees with *Hypoxylon* stem cankers died within six years of initial infection.

Despite the apparent resiliency of aspen, defoliation clearly suppresses its growth. During the peak of the Minnesota outbreak, Duncan and Hodson (1958) found that average basal area growth of defoliated aspens was as little as 13% of that of nondefoliated trees (Fig. 9.14). The year following the collapse of the infestation, previously defoliated trees produced smaller leaves and showed growth declines of approximately 15% (Fig. 9.14, year 4). The trees eventually recovered completely. In the second year after the infestation ended, there was no significant difference in growth between trees that had and those that had not been defoliated.

Up to the time that forest tent caterpillars reach the last instar, the larvae consume relatively small quantities of leaves, and there is no significant difference in radial growth between aspen that are insect-free

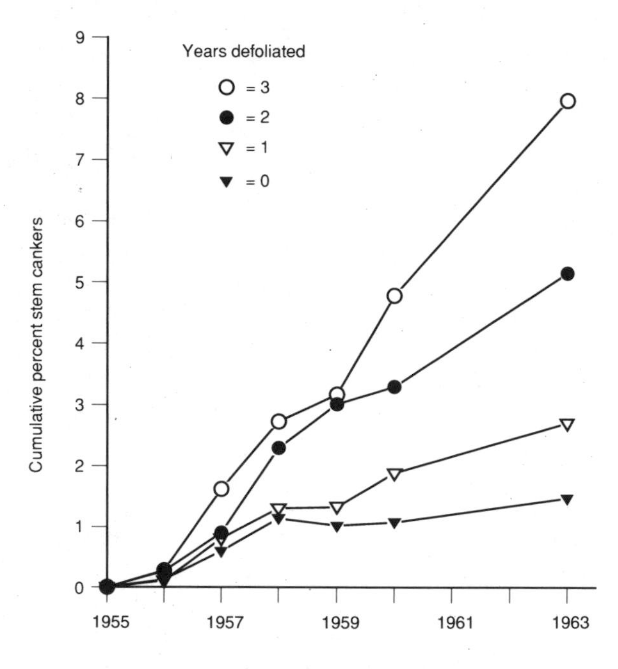

Figure 9.13. Cumulative percent of aspen trees infected with *Hypoxylon* stem cankers during the eight-year period following the collapse of an outbreak of forest tent caterpillars in 1955. The approximately 3500 trees included in the study experienced from 0 to 3 years of defoliation during the outbreak. (After Anderson and Martin 1981, by permission of the Society of American Foresters.)

and those that are infested. Massive defoliation during the last stadium, however, has a marked effect on radial growth (Rose 1958; Fig. 9.15). Three to four weeks after an aspen is completely defoliated newly formed buds that would otherwise not have opened until the following year become meristematically active and produce a new set of leaves. As shown in Figure 9.15, these new leaves do not contribute to new radial growth, and defoliated trees stop growing in early June, approximately two months before less-affected trees.

The ecological impact of severe outbreaks of the forest tent caterpillar extends well beyond direct damage to the ovipositional host and affects much of the plant community. Defoliation of aspen positively affects the growth of balsam fir, an understory species. On the basis of examination of core samples taken from trees, Duncan and Hodson (1958) found that in caterpillar-free years growth of aspen and balsam were positively correlated, but during periods of infestation a decline in the radial

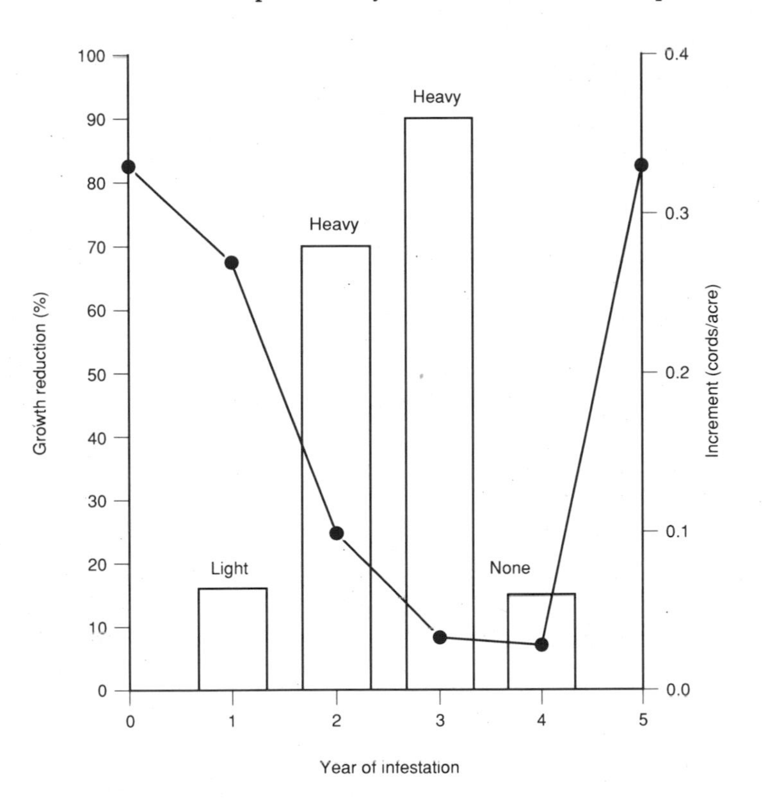

Figure 9.14. Average reduction in basal area growth (bars) and cords per acre during three successive years of light to heavy infestation of aspen by the forest tent caterpillar in Minnesota. Year 0 provides values for noninfested stands. (Redrawn, by permission of the Society of American Foresters, from Duncan and Hodson 1958.)

growth of aspen resulted in a significant increase in the radial growth of balsam. Defoliation of aspen appears to facilitate the growth of balsam by allowing more light to penetrate the canopy and by decreasing transpirational loss of soil water. Many understory species, however, are negatively affected. Duncan and Hodson (1958) found that 13 of the 17 brush species that commonly occur under aspen are eaten by the caterpillars. Some species, such as hazel (*Corylus* spp.), prickly ash (*Xanthoxylium americanum*), and species of cherry (*Prunus* spp.) are often completely defoliated.

Despite the negative impact of tent caterpillars on the growth of aspen, the economic significance of this insect to the forest industry has yet to be assessed. In Canada, the insect affects much of the hardwood re-

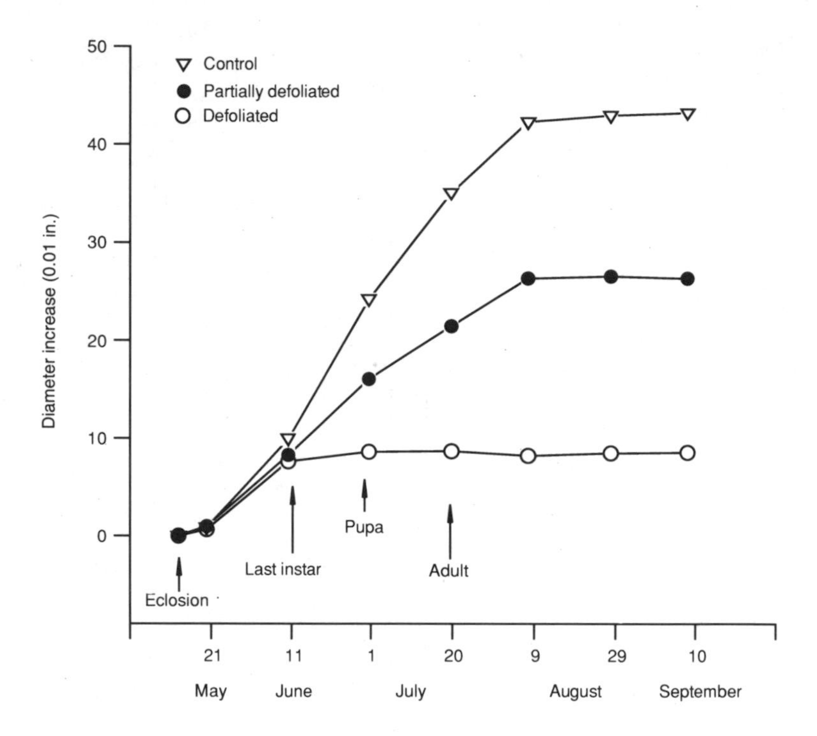

Figure 9.15. Seasonal growth of fully foliated aspen trees (control) and trees partially or completely defoliated by the forest tent caterpillar. Most of the loss in growth is attributable to defoliation by the last instar. Although defoliated trees produced new sets of leaves, they contributed little or nothing to the radial growth of affected trees. (After Rose 1958, with permission of the Society of American Foresters.)

source, yet only a small percentage is harvested each year and losses go largely unnoticed (Anonymous 1986a). Increased demand for pulpwood in the future, however, may change this situation.

Sugar maple. Sugar maple, another common host of the forest tent caterpillar, appears to be more susceptible to defoliation than aspen and may be irreversibly damaged. Field surveys to assess the effect of the forest tent caterpillar on maple indicate significant levels of dieback and whole-tree mortality following one to three years of repeated defoliation.

A major outbreak of the forest tent caterpillar occurred in Ontario during the 1970s, at one point affecting about 14 million hectares (Howse et al. 1981, Gross 1985). Some stands of sugar maple were defoliated three years in succession, and numerous pockets of dead and dying trees, some as large as 12 hectares, were observed a year or two afterward. Gross (1985) estimated that in southwestern Ontario 410,000 m^3 of maple timber were killed during the outbreak. Most of the loss occurred in 1977, the peak year of the infestation, and during the spring of the following year. Many trees that survived repeated defoliation suffered various degrees of branch loss, and many of those that lost over 40% of their branches were irreversibly damaged and died within three years of the end of the infestation. Less damaged trees, however, showed considerable resiliency. Those that sustained less than 40% branch dieback usually recovered completely and by 1980 were growing as well as trees that had not been defoliated. It was estimated that during the outbreak the growth increment of defoliated trees averaged 39.5% less than that of trees in nearby stands that escaped defoliation (Gross 1991).

During the peak of another infestation that occurred in the early 1980s in Vermont, 130,000 hectares were defoliated. In one area of heavy defoliation, 67% of sawlog trees and 79% of fuelwood trees were dead or had less than 10% of their crowns intact. Two years after the collapse of the infestation, 9000 hectares showed light to heavy mortality (Teillon et al. 1982, 1984). During the peak of a three-year infestation in New York State, 81,000 hectares were defoliated (R. Williams, cited in Allen 1987). In some stands, up to 95% of the defoliated trees subsequently died. Trees reportedly died after failing to produce new leaves following a single episode of defoliation by the caterpillar. Low soil water and high temperatures during the period of defoliation appear to have contributed to the sudden death of the trees.

In many parts of the northeastern United States and Canada, sugar maple trees are tapped commercially to obtain maple syrup. The sap sugar concentration is highest in the period just before leaf flush in the spring (Gregory and Wargo 1986), so trees are tapped before leaves are produced. Hence, defoliation has no effect on sap production during the initial year of an infestation. In the year after defoliation, however, significant reduction in sap flow and lower sugar sap content has been recorded (Winch and Morrow 1962).

The long-term impact of defoliation by the forest tent caterpillar on sugar production was assessed during a major outbreak of the caterpillar in New York State in 1953. Investigators established plots containing comparable numbers of trees in similar condition in each of two areas lying within a region heavily infested with caterpillars. The trees in half of the plots were protected from the caterpillar by spraying them with DDT. The other trees were repeatedly defoliated by the forest tent caterpillar during the course of the outbreak. The five-year study showed that defoliated trees had three times as much crown mortality as control trees. The loss of crown area resulted in reduced sap production; non-defoliated trees produced on average about twice as much sap as affected trees (Connola et al. 1965, Connola 1980)

Southern hardwoods. In the southern United States, the forest tent caterpillar has its greatest impact on swamp forests of tupelo. An estimated 202,350 hectares in Louisiana and 20,235 hectares in Alabama are completely defoliated each year (Abrahamson et al. 1982). Stands dominated by water tupelo that were defoliated yearly for at least 20 years had annual growth increments of approximately one-quarter that of non-infested stands. Sweet gum in nearby stands may also be affected during severe outbreaks, and in some cases trees have died after three successive years of defoliation (Batzer and Morris 1978). Infestations of forest tent caterpillars in mixed hardwood stands in southern Louisiana during 1986 involved 727,000 acres and were estimated to have resulted in a growth loss equivalent to 290,000 cords valued at $1.4 million (Anonymous 1986b).

Other Species

Other North American tent caterpillars are less widely distributed than the forest tent caterpillar and are of less overall economic importance. Before the widespread use of insecticides in commercial orchards, the eastern tent caterpillar was considered to be a serious pest of apple trees. For the most part, the caterpillar now attacks wild and neglected trees of little commercial value, but it also attacks ornamental rosaceous species and is consequently a major nuisance to homeowners throughout its range. Its principal host, black cherry, is widely distributed over the eastern half of the United States, but it is a commercially important member of the forest community only in areas of New York, Ohio, Pennsylvania, Maryland, and West Virginia. Under forest conditions, the trees rarely occur together in extensive stands as do aspen and sugar maple but are more likely to be found as scattered individuals. Moreover, the insect favors open-grown trees over forest trees throughout much of its range. Although repeated defoliation may result in some dieback, black cherry is particularly resilient and recovers rapidly after defoliation (Kulman 1965).

Populations of *M. californicum* and *M. incurvum* and their subspecies commonly cause extensive defoliation of cottonwood and aspen in the southwestern states. More than 400,000 hectares of aspen were reportedly defoliated in Colorado in 1934 (List 1934). Another outbreak that began in Colorado in 1977 persisted until 1986, affecting at one point 28,000 hectares of aspen (Anonymous 1986b). In New Mexico, 56,700 hectares of aspen were infested in 1929, and 30,000 in 1941 (Stelzer 1968).

Stelzer (1968) found that aspen in some stands in New Mexico were defoliated as many as four years in a row by *M. californicum*. Increment borings indicated that trees suffered annual radial growth losses of 28, 52, and 75% after one, two, and three successive years of complete defoliation. Moreover, repeated defoliation also appeared to cause overt tree mortality. During the five-year study, tree mortality in stands suffering four successive years of defoliation increased from 21 to 56% Ninety percent of the affected trees that were still alive at the end of the study showed significant death of branches in the crown of the tree.

Some populations of *M. californicum* are defoliators of bitterbush, *Purshia tridentata*, a shrub that is an important source of forage for domestic and big game animals. One particularly serious infestation occurred in California in 1944, involving approximately 29,000 hectares of rangeland. Nearly half of the bitterbush was completely defoliated. The carrying capacity of the infested range was reduced by an estimated 25%, and some mortality of plants was reported (Clark 1956b). Complete defoliation of bitterbush in Oregon over two successive seasons, however, reportedly did little noticeable damage to the plant despite heavy winter browsing by deer (Mitchell 1990).

Of the Eurasian species, *M. neustrium* and its subspecies *testacum* is the most economically important of the tent caterpillars. The insect ranges over all of Europe, the former Soviet Union, China, India, and Japan and attacks a wide array of trees wherever it is found. Bhandari and Singh (1991) investigated outbreaks of the Indian forest tent caterpillar *M. indicum*, a polyphagous species that attacks oaks, popular, willow, apple, and other broadleaf species. In the Himalayas at altitudes of 3500–4800 m the caterpillar attacks the bhojpatra tree (*Betula utilis*), a species that grows in association with other alpine species or in pure stands all the way to the treeline. Although the bhojpatra grows slowly and lacks a straight stem, it is considered a locally important species. The wood is harvested for lumber and fuel, and the bark is used for writing paper and as packing material. The bark is also used to make umbrellas, hookah tubes, and roofing material. The branches are used to make twig bridges on streams, and the leaves are used as fodder. Defoliation of *B. utilis* forests was reported for the first time in 1984, 1985, and 1986. The investigators reported finding patches of recently killed trees that they attributed to both defoliation and the habit of the caterpillar to feed on the green bark of the young shoots when the leaves are all eaten. Branches weakened by repeated defoliation are unable to withstand the weight of snow during the winter season, and massive loss of branches leads to eventual death of the tree.

10

Management of Populations

As curator of the Department of Entomology at the American Museum of Natural History in the 1930s, Frank Lutz was frequently besieged by people seeking advice on how to rid their trees of tent caterpillars. Lutz (1936) suggested that they do nothing, taking the position that to intervene interfered with the "balance of nature" and caused more harm than good. He argued that disease, natural enemies, and starvation would accomplish more in the long run. Although Lutz's ideas were not widely accepted at the time, the essential correctness of his position is now clear. Outbreaks of tent caterpillars are part of a natural cycle of the population dyamics of the insects, and they invariably run their course. Few species of forest trees suffer significant mortality, even when they are repeatedly defoliated, and the short-term loss of growth to all but highly valued trees rarely justifies costly containment efforts. Moreover, public concern for the environment has sharply curtailed the use of insecticides and has placed growing emphasis on biorational techniques that have high target-specificity. The current wisdom among many applied entomologists is that efforts to contain tent caterpillar outbreaks should be largely limited to high-use recreational areas or to areas where the potential loss of particularly valuable stands of trees justifies the economic and ecological costs of intervention. It has taken a long time to arrive at this consensus. Entomologists have waged intermittent warfare on tent caterpillars for the better part of two centuries, employing an arsenal of mechanical, chemical, and biological weapons.

Early Eradication Campaigns

In the 1800s and early 1900s, tent caterpillars were considered to be among the most insidious of insect pests, and whole communities were sometimes mobilized to combat the perceived threat of infestations. Residents of communities experiencing outbreaks were often issued emergency bulletins describing simple, labor-intensive procedures for combatting the insects. The entomologist T. W. Harris, credited with the founding of applied entomology in North America, was one of the earliest of a succession of entomologists to call for the mass destruction of the caterpillars. In his book *Report on Insects Injurious to Vegetation*, published in 1841, Harris wrote:

> [The larvae] may be effectually destroyed by crushing them by hand in the nests. A dried mullein head . . . will be useful to remove the nests . . . [or] . . . we may use, with nearly equal success, a small mop or sponge, dipped as often as necessary into a pailful of refuse soap-suds, strong white-wash, or cheap oil. The mop should be thrust into the nest and turned round a little, so as to wet the caterpillars with the liquid, which will kill every one that it touches. (Pp. 270–271)

Harris's call for a communitywide effort reflects the nearly religious fervor with which early entomologists often approached their work:

> I beg leave to urge the people of this Commonwealth to declare war against these caterpillars, a war of extermination, to be waged annually during the month of May and the beginning of June. Let every able-bodied citizen, who is the owner of an apple or cherry tree, cultivated or wild, within our borders, appear on duty, and open the campaign on the first washing-day in May, armed and equipped with brush and pail, as above directed, and give battle to the common enemy; and let every housewife be careful to reserve for use a plentiful supply of ammunition, strong waste soap-suds, after every weekly wash, till the liveried host shall have decamped from their quarters, and retreated for the season. If every man is prompt to do his duty, I venture to predict that the enemy will be completely conquered, in less time than it will take to exterminate the Indians in Florida. (P. 271)

During an outbreak of forest tent caterpillars in New York State, Slingerland (1899:557–558) wrote:

> An alarming state of affairs exists wherever this insect occurs. . . . Thousands of the shade trees in many New York villages are doomed

> unless prompt measures are taken to destroy the caterpillars. . . . Pay the boys and girls a few cents for each score or hundreds of the egg-masses they collect. . . . The rivalry between the children will soon spread to rivalries between schools and the result will be that the number of caterpillars will be reduced to the minimum by a single season's crusade of the children. . . . Begin the warfare in August or September . . . and keep it up until the last egg cluster is burned.

W. E. Britton (1935) reported the results of several campaigns by school children. In Newfields, New Hampshire, more than 8000 egg clusters of the eastern tent caterpillar were collected when school children were offered 10 cents for each hundred egg masses, resulting in the destruction of well over a million eggs. In 1913, the extension service of the Connecticut State College organized a contest for school children, offering a prize of $25 to the student who collected the largest number of egg masses and a scholarship to the college's summer school for a teacher from the school that collected the largest total number. The regionwide contest was reported to result in the collection and subsequent destruction of an estimated 10 million egg masses.

Despite the large numbers of egg masses destroyed during these campaigns, Sweetman (1940) reported that the results were often disappointing. Participants were surprised to see the extent of defoliation the following spring that was attributable to egg masses that had gone undiscovered. Sweetman's own diligent efforts at collecting eggs resulted in as many as 20% being overlooked, and he recommended that emphasis would be better placed on destroying the small colonies soon after they appeared in the spring. One particularly ardent school principal offered his students prizes ranging from $1 to $5 for the collection of tents, which he burned daily. During the campaign, the children collected nearly 17,000 tents, weighing over half a ton. In some villages, a reward of 10 cents per quart was offered for cocoons, and in the village of Glens Falls, New York, 1350 quarts were turned in (Felt 1899).

One manufacturer offered a specially designed brush for removing the tents from trees (Fig. 10.1), but colonies of tent caterpillars were often destroyed by setting fire to the inhabited tents. Swaine (1918) suggested that "asbestos fibre soaked in kerosene and placed in a tin can nailed to the end of a pole makes an excellent torch," and one manufacturer offered a kerosene-charged torch for use by orchardists (W. E. Britton 1935). Despite almost universal warnings against the procedure, due to the obvious threat to the tree, the sight of scorched and writhing caterpillars appears to satisify a desire for vengeance of a form reserved only

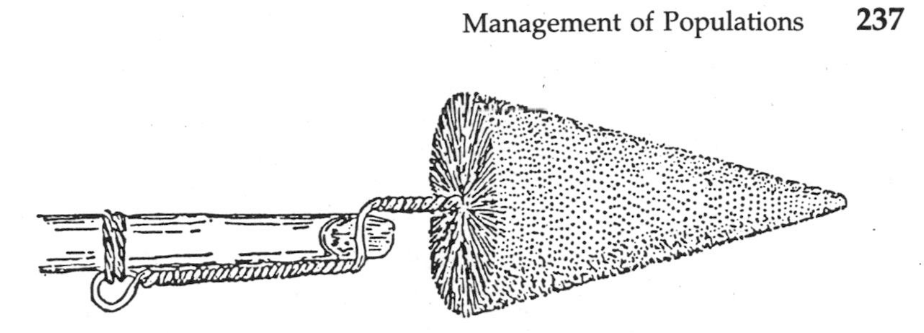

Figure 10.1. A brush manufactured in Connecticut in the 1930s to be used to remove the tents of the eastern tent caterpillar from infested trees. (From W. E. Britton 1935.)

for insects, and the procedure was widely applied. Indeed, scorn for tent catepillars persists unabated to the present. In the same volume in which Klein and Wenner (1991) suggest live-trapping of lawn-infesting moles so they can be safely relocated, they write, "The tent or webs (of tent caterpillars) can be torn out manually and viciously with a stick or long pole and then the caterpillars can be stomped upon."

Insecticides

Even though the simple mechanical expedients of destroying egg masses or tents were highly effective and ecologically desirable ways of suppressing small-scale infestations of tent caterpillars, early entomologists were quick to appreciate the advantages of insecticides. Before the advent of formulations based on synthetic chemicals, inorganic stomach poisons were widely used to kill tent caterpillars. One of the most popular was Paris green, a copper acetoarsenite originally used as a green pigment in paints. Its insecticidal properties were discovered serendipitously around 1870 (Whorton 1974). Although the compound had the distinct disadvantage of being a violent poison, affecting warm-blooded animals as well as insects, there were few chemical alternatives, and it was widely applied until it was replaced in the late 1800s by lead arsenate, a compound that proved less phytotoxic.

High concentrations of lead arsenate were required to control insects. Swaine (1918) recommended a mixture of 3 pounds of lead arsenate to 40 gallons of water for the control of young colonies of tent caterpillars but cautioned that "a very much stronger spray" was required to kill the larger caterpillars. Significant residues of lead arsenate eventually accumulated in the upper surface of the soil under sprayed trees, and some orchards, heavily sprayed for a variety of folivores, were reported to have 30–40 times the background level of arsenic and lead of unsprayed trees (Boswell 1952). After the insecticidal properties of chlorinated hydrocarbons were discovered, lead arsenate and other inorganic poisons were largely abandoned as insecticides.

The first of the synthetic insecticides, benzene hexachloride and DDT, were tested against the eastern tent caterpillar in small-scale field studies soon after they became generally available (Table 10.1; Manter 1945, Srivastava and Wilson 1947). They received high praise because of their remarkable ability to kill not only caterpillars that consumed sprayed foliage but also those whose bodies merely touched contaminated surfaces. Eight ounces of 50% wettable powder DDT proved to be more effective than 3 pounds of lead arsenate (Kerr 1952).

From 1945 to 1972, 88% of all the insecticide applied to forest lands was DDT (Anonymous 1975). Most, however, was applied before 1959, and its use against forest pests was largely phased out by 1969 because of growing evidence of the deleterious effects it had on nontarget organisms. Although approximately 12 million hectares of forest land in the United States were sprayed during this 27-year period, very little insecticide was directed against tent caterpillars. Most was used to combat the gypsy moth and lepidopteran pests of conifers. Only 22,000 hectares of forest land infested by either western tent caterpillars or the forest tent caterpillar were sprayed, despite the many extensive outbreaks of tent caterpillars that occurred during this period (Anonymous 1975).

When DDT first came into widespread use, its potential to cause delayed effects by accumulating in the food chain was unknown, and, on the basis of short-term studies, investigators at first concluded that it had little effect on warm-blooded animals. Fashingbauer et al. (1957) referenced numerous studies that showed that when applied at the nominal rate of 1 pound per acre, DDT had no apparent effect on woodland birds. His own studies showed that birds inhabiting a stand of aspen sprayed with DDT to control the forest tent caterpillar suffered no immediate ill effect and were as successful in rearing young as were birds in nearby untreated stands. In a separate note, however, Fashingbauer (1957) reported that populations of wood frogs inhabiting a pond in the sprayed area were decimated by the chemical. The frogs were bathed in residual spray that coated the surface of the ponds, and stomach analysis revealed they had gorged on the poisoned caterpillars that had fallen from the trees into the water. R. C. Morris and Orr (1962) also reported that DDT affected nontarget organisms after an experimental control program directed against the forest tent caterpillar in southwestern Alabama in 1961. Their investigation showed that, although one-half

Table 10.1. Experimental laboratory and field studies that have assessed the effectiveness of chemical insecticides, growth regulators, pathogens, and pheromones against tent caterpillars in North America

Year	Investigator	Target species	Agent	Site
1945	Manter	*Malacosoma americanum*	I	F & L
1947	Srivastava & Wilson	*M. americanum*	I	L
1952	Kerr	*M. americanum*	I	F
1954	Clark & Thompson	*M. californicum fragile*	NPV	F
1956	Clark & Reiner	*M. c. fragile*	NPV	F
1958	C. G. Thompson	*M. c. fragile*	NPV	F
1959	Angus & Heimpel	*M. americanum*	Bt	F
		M. disstria	Bt	F
1961	Jaques	*M. americanum*	Bt	F & L
1961a	Bucher	*M. americanum*	C	F
1962	R. C. Morris & Orr	*M. disstria*	I	F
1964	Oliver	*M. disstria*	I	F
1964	Stairs	*M. disstria*	NPV	F
1965	Angus	*M. disstria*	Bt	F
1965	Stelzer	*M. c. fragile*	NPV, Bt	F
1965	Stairs	*M. disstria*	NPV	F
1966	Stairs	*M. disstria*	NPV	F
1967	Stelzer	*M. c. fragile*	NPV, Bt	F
1969	Bird	*M. disstria*	NPV, CPV	L
1969	O. N. Morris	*M. c. pluviale*	Bt	F & L
1971	Wallner	*M. disstria*	I, Bt	F
1972	Lyon et al.	*M. disstria*	I	L
1972	R. F. Morris	*M. disstria*	Bt & I	L
1973	Abrahamson & Morris	*M. disstria*	I	F
1973	Abrahamson & Harper	*M. disstria*	I, NPV, Bt, FN	F
1973	Robertson & Gillette	*M. c. californicum*	I	L
1973	Page & Lyon	*M. c. lutescens*	I	L
1975	Frye & Ramse	*M. disstria*	I, NPV, Bt	L & F
1976	Nordin	*M. americanum*	M	F
1976	Retnakaran & Smith	*M. disstria*	G	L & F
1977	G. G. Wilson	*M. disstria*	M	L
1977	G. G. Wilson & Kaupp	*M. disstria*	M	F
1978	Ives & Muldrew	*M. disstria*	NPV	F
1979	Harper & Abrahamson	*M. disstria*	G, Bt	F
1979	Retnakaran et al.	*M. disstria*	G	F
1979	G. G. Wilson	*M. disstria*	M	L
1980	Milstead et al.	*M. c. californicum*	Bt	F
1980	Milstead et al.	*M. constrictum*	Bt	F
1981	Johnson & Morris	*M. disstria*	I, Bt	F
1982	Abrahamson et al.	*M. disstria*	I, Bt	F
1982	Chisholm et al.	*M. disstria*	P	F
1982	Ives et al.	*M. disstria*	NPV	F
1983	Palaniswamy et al.	*M. disstria*	P	F
1984	Pinkham et al.	*M. disstria*	Bt	L
1984	G. G. Wilson	*M. disstria*	M	L
1989	P. B. Schultz	*M. disstria*	I	L & F

Table 10.1—*cont.*

Year	Investigator	Target species	Agent	Site
1990	Nielsen	*M. americanum*	I, S	F
1991	Smitley et al.	*M. americanum*	I	F
1991	Nielsen & Dunlap	*M. americanum*	I, G, S	F

Note: BT = *Bacillus thuringiensis;* C = *Clostridium* spp.; CPV = cytoplasmic polyhedrosis virus; F = field; FN = fungus; G = growth regulator; I = chemical insecticide(s); L = laboratory; M = microsporidian; NPV = nuclear polyhedrosis virus; P = pheromone; S = insecticidal soaps and oils.

pound of DDT per acre gave highly effective control of the insect, fish and birds were also killed.

Indeed, in retrospect, it seems naive to have expected that broad-spectrum pesticides could be applied to forests and woodlands without significant, direct or indirect damage to nontarget organisms. Compared with single-crop agricultural farms, forests are elaborate ecosystems, populated by a broad diversity of interacting organisms. One inventory of the fauna of a mixed hardwood forest in southern Louisiana carried out in conjunction with an experimental spray program directed against the forest tent caterpillar gives some idea of the complexity of forest communities. The inventory of vertebrates included: green, spotted, yellow-belly, and warmouth sunfish; flagfin shiner; mullet; leopard frog; bullfrog; copperhead, cottonmouth, and water snakes; wood duck; white ibis; great blue, little blue, and green herons; cardinal; brown thrasher; blue jay; woodpeckers; nuthatches; warblers; muskrat; nutria; racoon; gray and fox squirrels; swamp rabbit; and deer (Oliver 1964). In addition, vastly larger numbers of invertebrates and microrganisms occur on the trees and in the forest soil. Abrahamson et al. (1982), for example, collected 112 taxa of invertebrates, mostly families of insects, from nets placed under the canopies of water tupelo and black gum trees infested with outbreak populations of the forest tent caterpillar.

It is of some historical interest to note that although benzene hexachloride and DDT were originally synthesized in 1825 and 1874, respectively (Bowen and Hall 1952), their insecticidal value was not recognized until the late 1930s, and they did not come into widespread commercial use until after World War II. It is sobering to contemplate the scale of environmental damage that might have occurred had they been widely disseminated before the development of sophisticated analytical tools that have made it possible to track the movement of toxic residues through communities.

When DDT was officially banned for use in the United States in 1973,

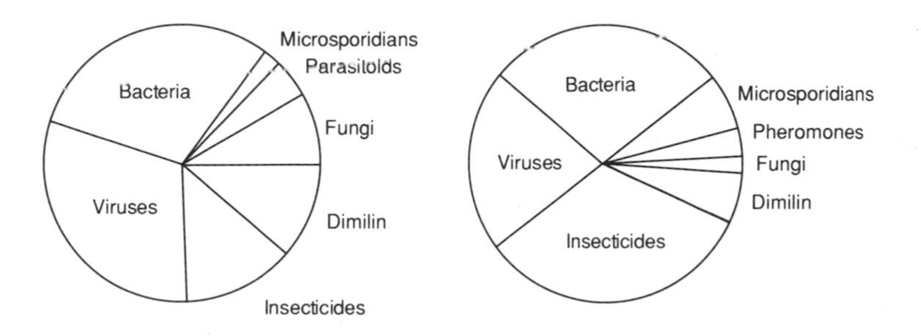

Figure 10.2. Proportions of experimental laboratory and field studies dealing with various chemical substances and biological organisms for the control of tent caterpillars in Eurasia (left) and North America (right). Data for Eurasia are for *Malacosoma neustrium* and are derived from Agricola and Common Wealth Agricultural Bureaux databases (1970–1991, $N = 46$ articles). Data for North America are for the period 1945 to 1991, and are based on Table 10.1.

it had been largely replaced for a number of years as the insecticide of choice for the control of forest defoliators. Chief among the insecticides to replace DDT was the carbamate carbaryl (also known as Sevin), which, while still toxic to a broad diversity of animals, persists for only weeks rather than years and does not accumulate in the food chain (Anonymous 1975). In response to both increased understanding of the ecological side effects of chemical insecticides and public opposition to them, both the United States and Canada have largely phased out their use for the control of tent caterpillars on forest lands.

Other Control Methods

Almost all experimental studies of tent caterpillar control before 1950 involved synthetic chemical insecticides, but most of the research effort since then has been directed at developing more biorational control techniques (Table 10.1, Fig. 10.2). These techniques include the propagation of viruses, bacteria, fungi, microsporidians, and entomophagous insects; the use of growth regulators and pheromones; and the development of trees that resist folivores.

The encouragement of entomophagous insects has achieved limited success with some other forest pests, but there have been no reported attempts at biological control by releasing non-native parasitoids or predators of tent caterpillars in North America. Some entomophages introduced into the United States for the control of the gypsy moth, however, have since been recovered from tent caterpillars (see Chapter 8). Limited studies in Russia (Romanova and Lozinsky 1958, Romanova

1972) indicate that the rate of parasitism of the eggs of tent caterpillars can be increased when wasp parasitoids are released in woodlands infested with the tent caterpillars, but there have been no systematic attempts at widespread control using this technique.

Since the 1960s, most of the laboratory and field research on the biological control of tent caterpillars in North America has been directed at developing bacterial and viral pathogens. A considerable effort to develop these pathogens for the control of *Malacosoma neustrium* has also occurred in eastern European countries and the former Soviet Union over this same period (Fig. 10.2). Pathogenic agents have the decided advantage over chemical insecticides of being relatively specific in their activity toward target insects. They are apparently harmless to vertebrates and have little direct impact on entomophagous parasitoids and predators. Principal drawbacks to their use as compared with traditional chemical approaches include lower cost-effectiveness and relative difficulty of application. Formulations of microbial agents are typically labile and must be consumed by the target insect soon after they are applied. In addition, because there is often a substantial period of time between application of some pathogens and the death of the victim, defoliation may continue for an unacceptably long time after the pathogen has been ingested.

Bacteria

Formulations based on the bacterium *Bacillus thuringiensis* (Bt), first isolated from the silkworm about 1900, have been commercially available since 1960. The facultative pathogen occurs worldwide, and more than 30 subspecies are now recognized. The subspecies *kurstaki* is toxic to approximately 200 species of Lepidoptera, and one particularly toxic isolate of this subspecies, HD-1, has been widely developed into a commerical product. Intensive efforts to develop the bacterium for commercial application have resulted in a marked decrease in the cost of producing and applying the pathogen, so that in some cases it is only slightly more expensive to use than an insecticide (Frankenhuyzen 1990). The major drawback to its effectiveness is its vulnerability to the elements and concomitant short residual life. Although dried Bt spores can retain their viability for several years in the soil (Krieg 1987), on exposed surfaces they typically retain their activity for only a few days and for even a shorter time in full sunlight. Thus, to be effective they must be eaten by the caterpillars soon after they are disseminated.

The active material is prepared by a fermentation process that induces the bacterium to produce spores and a crystalline, parasporal body. It is this crystalline material, the Bt delta endotoxin, that is toxic to the caterpillars. The formulation is sprayed on foliage and ingested by the caterpillars. The crystalline endotoxin is broken down by alkaline proteases in the gut, forming toxic products that destroy the epithelial cells of the gut wall, preventing further feeding. In some species, such as *Bombyx mori*, a general body paralysis occurs within an hour or so of ingesting the pathogen, but no such paralysis occurs in *Malacosoma* although there is a rapid inhibition of feeding (Angus and Heimpel 1959). Krywienczyk and Angus (1969) found that *M. disstria* larvae fed crystals of the endotoxin broken down by the gut juices of *B. mori* still did not develop a general paralysis, indicating that something other than differences in the gut enzymes of the caterpillars accounts for the paralysis.

Spores of Bt may also germinate in the gut creating a septicemia that contributes to the death of the insect. But the host environment does not typically support multiplication of the pathogen, as it does in viral diseases, so establishment of epizootics in populations is precluded. Caterpillars may die within several days, but caterpillars debilitated by sublethal doses may live for as long as a month after ingesting the endotoxin. Persistence of weakened larvae has the beneficial effect of providing a food base for parasitoids and predators for some time after the caterpillars are no longer a threat to the host tree.

Studies show that the interval between application of the material and mortality varies with the host plant of the caterpillar. In studies with *M. neustrium* and *M. paralellum*, Shcherbakova et al. (1981) found that preparations of Bt caused mass mortality of larvae feeding on poplar two to three days after treatment, on oak three to four days after treatment, and on cherry five to seven days after treatment. They attributed these differences to variation in the phytocidal activity of the host plants. In a separate study, Ovcharov (1984) found that leaves of cherry, poplar, willow, and apple, but not oak or alder, had an apparent antibiotic effect on bacterial preparations used for biological control of *M. neustrium*.

Bt is currently registered for use against tent caterpillars in North America and is commonly used to suppress small infestations. Several experimental studies have shown that the pathogen is also effective in controlling large outbreak populations of the forest tent caterpillar. One of the most extensive of these studies was undertaken to suppress infestations of forest tent caterpillar in tupelo stands in southwestern Alabama (Harper and Abrahamson 1979, Abrahamson et al. 1982). Formulations were applied by aircraft, and ground surveys were conducted after spraying in both treated and matched control plots to determine the effectiveness of the application. Moribund caterpillars began to drop from trees in treated plots within four hours of spraying, and those

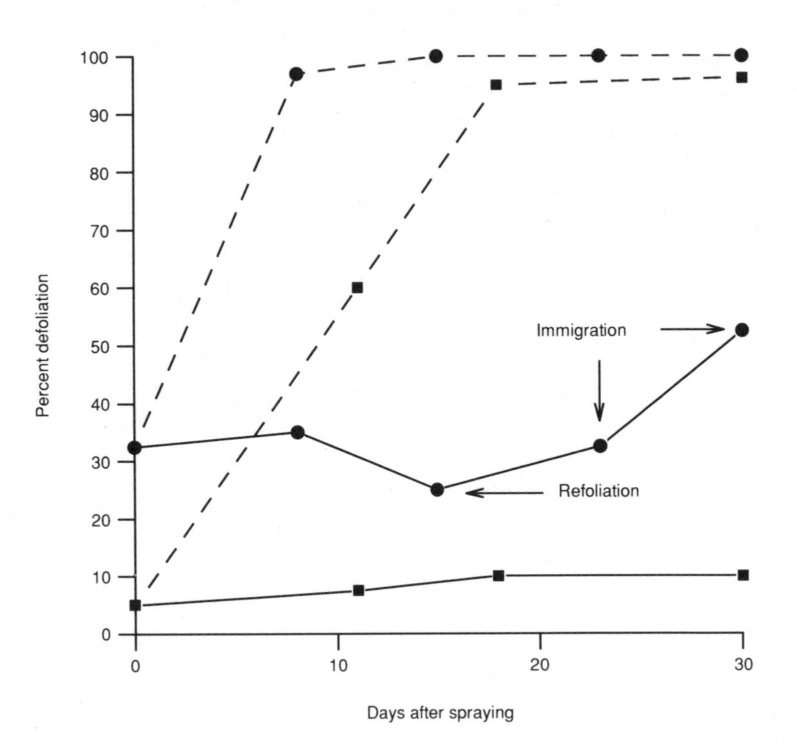

Figure 10.3. Loss of foliage to forest tent caterpillars in trees sprayed with *Bacillus thuringiensis* (solid lines) and in untreated trees (broken lines). The treatment was applied at the rate of 4.3 BIU/ha (circles) or 4.8 BIU/ha (squares) during two different years. (After Harper and Abramhamson 1979.)

remaining in the trees stopped feeding almost entirely shortly after ingesting the toxin. In contrast, trees in control plots were completely defoliated within 10–15 days after initiation of the study (Fig. 10.3). Follow-up studies to determine the impact of spraying on nontarget organisms were consistent with the known specificity of the material and indicated that that there was no apparent direct effect on wild bird populations or on the planktonic organisms in the river-swamp food chain. There also appeared to be little direct effect of the spray on arboreal invertebrates other than caterpillars.

Because of growing public pressure to limit the use of chemical insecticides, Bt is currently the method of choice for the control of many species of forest defoliators. Advances in the development of formulations and application techniques have made the use of Bt considerably more cost-effective than it was in the past. Undiluted formulations applied by air at the rate of 30 billion international units per hectare have

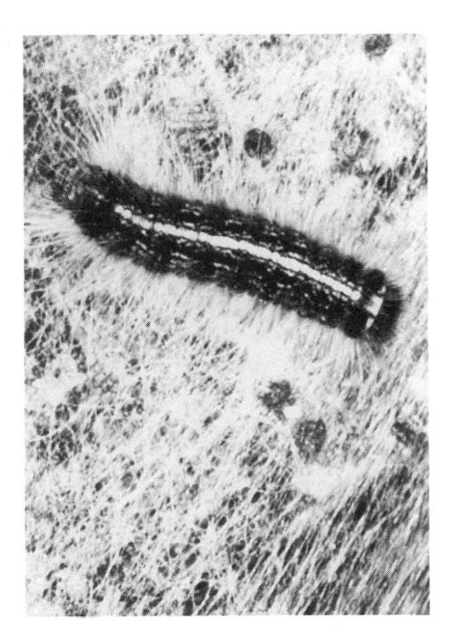

Figure 10.4. Characteristic shrunken appearance of an eastern tent caterpillar killed by brachyosis.

recently been shown to provide excellent control of the gypsy moth, but the use of similar high-potency formulations have not yet been reported for tent caterpillars (Frankenhuyzen 1990).

Tent caterpillars are also susceptible to the bacteria *Clostridium brevifaciens* and *C. malacosomae* (Bucher 1957, 1961a, 1961b). Originally isolated from *M. californicum pluviale,* they were subsequently shown to cause disease when fed to the larvae of *M. americanum* and *M. disstria,* although the latter is relatively resistant. After the spores are ingested, the bacteria multiply in the gut of the host. The course of the disease proceeds rapidly. Infested larvae of the western tent caterpillar become irritable and begin to regurgitate within 48 hours of infection. By the third day, the larvae feed less and void wet feces. In *M. americanum,* the rectum may be inverted through the anus. Contraction of the longitudinal muscles and the muscles of the gut wall cause the larvae to shrink markedly in length, the most characteristic symptom of the disease and the basis for the common name brachyosis (Fig. 10.4). By the fifth or sixth day, the larva evacuates most of the gut contents as a reddish brown liquid and becomes moribund. Death occurs one to four days later. The apparent immediate cause of death is water loss. Because the caterpillars regurgitate and void large numbers of spores on surfaces frequented by colony mates, the disease is highly infectious. The early instars are most susceptible, and brachyosis typically decimates any colony in which the symptoms develop early enough.

Bucher (1961a) conducted field tests against *M. americanum* to determine the feasibility of using *Clostridium* spp. as biological control agents. He found that the vegetative rods, which he was able to propagate on an artificial medium, were ineffective in establishing the disease, but spores obtained from sick caterpillars were highly infectious when sprayed on leaf surfaces or tents. The spray was most effective when it was applied to colonies before the caterpillars completed their third instar, and the disease spread from one caterpillar to another. Although the bacteria can survive the winter on contaminated egg masses or other surfaces, Bucher (1961b) found no evidence that the disease was carried over into the next generation.

Clostridium has been identified only from *M. californicum pluviale*, but it appears to occur naturally in other species as well. I observed symptoms identical to those described by Bucher in colonies of *M. americanum* reared in the laboratory from eggs collected in the field in Cortland, New York. The disease is particularly virulent under laboratory conditions and destroys whole colonies within a week or so after symptoms appear. I found it impossible to rear additional colonies in the laboratory in the same season that the disease appeared, despite a diligent effort to disinfect all surfaces and prescreen egg masses for disease symptoms. Stehr and Cook (1968) also reported that their mobile insectary was so contaminated that it became impossible to rear additional insects. Although they did not identify the organism responsible, they describe symptoms identical to those caused by *Clostridium*.

Viruses

Tent caterpillars are susceptible to nuclear polyhedrosis virus (NPV) and granulosis virus, both of which are baculoviruses. They also harbor cytoplasmic polyhedrosis virus (CPV) and various nonoccluded viruses (Martignoni and Iwai 1986). Of these, the nuclear and cytoplasmic polyhedrosis viruses are the best studied and the most important viral pathogens of tent caterpillars. Nuclear polyhedrosis virus contains DNA and infects the nuclei of the tracheae, blood, muscle, fat body, ganglia, pericardial cells, and epidermis of the host (H. F. Evans and Entwistle 1987). Larvae infected early enough are completely overtaken by the pathogen. The body contents turn into a liquid mass of viruses that spills from the cadaver when the cuticle eventually ruptures (Fig. 10.5).

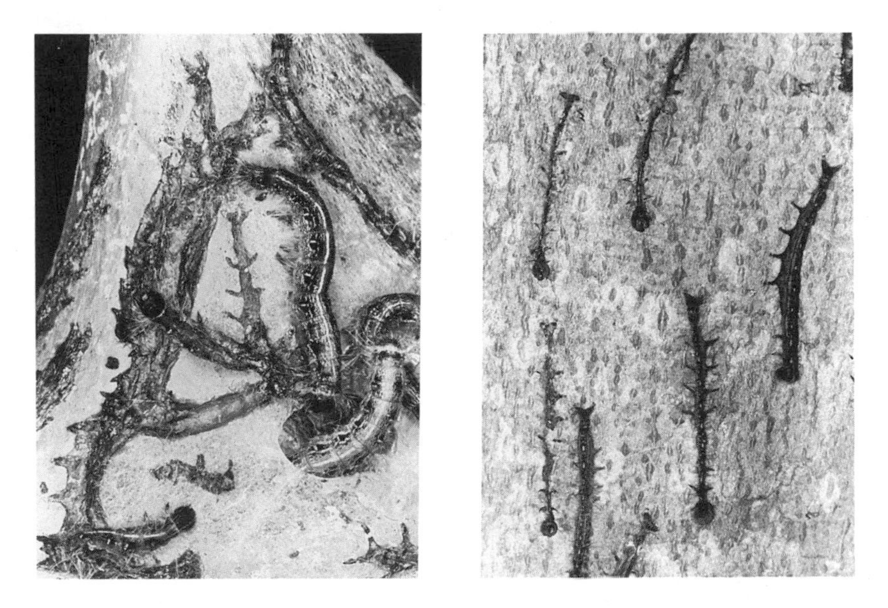

Figure 10.5. Eastern (left) and forest (right) tent caterpillars killed by a viral disease. Flattened cuticles remain after the virus has spilled from the diseased caterpillar.

Cytoplasmic polyhedrosis virus contains RNA and multiplies only in the cytoplasm of the midgut epithelium (Bird 1969). CPV is comparatively rare in natural populations of North American species of *Malacosoma* and is not considered as promising a potential control agent as NPV. Laboratory tests involving the forest tent caterpillar show that, although CPV is somewhat more infectious than NPV, it is significantly less lethal (Bird 1969; Fig. 10.6). Moreover, when larvae are fed both CPV and NPV, CPV tends to interfere with the development of NPV and prolongs the survival of infected larvae. Studies in Russia (Golosova 1986), however, showed that even though infected *M. neustrium* caterpillars lived 20–30 days, infection of the gut with CPV caused larvae to stop feeding after three to five days, and the caterpillars caused less damage than those infected with either NPV or granulosis viruses. Moreover, the investigators found that CPV in fecal material survived in the soil for as long as three years.

Although the susceptibility of tent caterpillars to viral diseases has been known for many years, their use to suppress infestations of tent caterpillars has lagged behind the development of Bt. Commercial formulations of viruses are registered for use against the gypsy moth and several other forest pests in North America, but none is currently reg-

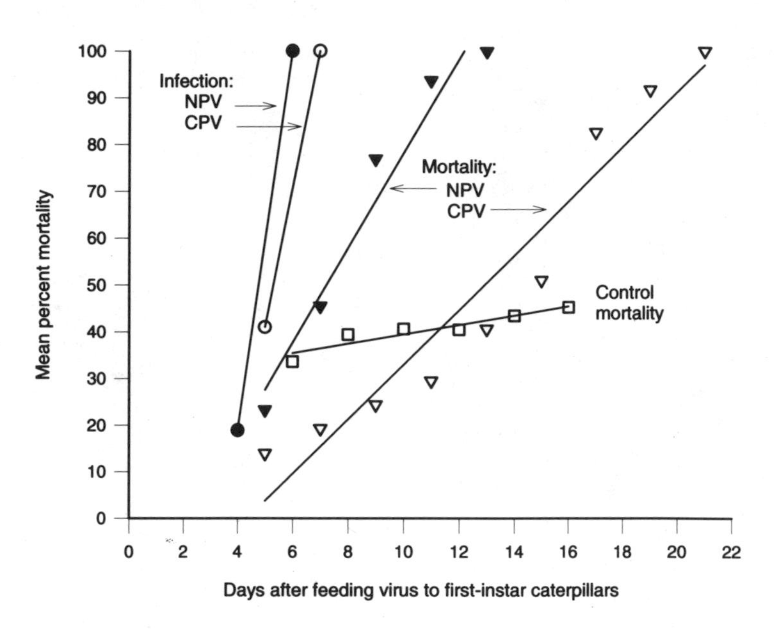

Figure 10.6. Percent infection and mortality of forest tent caterpillars fed nuclear polyhedrosis virus (NPV), cytoplasmic polyhedrosis virus (CPV), and untreated controls. (After Bird 1969, *The Canadian Entomologist* 101[1969]:1279, by permission of the Entomological Society of Canada.)

istered for tent caterpillars, for several reasons. Unlike Bt, viruses are relatively host-specific. Because tent caterpillars are readily controlled with Bt and are not often perceived as a serious threat, there has been little economic incentive to develop viral diseases of tent caterpillars. In addition, viruses are relatively difficult to culture in the numbers required for control programs, because they must be grown in the target insect. It takes about 1000 infested gypsy moth larvae to produce adequate virus to treat 1 hectare of forest (Cunningham et al. 1991). In vivo culturing has proved manageable in the case of insect pests readily reared on artificial diets, but most species of tent caterpillars do poorly on such diets.

A more serious consideration is that, unlike Bt, which acts much like an insecticide, viruses act as pathogens and are slow to kill the host. The incubation period of tent caterpillar virus as determined under laboratory conditions is 8–20 days, depending on temperature, so infected caterpillars can be expected to continue to damage host trees before they eventually succumb (Fig. 10.7). Applications must also be timed to correspond to the initial stages of host development. Stairs (1964) found that when an NPV spray containing 1 to 10 million per milliliter of water

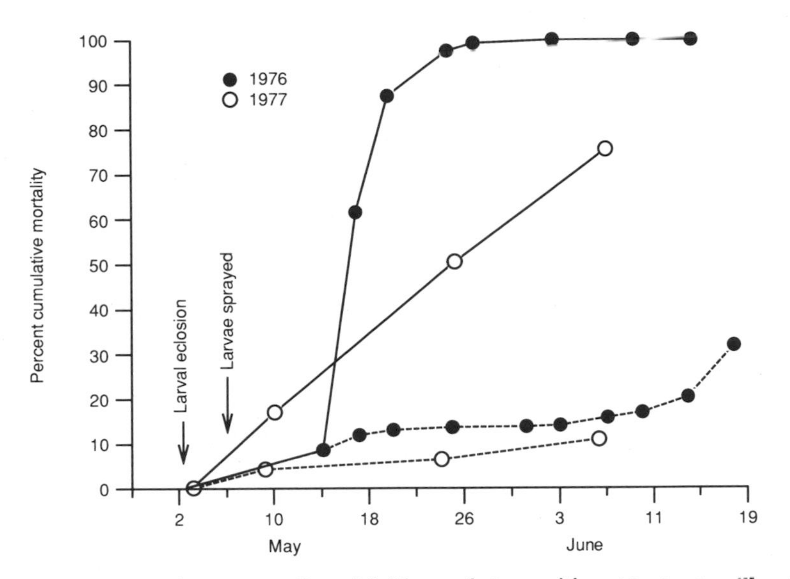

Figure 10.7. Cumulative mortality of field populations of forest tent caterpillars in plots sprayed with nuclear polyhedrosis virus in 1976 (solid lines) and in control plots (broken lines). Mortality in 1977 was attributable to carry over of the virus from the previous year. (Data from Ives and Muldrew 1978.)

was applied to the first-instar larvae of the forest tent caterpillar, 92% of the larvae subsequently died; when the third instar was sprayed only 14% died. No mortality attributable to the virus occurred when the fourth instar was sprayed. Magnoler (1985) also found that when NPV was applied to the third instar at the rate of 1 million polyhedra per milliliter, the virus killed 99% of a field population of *M. neustrium,* but much higher concentrations were needed to suppress larger caterpillars.

Viruses are highly labile when sprayed on foliage. Exposure of the nuclear polyhedrosis virus of *M. neustrium* to concentrated ultraviolet light for 20 minutes causes a 40–50% reduction in activity. On the upper surface of a leaf in full sunlight, unprotected deposits may have a half-life of only a few hours (Zarin'sh and Eglite 1985). Broome et al. (1974) found that an NPV of *M. disstria* applied to the foliage of gum leaves was completely deactivated when exposed to sunlight for 10 hours. Thus, to be effective the virus must be consumed soon after it is applied. The viability of the virus can be enhanced to some extent by incorporating chemicals that protect the preparation from damage due to exposure to UV radiation, but careful timing has proved critical to the success of experimental spray programs.

One potential advantage viruses hold over chemical agents and Bt for the control of pest insects is their ability to multiply and spread through the population. Such potentiality would be expected to be particularly high in the case of tent caterpillars because of the gregarious habits of the larvae. Thus, Stairs (1965, 1966) found that an epizootic of an NPV of *M. disstria* he initiated not only spread from caterpillar to caterpillar but also carried over into the next generation. Ives and Muldrew (1978) and Ives et al. (1982) also found evidence that populations of forest tent caterpillars in areas that had been sprayed with the virus one to three years earlier were more heavily infected with the virus than control populations in nearby areas. This cross-generational transmission is attributable primarily to contaminated adults that passed the virus to their egg masses. Some contamination of egg masses is also attributable to infected larvae that survived long enough to transfer the virus to the newly deposited egg masses they crawled over and to the adult of the common parasitoid *S. aldrichi* which feeds on diseased larvae (Stairs 1965, 1966).

Almost all attempts to assess the potential of viruses to suppress populations of tent caterpillars have involved nuclear polyhedrosis viruses. In one of the earliest records of microbial control, Davis (1903) reported that he was able to infect colonies of *M. americanum* by spraying leaves near the tent with water in which diseased caterpillars had been mixed. The identity of the infectious agent, while not clearly established, was probably a nuclear polyhedrosis virus. Subsequent investigations (reviewed in Clark and Thompson 1954) showed that *M. disstria* and *M. californicum* were susceptible to NPV, and it is probable that all North American species of *Malacosoma* show cross-infectivity to viruses collected from any one species. Indeed, Stairs (1964) demonstrated that the forest tent caterpillar was susceptible to a virus isolated from the European species *M. alpicolum,* though the pathogen was less virulent than native strains.

Experimental studies to control field populations of tent caterpillars with NPV have given mixed results. Clark and Thompson (1954) and Clark and Reiner (1956) conducted the first extensive field studies in North America. They collected the virus from diseased larvae and sprayed suspensions of the pathogen on field populations of the Great Basin tent caterpillar infesting bitterbush. Ground applications at concentrations of a million polyhedra per milliliter of water applied at the rate of 10 gallons (38 liters) per acre resulted in the apparent establishment of an epizootic and significant mortality in the populations. Large-scale aerial applications involving lower concentrations of polyhedra, however, were ineffective.

Field studies conducted by Stelzer (1967) involved the aerial application of a formulation containing both *B. thuringiensis* and NPV, the latter collected from a natural field epizootic. The material was applied by helicopter to plots set up within a 20-hectare area infested by *M. californicum.* Follow-up surveys indicated that 16–24 days after spraying from 31 to 92% of the colonies in the plots were infected with the virus and only 1% of the colonies in an adjacent control area were infected. Larva-to-larva transmission both within and between colonies was observed. One year later the tent caterpillar infestation in the treated area was only 5% less than that in the control area.

Studies show that timing of NPV application is critical, and the pathogen may have little or no effect if applied to larger caterpillars. Abrahamson and Harper (1973) applied NPV from the air at the rate of 10 billion polyhedral bodies per acre to control populations of the forest tent caterpillar. The material was not applied until the insects were in the third to fourth larval instar, and the caterpillars were not noticeably affected by the treatment. The plots treated with NPV experienced 95–100% defoliation, whereas plots treated at the same time with the insecticide trichlorfon or Bt experienced only 30–40% defoliation, little different from the extent of defoliation before the plots had been sprayed.

The most ambitious attempt to assess the potential for NPV to suppress outbreak populations of tent caterpillars was undertaken in Alberta, Canada, from 1976 to 1980 (Ives and Muldrew 1978, Ives et al. 1982). Preliminary studies to assess the virulence of the viral preparation were carried out by hand spraying egg masses or early-instar larvae under field conditions. Egg masses were saturated with a preparation containing 100 million polyhedral bodies per milliliter just before the date of hatching. Newly emerged caterpillars were sprayed with a mixture containing 10 million polyhedral bodies per milliliter. Both treatments resulted in massive mortality of the insect (Fig. 10.7). Moreover, the disease carried over into the subsequent generation, and evidence of its persistence in the treated area was found in the third generation posttreatment.

Encouraged by these preliminary studies, the investigators initiated a three-year program to assess the effectiveness of aerial applications of the pathogen, but the results were less promising. The viral preparation was applied to experimental plots by helicopter at rates as high as 10 trillion polyhedra per hectare. Although the study was somewhat complicated by the occurrence of a natural epizootic of the pathogen during the course of the investigation, the applications appeared to increase infection rates above the background level. The treatments, however, failed to cause population collapse, and subsequent generations of caterpillars rebounded. Moreover, there was little difference in the extent to which trees in control and sprayed stands were defoliated. The investigators noted that the natural epizootic of the NPV had little apparent impact on populations of the caterpillar and concluded that unless a more virulent strain of the virus was developed, aerial application of NPV held little future promise for controlling outbreaks of the forest tent caterpillar.

Apparently disheartened by the mixed results of field studies and an increasing emphasis on the use of Bt for caterpillar control, entomologists have shown relatively little interest in developing viruses for the control of North American tent caterpillars since the 1980s (see Table 10.1). Viruses, however, have the advantage over control agents that are currently operational of being both highly host-specific and transmissible, and it is likely that they eventually will be used to manage tent caterpillar populations. Techniques now at hand that enable the insertion of genes into viruses hold much promise for the future of control strategies based on the pathogen. Such procedures are expected to lead to the development of strains of viruses that act faster and have increased transmissibility. It may also prove possible to make viruses more virulent by inserting genes that induce the pathogens to produce toxins, hormones, or enzyme inhibitors in the infected host.

Fungi

The most significant fungal pathogen of tent caterpillars in North America is *Furia crustosa* (Entomophthorales). The widespread fungus has been isolated from *M. americanum, M. disstria,* and *M. californicum pluviale* (MacLeod and Tyrrell 1979, Tyrrell and Ben-Ze'ev 1990, Sampson and Nigg 1992). The pathogen overwinters as a resting spore (zygospore) in the soil. Spores that occur above the snowline do not survive the winter (D. Tyrrell, pers. comm., 1993). Laboratory studies show that the resting spore germinates when held at 8–28°C after storage in the soil at 4°C for six months (Perry and Fleming 1989). In the field, the spore germinates in the spring and produces germ conidia, an infectious stage of the pathogen. Tent caterpillars come into contact with the conidia when they disperse over the ground after defoliation of their natal tree. Forest tent caterpillars typically disperse in the fifth instar, although under outbreak conditions they may disperse earlier.

Unlike bacteria or viruses, the fungal conidia do not need to be ingested for the insect to be infected. Germ tubes produced by germinating spores may penetrate the integument or spiracular openings of the insect. In the infected caterpillar, the mycelium grows to eventually fill the body cavity. After the host dies, the hyphae emerge and the mycelium

Figure 10.8. (Above) Caterpillar infected with *Furia crustosa.* (Below) Resting spores of *F. crustosa.* (Reprinted, courtesy of H. N. Nigg, from Sampson and Nigg 1992.)

grows over the outside of the caterpillar's body, eventually completely covering it with a light to dark brown, crustose coating (Fig. 10.8). Once the insect is infected, the course of the disease is rapid, and the insect typically dies four to five days after first contacting the conidia. The disease spreads from caterpillar to caterpillar when conidia are actively ejected from the condiophores that cover the outside of the cadaver. Sarcophagid flies, which frequently attend outbreaks of the forest tent caterpillar in huge numbers, may also be a significant factor in the spread of the conidia.

For reasons that are unclear, some infected larvae fail to develop the external mycelial coat, and the hyphae that pack the body cavity give rise to resting spores. When these caterpillars eventually break apart, the resting spores are dispersed, and those falling to the soil are capable of reinitiating the epizootic the following year. Although there are no data documenting the impact of the fungus on field populations of tent caterpillars, observations indicate that the pathogen can become a significant mortality factor in the later stages of an outbreak.

F. crustosa can be cultured on coagulated egg yolk to produce both conidia and resting spores (MacLeod and Tyrrell 1979), greatly facilitating its potential use as a control agent. Indeed, the gregarious habit of tent caterpillars and the tendency of many species to aggregate in humid tents is highly favorable to the establishment of epizootics. Abrahamson and Harper (1973) reported the only attempt to disseminate the pathogen to suppress a population of tent caterpillars in North America. They recovered resting spores from diseased caterpillars and distributed them at the rate of 4.2 million spores per acre but failed to establish an epizootic in populations of the forest tent caterpillar. Failure of the application may have resulted from inadequate cold treatment of the spores before dissemination or the fact that the caterpillars were in the third to fourth instar at the time the pathogen was applied.

Other less host-specific species of entomophagous fungi have also been tested in experimental control programs. Joshi and Agarwal (1987) reported that *Aspergillus flavus* caused losses of caterpillars in laboratory cultures. Machowicz-Stefaniak (1979) tested various mixtures of the fungi *Beauveria bassiana, Verticillium lecanii, Metarhizium anisopliae,* and *Paecilomyces farinosus* against *M. neustrium* in Poland. Mixtures containing *B. bassiana* reportedly gave the best results under both laboratory and field conditions. Leathers and Gupta (1993) tested strains of *B. bassiana* against *M. americanum* larvae in laboratory experiments. Larvae inoculated with spores developed symptoms as early as six hours after exposure, and the caterpillars in all treatment groups were dead within four days. The rapid onset of symptoms was consistent with the production by the fungus of both cuticle-degrading enzymes that facilitate the invasion of the hyphae and insecticidal toxins. The rapid action of the pathogen makes it a promising candidate as a control agent, but the fungus has yet to be tested under field conditions.

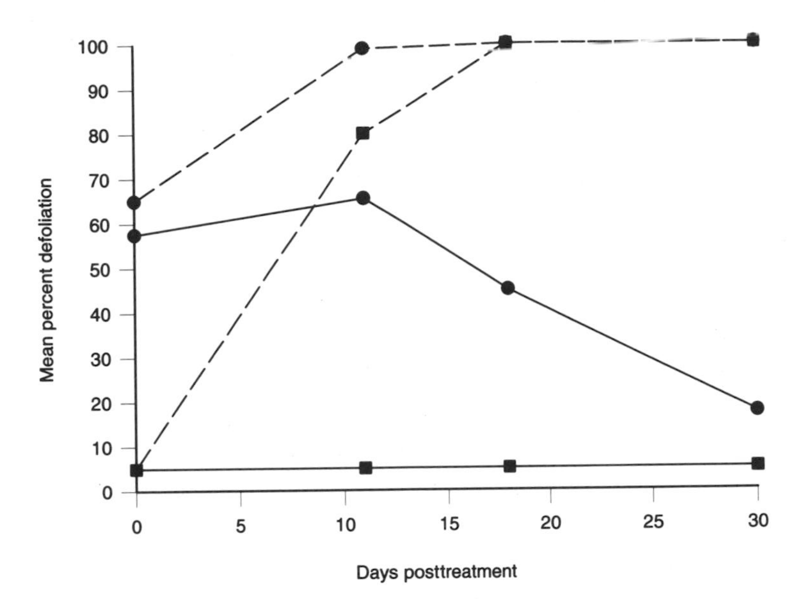

Figure 10.9. Loss of foliage to forest tent caterpillars in control stands of water tupelo (broken lines) and stands treated with the insect growth regulator Dimilin (solid lines) applied at rates of 67 g/ha (circles) and 108 g/ha (squares). (After Harper and Abrahamson 1979.)

Growth Regulators

The insect growth regulator Dimilin inhibits molting and has been used in pilot projects to control the forest tent caterpillar. Field testing of the compound was conducted in Ontario in 1978 (Retnakaran et al. 1979). The material was applied by aircraft at the rate of 70 g/ha. The effect of the treatment was apparent as early as four days after application. Caterpillars in treated areas ceased feeding, and many showed molting abnormalities. Postspray sampling of trees in control and treatment plots showed marked differences in population densities. Caterpillars were virtually absent from sprayed trees; trees in control plots were heavily defoliated and had average populations of more than 200 caterpillars.

Another field study conducted in Alabama in water tupelo stands infested with forest tent caterpillar gave similiar results. Harper and Abrahamson (1979) found that when Dimilin was applied at rates as low as 34 g/ha, additional foliage losses in treated plots could be limited to 20% or less. At higher application rates, losses stabilized or decreased in treated stands at the same time that control stands were experiencing increased losses (Fig. 10.9). Dimilin has also been applied to control *M.*

neustrium. Aerial application of the material at the rate of 113–150 g/ha in China resulted in the reduction of caterpillar populations by over 95% (Miao et al. 1986).

Pheromones

Female tent caterpillar moths produce pheromones that attract males. The sex pheromones of some species have been identified and synthesized (Chapter 4). Field experiments show that at least one of these pheromones, that produced by the forest tent caterpillar, can be used to trap males as a means of monitoring the growth of a population (Chisholm et al. 1981, 1982). Moreover, a blend of the compounds (*Z*)-5,(*E*)-7-dodecadienal, (*Z*)-5,(*Z*)-7-dodecadienal, and (*Z*)-7-dodecenal at a ratio of 10:1:1, respectively, has been shown to be as effective as live virgin females in luring males to traps (Palaniswamy et al. 1983).

Palaniswamy et al. (1983) investigated the possibility of disrupting mating of the forest tent caterpillar by liberating synthetic pheromone in quantities that interfere with the ability of the males to locate females. One component of the forest tent caterpillar pheromone, (*Z*)-5,(*E*)-7-dodecadienal, was formulated in rubber or plastic dispensers to release from 100 to 1500 μg of pheromone over a 24-hour interval. Dispensers were distributed around experimental plots by placing them on the ground or by tying them to trees. Sticky traps bearing live females were set up in both treatment and control plots, and the number of males they captured was compared as an index of the success of the procedure in preventing males from finding females. Success of males in finding the caged females in treated plots was less than 15% of that of males in control plots when release rates equaled or exceeded 300 μg per plot per 24 hours.

Field observations of flying males indicated that their limited success in locating females in treatment plots was due both to the confusion caused by the dispersed pheromone and to the fact that males spent less time in plots permeated with synthetic pheromone. Although the results of these experimental studies were encouraging, it is unknown whether the extent of mating disruption would be adequate to contain outbreaks of tent caterpillars over larger areas. Scaled-up experimental studies with the gypsy moth showed that microencapsulated pheromone dispersed by airplane over a 60-km^2 area significantly reduced the level of mating and subsequent production of egg masses (Beroza et al. 1974). The gypsy moth researchers suggested that the use of pheromone as a confusant is likely to be most effective during the early stages of an infestation, when population levels are low.

Host Tree Resistance

Still another means of combatting phytophagous insects pests of agricultural plants involves the development of plants that resist insect attack. Most of the work in this field has thus far been directed at deterring herbivores of agricultural plants. Many fewer attempts have been made to develop resistance to herbivores in long-lived woody plants. Recent preliminary studies show, however, that it may be possible to alter the resistance of host tree foliage to forest tent caterpillars either by inducing short-term changes in leaf quality or by breeding plants that are naturally resistant.

Experimental studies by Haanstad and Norris (1992) show that Siberian and American elm saplings treated with either *N*-ethylmaleimide in paraffin oil or heat-killed microbial suspensions produce leaves with altered palatability for the forest tent caterpillar. When applied topically to the stem or introduced into a shallow hole, the treatments appear to stress the plant and induce it to produce foliar chemicals that act as antifeedants. The results of this preliminary study were highly variable, but all treatments significantly reduced feeding below control levels on some leaves during at least one of the posttreatment sampling intervals. Despite the promising results of these preliminary studies, it seems unlikely that such procedures could provide cost-effective alternatives to other established control techniques. Moreover, we presently know too little of the effects of induced phytochemicals to state that such compounds have significant impact on field populations of tent caterpillars.

A more promising approach to the development of resistant trees involves the propagation of constitutionally resistant clones. To date, the species best suited to this technique are hybrid willows, cottonwoods, and poplars. Hybrids show high phenotypical variability, are easily propagated vegetatively, and grow rapidly. Pure stands of hybrid trees are highly productive and can be grown much like agricultural crops. They are likely to become increasingly important sources of wood products. Because the trees are attacked by tent caterpillars, there is incentive to propagate trees that show maximal resistance to the insects.

Recent studies demonstrate that clones of hybrid poplars vary markedly in their resistance to forest tent caterpillars. Robison (1993) conducted growth studies in the laboratory and field to assess the relative palatability of 15 clones of poplar to the caterpillar. Larval growth rate and survival varied significantly among clones (Fig. 10.10). For all 15

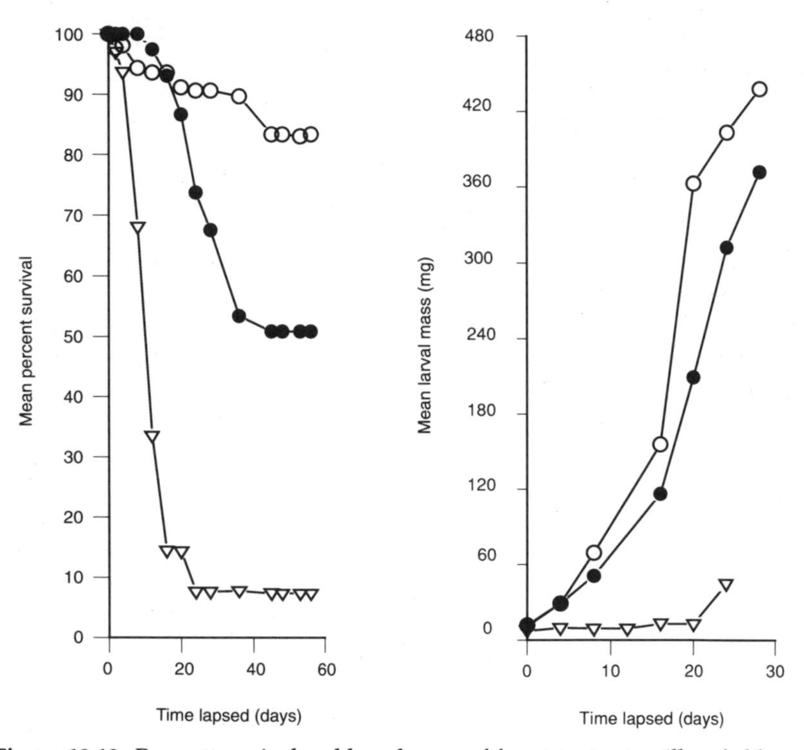

Figure 10.10. Percent survival and larval mass of forest tent caterpillars fed leaves of three different clones of hybrid poplar (*Populus* spp.). (Data from Robison 1993.)

clones, mean survival varied from 7 to 83%. The mean time lapsed between the second larval stadium and pupation varied from 30 to 52 days, and the mean final mass of female pupae varied from 166 to 446 mg. In free-foraging experiments, caterpillars were restless on unsuitable clones and left the trees after feeding little or not at all. Moreover, Robison provided preliminary evidence, based on a small-scale pilot study, that some clones will not support a buildup of tent caterpillar populations. When caged with whole trees of a clone shown to promote growth, field populations of caterpillars increased by a factor of 9.7 over two years. But caterpillars caged with trees of an unsuitable clone declined by a factor of 2.7 over the same period.

Although the specific basis for the variation in acceptability among the clones was not determined, Robison found no significant correlation between larval development and leaf toughness. Similiarly, there was no correlation between larval development and the water and nitrogen content of leaves. The concentrations of the phenoloic glycosides tremuloidin, tremulacin, salicin, salicortin, and populin were not measured, but variation in those compounds, which are known feeding deterrents (see Chapter 5), is likely to account, to some degree, for the differential response of the caterpillars to the poplar clones. Growth and survival curves for the caterpillars suggested marked variation among the clones in toxicity, deterrency, digestibility, and nutritional quality (Robison 1993).

The potential for the forest tent caterpillar to respond to resistant host plants by evolving countermeasures is completely unknown. The insect has a high reproductive potential and a short generation time. Moreover, Robison has shown that the females do not discriminate among clones when ovipositing. Taken collectively, these factors favor the rapid evolution of larval adaptations, and it is likely that there will be no shortage of work for future generations of plant geneticists.

Still another technique that holds promise for the development of trees that are resistant to caterpillars involves gene transfer technology. The incorporation of the Bt delta-endotoxin gene into the genome of trees, for example, would render foliage toxic to tent caterpillars and, in theory, would decimate incipient colonies before they could seriously damage trees (Robison et al. 1994). Klopfeinstein et al. (1993) also reported preliminary experiments to transfer the proteinase inhibitor II gene from potato into hybrid poplar, but the potential effect on folivores has not yet been assessed. As with other attempts to develop resistant plants, the evolutionary response of the caterpillars to genetically altered trees remains to be determined.

11

Maintaining Colonies and Suggestions for Classroom Studies

The studies suggested in this chapter are designed to demonstrate some of the more significant social behaviors of tent caterpillars: group foraging, tent building, trail marking, recruitment, and thermoregulation. The general procedures discussed here will enable interested individuals to conduct more extended investigations. Reference is made mainly to the eastern tent caterpillar, but the basic features of the life histories of all species of tent caterpillars are similar, and other species can be reared and studied in a similar manner.

Rearing Tent Caterpillars

Food Sources

Studies of tent caterpillars necessitate an adequate supply of host material. Tent caterpillars are typically adapted to feed on the young leaves of their host plants, and their foraging behavior is strongly affected when they are forced to feed on foliage of suboptimal quality. The best procedure when selecting food for colonies is to provide them with host plants that are in the same stage of development as plants eaten by naturally occurring colonies.

Tent caterpillars are readily reared in the laboratory on the spring foliage of their host plants. Although such foliage is abundantly available only in the spring, branches or small specimens of many trees can be brought into the laboratory in late winter to induce early leaf flush (Bucher 1959). Some investigators have also maintained colonies of tent caterpillars out of season by culturing the host tree under greenhouse conditions (Fitzgerald and Gallagher 1976). Robison (1993) produced fresh foliage for his forest tent caterpillar colonies by planting cuttings from poplar clones kept frozen until needed. When only small numbers of early instars are needed for studies, caterpillars can be fed previously frozen leaves. Schroeder and Lawson (1992) conducted studies of caterpillar growth using spring leaves of black cherry stored at −20°C.

Synthetic diets have been used to rear the forest tent caterpillar, but there have been no successful attempts to rear other species of tent caterpillars on such diets. Addy (1969) and Grisdale (1985) provide a list of ingredients for preparing a diet and a procedure for mass-rearing the forest tent caterpillar. The insect can be reared from the egg to the adult stage in approximately 45 days when maintained on the diet (Table 11.1). Caterpillars reared in this manner have been used in studies of the insect's physiology and for the bioassay of potential control agents. It would not be appropriate, however, to use synthetic diets when studying the behavior of free-foraging colonies, most obviously because their use precludes dynamic interactions between the host and the caterpillar, the basis of a significant fraction of caterpillar's adaptive behavior. In addition, tent caterpillars are finicky feeders, and larval behavioral patterns may be strongly affected when they are reared on a suboptimal food supply. The fact that synthetic diets are nearly always less attractive to caterpillars can easily be demonstrated by noting the choice made by caterpillars when they are allowed to forage freely in a container having both host leaves and artificial diet.

Breeding Moths

Tent caterpillars are easily bred. Moths for breeding can be obtained by collecting pupae. Pupae, in turn, can be obtained from laboratory rearings or by collecting nearly mature caterpillars from the field. Although it is not necessary to free the pupae from their cocoons, surface sterilization of the pupae can reduce the possibility that pathogens will be transferred to the egg masses. Cocoons can be removed and the pupae surfaces sterilized by soaking them in a solution of equal parts of 5% sodium hypochloride (common houshold bleach) and water until the pupae are free of the silk. The pupae should then be rinsed in running water and allowed to dry (Grisdale 1985).

Grisdale (1985) recommended that tent caterpillar moths be bred by housing two pairs in pint-size cardboard containers containing a twig of the host tree. Oviposition typically takes place within 24 hours. Moths can also be kept in larger breeding cages. I have found that a 50-gallon galvanized trash container makes an excellent and inexpensive breeding

Table 11.1. Duration of the larval instars and pupal stage of the forest tent caterpillar when reared on a synthetic diet

Instar	Duration of stadium (days)
1	3
2	5
3	5
4	5
5	6
6	10
Pupa	10

Source: Data from Grisdale 1985.
Note: Caterpillars maintained under a photoperiod of 18 hours light : 6 hours dark at 24 ± 1°C and 50–60% relative humidity.

cage for tent caterpillars. A 10- to 12-inch-diameter opening can be cut in the lid and covered with screening to provide ventilation. The screen is fixed to the metal with hot glue. Cut branches of the host tree should be placed in the container to provide an ovipositional substrate. Fifty or more pairs of moths can be transferred to the cage while still in the pupal stage. It is a good policy to stock the cage with pupae obtained from several different colonies.

Storing and Sterilizing Eggs

Whether eggs are collected from the field or obtained by breeding adults, they need to undergo a period of cold treatment before they will hatch. This requirement is already satisfied if egg masses are collected in the spring. When eggs are collected in the fall, the eggs of both the forest tent caterpillar and the eastern tent caterpillar need to be stored at 2°C for at least 15 weeks to satisify the requirements of the pharate larva. The eggs must be kept in a sealed container at elevated humidity during the period of cold storage to prevent desiccation. A saturated solution of potassium chloride will produce an ideal relative humidity of 85% when the eggs are stored at 2°C (Bucher 1959). The exact cold-storage requirements of other species have not been determined but are likely to be similar.

Grisdale (1985) determined that eggs of the forest tent caterpillar collected in the fall and stored under refrigeration will hatch in 5–6 days when removed from cold storage in January and held at room temperature, and in 2–3 days when removed in June or July. Egg masses can be kept under cold storage for even longer periods, but the proportion of eggs that hatch declines markedly. If eggs are obtained by breeding adults in the laboratory, they must be kept for at least three weeks at room temperature before refrigerating them to allow embryogenesis to proceed. Although the information given here is based on work with the forest and eastern tent caterpillars, other species of tent caterpillars are likely to have similar requirements.

Before egg masses are set out for hatching, they should be surface-sterilized to reduce the possiblity of viral contamination (Grisdale 1985). The eggs can be rinsed with agitation in a 5% solution of sodium hypochloride for as long as is needed to remove the spumaline coating. Typically, this procedure takes only a minute or two. Care should be taken not to leave them in the solution much longer than is needed to remove the spumaline, because prolonged exposure to the solution will cause the chorion to erode. The eggs need to be thoroughly rinsed in running water and allowed to dry before being set out to hatch.

Maintaining Larvae

Newly emerged colonies of tent caterpillars may be fed new, partially unfolded leaves in Petri dishes and maintained through several instars if desired. The larvae can be maintained under more natural conditions if they are established on branches of their host trees or on stands made from wooden dowels (Fig. 11.1). Colonies and their accumulated silk can be transferred from the Petri plates to the stands. Tent caterpillars that build large, permanent tents show strong attachment to their silk and will usually construct a tent at the transfer site. A bridge (Fig. 11.1) can be used to connect the tent site to branches, which can be replaced as needed. Colonies of eastern tent caterpillars typically build only a single tent during their larval life and can be reared through the entire larval stage using this procedure. The technique may need to be varied somewhat for species that build more than one tent. Forest tent caterpillars, which build no tent at all, have been maintained through several instars on a flat platform such as that illustrated in Figure 11.2. An alternative and simpler approach to studying tent-building species is to cut an established tent from a tree in the field, leaving adequate branch area for future tent expansion. These collected tents can be set up in the same way as described above.

Studies with Tent Caterpillars

Tent caterpillars are conspicuous features of forest and field communities wherever they occur and lend themselves well to educational pro-

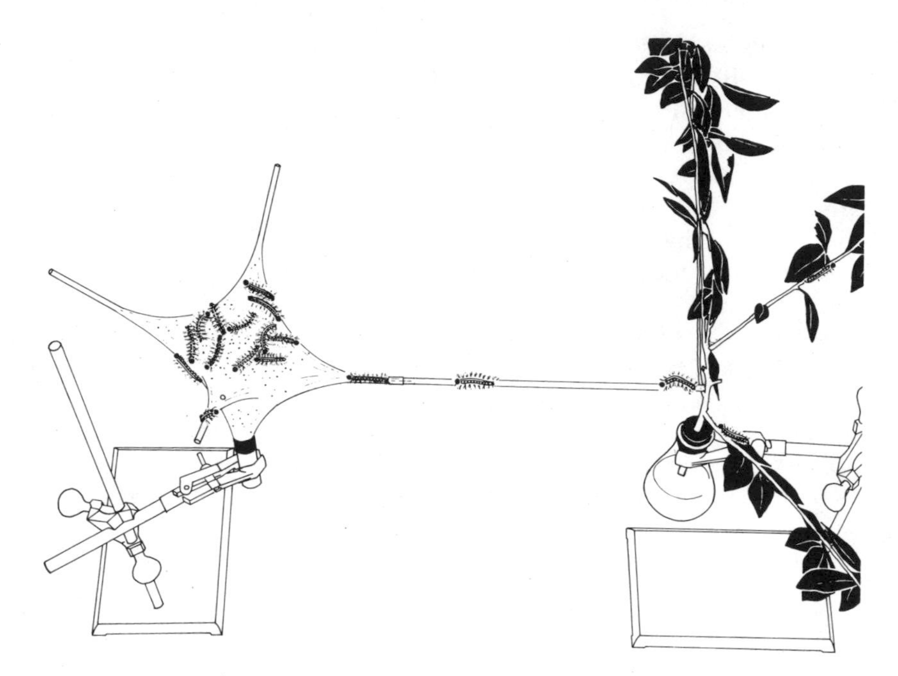

Figure 11.1. A technique for maintaining laboratory colonies of the eastern tent caterpillar to study colony foraging behavior.

grams that involve the study of ecology and behavior (Fitzgerald 1982). They are particularly well suited to laboratories in graduate and undergraduate courses in college programs but have also been used successfully to demonstrate general principles of life history and behavior in high schools and grade schools. Many students will readily recognize tent caterpillars and will often have made negative associations between the caterpillars and their leaf-feeding habits. Few, however, will have any detailed knowledge of the life history of the colony or of the relatively sophisticated communication system the caterpillars have evolved to facilitate foraging. Students typically find their behavior fascinating.

Although tent caterpillars can be used to demonstrate general principles of insect growth and development, it is their social behavior that makes them particularly attractive subjects. Tent caterpillars have some distinct advantages over other social species for classroom study: colonies are readily reared in the open, where their behavior can be directly observed and subjected to experimentation; naturally occurring colonies are easy to locate in the field; the entire life cycle is completed in six to seven weeks in the field and in fewer than four weeks under laboratory

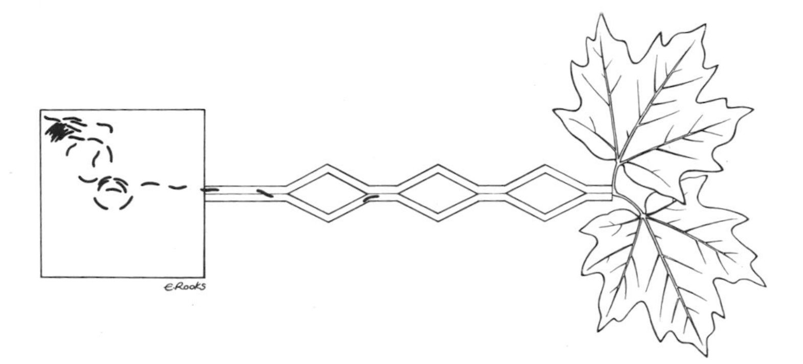

Figure 11.2. A technique for maintaining laboratory colonies of the forest tent caterpillar, a species that does not build a tent. (Drawing by Edward Rooks, used with permission of the Entomological Society of America.)

conditions; and colonies are easily moved about and adapt readily to indoor rearing. The main disadvantages of studying tent caterpillars are that they have only one generation each year and they are conveniently studied only in the spring.

It is most interesting and profitable to study caterpillar behavior in the day-to-day life of the colony under field conditions, but doing so is often impractical because of the caterpillar's brief period of seasonal activity and classroom time constraints. Fortunately, tent caterpillar colonies can easily be reared indoors through their entire life cycle under conditions simulating natural conditions.

Whole Colonies

Life History

Colonies of eastern tent caterpillars established as illustrated in Figure 11.1 have been used in the classroom to demonstrate many aspects of the life history of insects. When allowed to forage around the clock, tent caterpillars grow rapidly and achieve the last instar within three weeks of eclosion. This rapid growth can be studied by measuring or weighing a representative sample of the larvae at intervals of several days, then plotting the resulting values over time.

The amount of food eaten by the larvae increases with age, fueling their rapid growth. The amount of food eaten in a single foraging bout can be assessed by weighing samples of caterpillars just before and just after a foraging bout. If samples are weighed at different points in the life cycle, comparisions can be made of the relationship between larval size and the rate of food intake. Selectivity in feeding can be studied by placing new and aged leaves of the host plant or nonhost species within reach of the foraging caterpillars, then assessing feeding damage after a foraging bout.

The discrete nature of colony feeding bouts is readily apparent when the colonies are allowed to forage ad libitum. This aspect of colony behavior can be studied in more detail with a videocamera. Many inexpensive home videocameras have time-lapse functions that make it possible to monitor foraging behavior around the clock for several days at a time. A 5-W red lamp placed just near enough to the tent to provide threshold illumination will provide adequate light to monitor nighttime activity without interfering with foraging behavior. A small timing device placed in the field of view can provide a measure of the time of onset and the duration of feeding bouts if one cannot be generated by the videocamera.

When the caterpillars reach the end of the last instar they will need to be removed from their tent or they will disperse in search of pupation sites. Pupation can be readily observed by placing some of these larvae in vented, glass-sided containers such as small, screen-topped aquaria. The larvae should be provided with folded pieces of cardboard to which they can attach their cocoons. Adults allowed to emerge in these aquaria will usually mate and, if provided with branches of the host tree, oviposit, completing the life cycle. Indoors, the entire process, from egg hatch to adult eclosion, is completed in about five weeks in the case of the eastern tent caterpillar.

Tent-Building Behavior and Thermoregulation

Tent-building behavior has been studied extensively only in the eastern tent caterpillar. Yet, the behavior is well developed in many of the common species and can be readily observed in the laboratory. Eastern tent caterpillars add to their tent just before each bout of foraging. The caterpillars spin en masse and add one or more new layers to the tent in the process. Tent-building behavior is particularly fascinating when viewed as a time-lapse videorecording.

Light plays a dominant role in the tent-building process. In general, the caterpillars are strongly attracted to the most highly illuminated areas of the tent when spinning. This effect of a light source on the tent-building process can be dramatically demonstrated by placing the major source of illumination under the structure. The caterpillars will concentrate most of their spinning effort in the illuminated area and extend the tent down toward the light, in effect, building it upside down. Outdoors, this photopositive behavior causes the colony to develop their tent toward the south, facilitating thermoregulation. Although caterpillars in field colonies typically leave and enter the tent at its highest point, this behavior is a response to light: the caterpillars are attracted by skylight, causing the holes in the tent that serve as both exit and entrance sites to be formed at the top of the structure. When light is directed from below, the caterpillars exit and enter the tent from the underside of the structure.

Thermoregulation can be demonstrated in the laboratory with the aid of an incandescent lamp. High body temperatures facilitate digestion, and the caterpillars will bask en masse after feeding if a radiant heat source is provided. The effects of thermoregulation on the growth of the colony can be shown by dividing a colony in half and maintaining one half without a radiant light source at room temperature and the other under an incandescent lamp. Although studies have shown that the frequency of foraging bouts is not affected by these different regimens, the rate of growth is significantly faster when a colony is exposed to radiant light because the caterpillars process the food in their guts faster. Tent caterpillars are early spring foragers and must deal with low ambient temperatures; thermoregulation is essential to promote growth under these otherwise adverse conditions.

Tent caterpillars bask both on and in the tent. The layers of the tent trap infrared radiation, and the structure functions like a minature greenhouse. This effect can be demonstrated by placing a thermometer in the tent (or by fastening a sufficiently long one to a branch that supports the tent and allowing the caterpillars to gradually web it into the structure). The difference in temperature attributable to the trapping of infrared radiation can be measured by placing a second thermometer in the path of the external light source in the air near the tent. The lack of significant insulating capability of the silk can also be demonstrated by recording the rate at which the tent returns to ambient temperature after the external heat source is removed.

On hot days, tent caterpillars risk overheating when the tent is exposed to direct sunlight. To demonstrate the evasive action that the caterpillars take when exposed to direct sunlight, bring an incadescent lamp gradually closer to the tent. Proceed a step at a time to avoid too sudden a heat load. The caterpillars will emerge from the tent and then begin to move away from the light, clustering on the shaded side of the tent. The lamp can be gradually moved closer to demonstrate how the caterpillars hang from their prolegs to maximize convective heat loss when the temperature approaches the maximium they can tolerate. This critical temperature can be measured with a thermometer placed near the caterpillars.

Trail Marking, Exploration, and Recruitment

One of the most fascinating aspects of tent caterpillar behavior is their ability to communicate by means of chemical trails. Chemical trail marking can be studied using either individual caterpillars isolated from the colony or whole colonies. The trail pheromone 5β-cholestane-3-one is inexpensive and can be obtained from Sigma Chemical Company or other suppliers of organic chemicals. This compound has been shown to elicit trail following from the eastern tent caterpillar, the forest tent caterpillar, and the lackey moth *Malacosoma neustrium*. It has not yet been tested against other species, but it is likely that the compound is effective with other species as well. The material must be dissolved in a solvent such as hexanes, pentane, or ethyl alcohol to achieve a proper dilution. The solution should be laid down in a narrow line with a micropipette at the rate of 10^{-9} g/mm of trail.

Both the importance of trails to foraging behavior and trail-marking behavior can be demonstrated by placing a small sheet of clean paper over a section of the trail leading from a tent to a food source (Fig. 11.1). When caterpillars encounter this trail discontinuity they will stop abruptly and turn back to the tent. The gap will be gradually repaired by the caterpillars as successive groups of foragers advance hesistantly onto the paper and mark the substrate. That trail following in tent caterpillars is dependent on a chemical marker can be demonstrated by placing either a blank sheet of paper or a sheet of paper marked with the synthetic trail pheromone over the trail and comparing the response of the caterpillars to the treatments. A convincing display is shown in Figure 6.8. The chemical trail is laid out on a sheet of cardboard, and the card is placed between two sections of the bridge joining the tent and the food supply. The larvae will readily follow such a trail as they move between their tent and feeding sites. Caterpillars will build their silk trails over the synthetic trail and reinforce it with additional pheromone during their foraging bouts. Once established, the trails will be persistently followed for the life of the colony. The lasting nature of these trails is demonstrated by setting aside such an established trail for a week or so then placing it back into the path of the colony.

Individual Caterpillars

Trail-following behavior can be studied with individual caterpillars as well as with whole colonies. Second- or early-third-instar caterpillars make the best subjects for such studies. Synthetic pheromone trails can

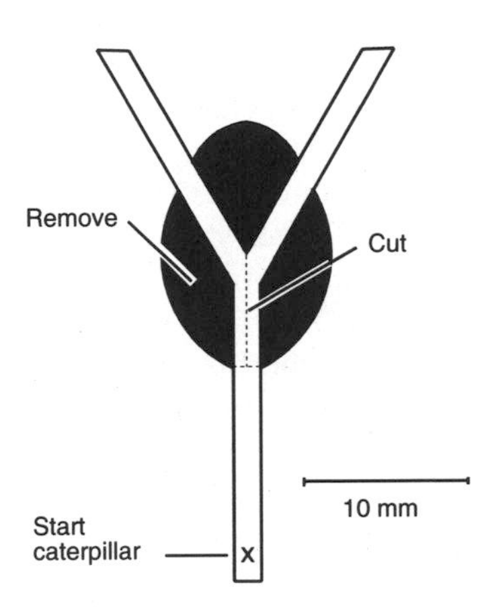

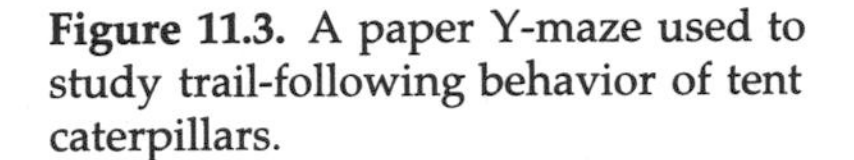

Figure 11.3. A paper Y-maze used to study trail-following behavior of tent caterpillars.

be laid out on paper cards or on Y-mazes cut from paper. Both the forest and eastern tent caterpillars will readily follow trails laid out at the rate of 10^{-9} g/mm of trail. The solution should be deposited in a narrow line to simulate the trail that the caterpillars create.

The branching pattern of the trees on which caterpillars naturally forage creates a series of Y-choices. At each point the caterpillar typically swings its head from one branch to the other, and, if trails are present on both, it assesses their strength and chooses the stronger of the two. This process is readily simulated in the laboratory with Y-mazes cut from paper as shown in Figure 11.3. Trails of different strength can be laid out on the alternate arms of the mazes. The caterpillar is placed at the end of the stem of the maze and observed to determine which of the alternate arms it takes when it reaches the choice point.

Field Studies

Field studies will usually require more patience and time commitment than laboratory studies. Interesting colony behaviors will often occur at hours that are not convenient for whole-class participation. Such field projects can be carried out by smaller groups of students or individuals who can later report the results of their observations to the class as a whole. Colonies can be studied where they occur naturally, or they can be established outdoors on trees near the classroom. Students may also be given egg masses to place on trees near where they live to facilitate observations early in the morning or later at night.

Thermoregulation

Field colonies spend a considerable part of each day basking in the sun. On cool mornings, the caterpillars cluster together just under the surface of the tent after their early morning foraging bout. The effectiveness of this tactic in elevating their temperature is readily investigated by comparing air temperature with the temperature of the aggregate. Readings of colony temperature should be made by placing a thermometer in the midst of the basking aggregate. Air temperature should be measured in the shade. Measurements can be made at 15-minute intervals from just before sunrise to midmorning or later. If measurements are made on several days, under different conditions of cloud cover, a range of values can be generated.

It is also interesting to determine whether the tent has any insulating value, allowing the caterpillars to retain their acquired body heat after the sun sets. This determination can be made by measuring air and cluster temperatures in the early evening just before and after sunset.

Foraging and Recruitment

Recruitment behavior and chemical trail following are easily demonstrated under field conditions. The chemical trail marker can be laid down on branches radiating from the tent that have not been previously marked by the colony. The synthetic chemical trail will simulate a newly deposited recruitment trail. If the synthetic trail is laid just before a colony's activity period, the caterpillars will typically follow it in preference to their established trail system during their next foray en masse.

Recruitment by successful foragers can be demonstrated as well. This demonstration is most easily accomplished by selecting a colony that is established in a smaller tree. The tree can be manually defoliated during the colony's rest period to elicit intense exploratory behavior during the next foraging bout. Pin a fresh leaf in the path of a caterpillar that is exploring at some distance from the aggregate and carefully note the pathway it follows back to the tent after feeding. Hungry tentmates that discover the trail left by their fed sibling will follow the exact pathway to the food find. Various alternative procedures can be followed to quantify the results.

12

Epilogue

Interest in tent caterpillars has increased steadily through each decade of this century. Nearly 40% of the papers on tent caterpillars cited in this book were published since 1980 (Fig. 12.1). If this trend continues we can expect a flood of new studies to appear during the next decade. What future trends in research are suggested by the studies reviewed in this book? There can be little doubt that tent caterpillars will continue to serve in their role as "white mice" and as such to contribute to our knowledge of the behavior, physiology, and ecology of insects. The interest that developed over this past decade in the interactions that occur among tent caterpillars and their host trees will likely be sustained. Research into the means of containing tent caterpillar populations can be expected to continue to focus on nonchemical approaches and to take increasing advantage of the techniques of molecular biology. One hopes that progress will also be made in expanding our understanding of the phylogenetic relationships of tent caterpillars so that we will have more insight into the evolutionary origins of the many patterns of behavior that are found among these insects. It is my hope that the synthesis of some 150 years of research on tent caterpillars presented in this book will serve in some small way to both stimulate and facilitate those studies. But to readers just starting out, or otherwise unconstrained by a research protocol, I would like to go further and suggest a specific area of need.

In Chapter 2 I addressed the question of whether we can expect to find other species of caterpillars that are as social as the tent caterpillars and presented a tentative list of candidate species. Some of these, such as *Eucheria socialis* and *Hylesia acuta,* not only forage from a central place

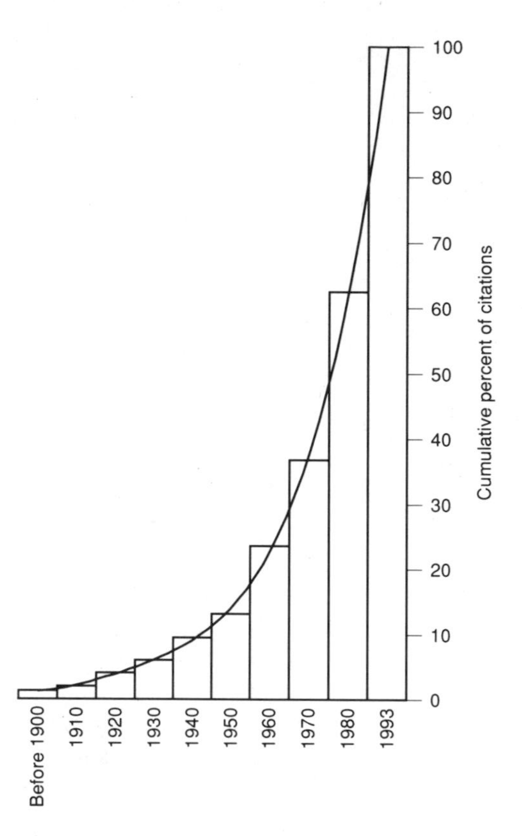

Figure 12.1. Cumulative percentage of articles on tent caterpillars cited in this book, arranged by decade of publication.

and stay in close contact for their entire larval lives, but they also pupate gregariously within the confines of a communal shelter. On the surface they would appear to be very social caterpillars. Clearly, there is a communal effort among the larvae in shelter building, and the caterpillars probably engage in some collective patterns of behavior that deter predators. But for these and other species listed here and elsewhere (Fitzgerald 1993a), and, indeed, for most species of tent catepillars, we have almost no knowledge of larval foraging behavior, the most likely arena within which cooperative interactions can be expected to evolve. What is needed are studies of the foraging behavior of a broad diversity of species, sufficiently detailed to allow us to begin to assess the range of communicative interactions that occur among social caterpillars.

References

Abrahamson, L.P., and J.D. Harper. 1973. Microbial insecticides control forest tent caterpillar in southwestern Alabama. U.S. Dep. Agric. For. Serv. Res. Note SO-157.

Abrahamson, L.P., and R.C. Morris. 1973. Forest tent caterpillar: control with ULV trichlorfon in water tupelo ponds. J. Econ. Entomol. 66:574.

Abrahamson, L.P., J.D. Harper, I.R. Ragenovich, and J.R. Hyland. 1982. Pilot control project using trichlorforn and *Bacillus thuringiensis* against the forest tent caterpillar in southwest Alabama. U.S. Dep. Agric. For. Serv. Tech. Publ. SA-TP 18.

Adams, A.B. 1989. The relationship between the leaf nutritional quality of *Alnus rubra* Bong., and the relative fitness of *Malacosoma californicum pluviale* (Dyar). Ph.D. dissertation, University of Washington, Seattle.

Addy, N.D. 1969. Rearing the forest tent caterpillar on an artificial diet. J. Econ. Entomol. 62:270–271.

Aidley, D.J. 1976. Increase in respiratory rate during feeding in larvae of the armyworm, *Spodoptera exempta*. Physiol. Entomol. 1:73–75.

Allen, D.A. 1987. Insects, declines and general health of northern hardwoods: issues relevant to good forest management in managing northern hardwoods. SUNY Coll. Environ. Sci. For. Misc. Pub. 13. 430 pp.

Ancona, L. 1930. Biología de la *Clisiocampa azteca* Neum. Ann. Inst. Biol. Univ. México 1:215–225.

Anderson, G.W., and M.P. Martin. 1981. Factors related to incidence of *Hypoxylon* cankers in aspen and survival of cankered trees. For. Sci. 27:461–476.

Angus, T.A. 1965. Mortality due to *Bacillus thuringiensis* in post-larval stages of some Lepidoptera. Proc. Entomol. Soc. Ont. 95:133–134.

Angus, T.A., and A.M. Heimpel. 1959. Inhibition of feeding, and blood pH changes, in lepidopterous larvae infected with crystal-forming bacteria. Can. Entomol. 91:352–358.

Anonymous. 1960–1990. Forest insect and disease conditions in Canada. Ottawa: Canadian Forest Service.

Anonymous. 1975. DDT: a review of scientific and economic aspects of the decision to ban its use as a pesticide. EPA-540/1–75–022. Washington, D.C.: U.S. Environmental Protection Agency.

Anonymous. 1986a. Forest insect and disease conditions in Canada. Ottawa: Canadian Forest Service.

Anonymous. 1986b. Forest insect and disease conditions in the United States. Washington, D.C.: U.S. Department of Agriculture, Forest Service.

Atwood, C.E. 1943. A third tent caterpillar in eastern Canada (Lepidoptera: Lasiocampidae). Can. Entomol. 75:203–205.

Ayre, G.L., and D.E. Hitchon. 1968. The predation of tent caterpillars, *Malacosoma americanum* (Lepidoptera: Lasiocampidae), by ants (Hymenoptera: Formicidae). Can. Entomol. 100:823–826.

Bacot, T.T. 1902. Current notes. Entomol. Rec. 14:106.

Baerg, W.J. 1935. The tent caterpillar. Arkansas Agric. Exp. Stn. Bull. 317:20–27.

Baird, A.B. 1917. An historical account of the forest tent caterpillar and of the fall webworm in North America. 47th Ann. Rep. Entomol. Soc. Ont. 1916: 73–87.

Baker, B.H. 1969. Larval instar determination of *Malacosoma californicum fragile* on bitterbush in southern Utah. J. Econ. Entomol. 62:1511.

Baker, B.H. 1970. Occurrence of *Malacosoma incurvum discoloratum* in Zion Canyon, Utah. Pan-Pac. Entomol. 46:27–33.

Baldwin, I.T., and J.C. Schultz. 1983. Rapid changes in tree leaf chemistry induced by damage: evidence for communication between plants. Science 221:277–279.

Balfour-Browne, F. 1925. The evolution of social life among caterpillars. *In* Proc. Third Int. Cong. Entomol., Zurich, pp. 334–339.

Balfour-Browne, F. 1933. The life-history of the "small eggar moth," *Eriogaster lanestris* L. Proc. Zool. Soc. Lond.:161–180.

Batzer, H.O., and R.C. Morris. 1978. Caterpillar forest tent. U.S. Dep. Agric. For. Serv. For. Insect Dis. Leafl. 9. 8 pp.

Behr, H. 1869. Description of a new species of Pieridae, and certain new species of butterflies from California. Trans. Am. Entomol. Soc.:303–304.

Bent, A.C. 1940. Life histories of North American cuckoos, goatsuckers, hummingbirds and their allies. Smithson. Inst. Bull. 176.

Beroza, M., C.S. Hood, D. Trefrey, D.E. Leonard, E.F. Knipling, W. Klassen, and L.J. Stevens. 1974. Large field trial with microencapsulated sex pheromone to prevent mating of the gypsy moth. J. Econ. Entomol. 67:659–664.

Bess, H.A. 1939. The biology of *Leschenaultia exul* Townsend, a tachinid parasite of *Malacosoma americanum* Fab., and *Malacosoma disstria* Hübner. Ann. Entomol. Soc. Am. 29:493–613.

Bhandari, R.S., and P. Singh. 1991. Tent caterpillar epidemic in *Betula utilis* forest in Pithoragarh district of Uttar Pradesh and its control. Indian For. 117:267–273.

Bieman, D.N. 1980. An evolutionary study of mating of *Malacosoma ameri-*

canum (Fabricius) and *Malacosoma disstria* Hübner (Lepidoptera: Lasiocampidae). M.S. thesis, University of Michigan, Ann Arbor.

Bieman, D.N. 1982. Mating wounds in *Malacosoma*: an insight into bed bug mating behavior. Fla. Entomol. 65:377–378.

Bieman, D.N., and J.A. Witter. 1981. Differences in emergence date and size between the sexes of *Malacosoma americanum*, the eastern tent caterpillar (Lepidoptera: Lasiocampidae). Great Lakes Entomol. 14:51–53.

Bieman, D.N., and J.A. Witter. 1983. Mating behavior of *Malacosoma disstria* at two levels of mate competition. Fla. Entomol. 66:72–279.

Bird, F.T. 1969. Infection and mortality of spruce budworm, *Choristoneura fumiferana*, and forest tent caterpillar, *Malacosoma disstria*, caused by nuclear and cytoplasmic polyhedrosis viruses. Can. Entomol. 101:1269–1285.

Blackman, M.W. 1918. Apple tent caterpillar. J. Econ. Entomol. 11:432–433.

Blais, J.R., R.M. Prentice, W.L. Sippell, and D.R. Wallace. 1955. Effects of weather on the forest tent caterpillar *Malacosoma disstria* Hbn. in central Canada in the spring of 1953. Can. Entomol. 87:1–8.

Bloome, C.G. 1991. Ring-billed gulls, *Larus delawarensis*, feeding in flight on forest tent caterpillar, *Malacosoma disstria*, cocoons. Can. Field-Nat. 105:280–281.

Borah, W. 1943. Silk raising in colonial Mexico. Berkeley: University of California Press.

Boswell, V.R. 1952. Residues, soils, and plants. *In* A. Stefferud, ed., Insects: the yearbook of agriculture, pp. 284–297. Washington, D.C.: U.S. Department of Agriculture.

Bowen, C.V., and S.A. Hall. 1952. The organic insecticides. *In* A. Stefferud, ed., Insects: the yearbook of agriculture, pp. 284–297. Washington, D.C.: U.S. Department of Agriculture.

Boyden, T.C. 1976. Butterfly palatability and mimicry: experiments with *Ameiva* lizards. Evolution 30:73–81.

Brattsten, L. 1979. Biochemical defense mechanisms in herbivores against plant allelochemicals. *In* G.A. Rosenthal and D.H. Janzen, eds., Herbivores: their interaction with secondary plant metabolites, pp. 199–270. New York: Academic Press.

Britton, E.B. 1954. Report by P.J.L. Roche. Proc. Entomol. Soc. Lond. Ser. C 19: 39.

Britton, W.E. 1935. The eastern tent caterpillar. Conn. Agric. Exp. Stn. Bull. 378:65–82.

Broome, J.R., P.P. Sikorowski, and W.W. Neel. 1974. Effect of sunlight on the activity of nuclear polyhedrosis virus from *Malacosoma disstria*. J. Econ. Entomol. 67:135–136.

Brown, C.E. 1965. Mass transport of forest tent caterpillar moths, *Malacosoma disstria* Hübner, by a cold front. Can. Entomol. 97:1073–1075.

Bucher, G.E. 1957. Diseases of the larvae of tent caterpillars caused by a spore-forming bacterium. Can. J. Microbiol. 3:695–709.

Bucher, G.E. 1959. Winter rearing of tent caterpillars, *Malacosoma* spp. (Lepidoptera: Lasiocampidae). Can. Entomol. 91:411–416.

Bucher, G.E. 1961a. Control of the eastern tent caterpillar, *Malacosoma ameri-*

canum (Fabricius), by distribution of spores of two species of *Clostridium*. J. Insect Pathol. 3:439–445.
Bucher, G.E. 1961b. Artificial culture of *Clostridium brevifaciens* n. sp., and *C. malacosomae* n. sp., the causes of brachyosis of tent caterpillars. Can. J. Microbiol. 7:641–655.
Burgess, A.F., and C.W. Collins. 1915. The *Calosoma* beetle (*Calosoma sycophanta*) in New England. U.S. Dep. Agric. Bull. 251:1–40.
Byers, J.R. 1975. Tyndall blue and surface white of tent caterpillars, *Malacosoma* spp. J. Insect Physiol. 21:401–415.
Byers, J.R., and C.F. Hinks. 1973. The surface sculpturing of the integument of lepidopterous larvae and its adaptive significance. Can. J. Zool. 51:1171–1179.
Capinera, J.L. 1980. A trail pheromone from the silk produced by larvae of the range caterpillar *Hemileuca olivae* (Lepidoptera: Saturnidae) and observations on aggregation behavior. J. Chem. Ecol. 3:655–664.
Carlberg, U. 1980. Larval biology of *Eriogaster lanestris* (Lepidoptera: Lasiocampidae) in S.W. Finland. Not. Entomol. 60:65–72.
Carmona, A.S., and P. Barbosa. 1983. Overwintering egg mass adaptations of the eastern tent caterpillar, *Malacosoma americanum* (Fab.) (Lepidoptera: Lasiocampidae). J. N.Y. Entomol. Soc. 91:68–74.
Carter, D.J., and B. Hargreaves. 1986. A field guide to the butterflies and moths in Britain and Europe. London: Collins.
Casey, T.M. 1981. Energetics and thermoregulation of *Malacosoma americanum* (Lepidoptera: Lasiocampidae) during hovering flight. Physiol. Zool. 54:362–371.
Casey, T.M. 1983. Metabolism of crawling caterpillars–soft bodied insects. Am. Zool. 23:112–114.
Casey, T.M. 1988. Thermoregulation and heat exchange. Adv. Insect Physiol. 20:119–146.
Casey, T.M. 1991. Energetics of caterpillar locomotion: biomechanical constraints of a hydraulic skeleton. Science 252:112–114.
Casey, T.M., and J.R. Hegel. 1981. Caterpillar setae: insulation for an ectotherm. Science 214:1131–1133.
Casey, T.M., and J.R. Hegel-Little. 1987. Instantaneous oxygen consumption and muscle stroke work in *Malacosoma americanum* during pre-flight warm-up. J. Exp. Biol. 127:389–400.
Casey, T.M., J.R. Hegel, and C.S. Buser. 1981. Physiology and energetics of pre-flight warm-up in the eastern tent caterpillar moth *Malacosoma americanum*. J. Exp. Biol. 94:119–135.
Casey, T.M., B. Joos, T.D. Fitzgerald, M.E. Yurlina, and P.A. Young. 1988. Synchronized group foraging, thermoregulation, and growth of eastern tent caterpillars in relation to microclimate. Physiol. Zool. 61:372–377.
Casson, E.M. 1979. An invasion of forest tent caterpillars. Blue Jay 37:198–199.
Chadab, R., and C.W. Rettenmeyer. 1975. Mass recruitment by army ants. Science 188:1124–1125.
Chisholm, M.D., E.W. Underhill, W. Steck, K.N. Slessor, and G.G. Grant. 1980. (*Z*)-5,(*E*)-7-dodecadienal and (*Z*)-5,(*E*)-7-dodecadien-1-ol, sex pheromone

components of the forest tent caterpillar, *Malacosoma disstria*. Environ. Entomol. 9:278–282.

Chisholm, M.D., W.F. Steck, B.K. Bailey, and E.W. Underhill. 1981. Synthesis of sex pheromone components of the forest tent caterpillar *Malacosoma disstria* (Hübner) and of the western tent caterpillar, *Malacosoma californicum* (Packard). J. Chem. Ecol. 7:159–164.

Chisholm, M.D., P. Palaniswamy, and E.W. Underhill. 1982. Orientation disruption of male forest tent caterpillar, *Malacosoma disstria* (Hübner) (Lepidoptera: Lasiocampidae), by air permeation with sex pheromone components. Environ. Entomol. 11:1248–1250.

Churchill, G.B., H.H. Duncan, D.P. Duncan, and A.C. Hodson. 1964. Long-term effects of defoliation of aspen by the forest tent caterpillar. Ecology 45:630–633.

Clark, E.C. 1955. Observations on the ecology of a polyhedrosis of the Great Basin tent caterpillar *Malacosoma fragilis*. Ecology 36:373–376.

Clark, E.C. 1956a. Survival and transmission of a virus causing polyhedrosis in *Malacosoma fragile*. Ecology 37:728–732.

Clark, E.C. 1956b. The Great Basin tent caterpillar in relation to bitterbush in California. Calif. Fish Game 42:131–142.

Clark, E.C., and C.E. Reiner. 1956. The possible use of a polyhedrosis virus in the control of the Great Basin tent caterpillar. J. Econ. Entomol. 49:653–659.

Clark, E.C., and C.G. Thompson. 1954. The possible use of microorganisms in the control of the Great Basin tent caterpillar. J. Econ. Entomol. 47:268–272.

Clausen, C.P. 1972. Entomophagous insects. New York: Hafner.

Clausen, T.P., P.B. Reichardt, J.P. Bryant, R.A. Werner, K. Post, and K. Frisby. 1989. Chemical model for short-term induction in quaking aspen (*Populus tremuloides*) foliage against herbivores. J. Chem. Ecol. 15:2335–2345.

Collier, W.A. 1936. *Malacosoma. In* E. Strand, ed., Lepidopterorum catalogus, part 73. Lasiocampidae, pp. 94–141. The Hague: Dr. W. Junk.

Connola, D.P. 1980. Permanent damage to sugar maple groves by forest tent caterpillar in New York. N.Y. State Department of Environmental Conservation. Unpublished report.

Connola, D.P., R.C. Sweet, and J. Lev. 1965. Report on a six year study of forest tent caterpillar damage in eight sugarbushes in New York. State Museum and Science Service, Albany, N.Y. Unpublished report.

Coroiu, I., N. Tomescu, and G.H. Stan. 1984. Comportamentul de reproducere la *Malacosoma neustria* L. (Lepidoptera, Lasiocampidae) in conditii de laborator. Stud. Cercet. Biol. Ser. Biol. Anim. 36:57–62.

Coroiu, I., C.M. Roman, N. Tomescu, and G. Stan. 1986. Determination of the sexual pheromone of the female *Malacosoma neustria* (Lepidoptera: Lasiocampidae) and its role in reproduction. Stud. Cercet. Biol. Ser. Biol. Anim. 38(1):11–16.

Costa, J.T., III, and K.G. Ross. 1993. Seasonal decline in intracolony genetic relatedness in eastern tent caterpillars: implications for social evolution. Behav. Ecol. Sociobiol. 32:47–54.

Costa, J.T., III, and K.G. Ross. 1994. Hierarchical genetic structure and gene flow in macrogeographic populations of the eastern tent caterpillar (*Malacosoma americanum*). Evolution 48:1158–1167.

Courtney, S.P. 1984. The evolution of egg clustering by butterflies and other insects. Am. Nat. 1233:276–281.

Crump, D., R.M. Silverstein, H.J. Williams, and T.D. Fitzgerald. 1987. Identification of the trail pheromone of the eastern tent caterpillar, *Malacosoma americanum* (Lepidoptera: Lasiocampidae). J. Chem. Ecol. 13:397–402.

Culver, J.J. 1919. A study of *Compsilura concinnata,* an imported tachinid parasite of the gipsy moth and the brown-tail moth. U.S. Dep. Agric. Bull. 766. 27 pp.

Cunningham, J.C., W.J. Kaupp, and G.M. Howse. 1991. Development of nuclear polyhedrosis virus for control of gypsy moth (Lepidoptera: Lymantriidae) in Ontario. I. Aerial spray trials in 1988. Can. Entomol. 123:601–609.

Curtis, J. 1828. British entomology. 5: Insect Number 229.

Darling, C.D., and N.F. Johnson. 1982. Egg mortality in the eastern tent caterpillar, *Malacosoma americanum* (Lepidoptera: Lasiocampidae): the role of accessory gland secretion and egg mass shape. Proc. Entomol. Soc. Wash. 84:448–460.

Davis, N.F. 1903. The apple-tree tent caterpillar and its life history. Pa. Dep. Agric. Bull. 120:1–40.

de Freina, J.J., and T.J. Witt. 1987. Die Bombyces und Sphinges der Westpalaearktis (Insecta, Lepidoptera). Band 1. Munich: Edition Forschung & Wissenschaft.

Delrio, G., P. Luciano, and R. Prota. 1983. The parasites of *Malacosoma neustria* L. in Sardinia. Atti XIII Congr. Nazionale Ital. Entomol. 237–244.

Dethier, V.G. 1942. The dioptric apparatus of lateral ocelli. I. The corneal lens. J. Cell. Comp. Physiol. 19:301–313.

Dethier, V.G. 1943. The dioptric apparatus of lateral ocelli. II. Visual capacities of the ocellus. J. Cell. Comp. Physiol. 22:115–126.

Dethier, V.G. 1972. Electrophysiological studies of gustation in lepidopterous larvae. J. Comp. Physiol. 82:103–134.

Dethier, V.G. 1980a. Responses of some olfactory receptors of the eastern tent caterpillar (*Malacosoma americanum*) to leaves. J. Chem. Ecol. 6:213–220.

Dethier, V.G. 1980b. The world of the tent-makers: a natural history of the eastern tent caterpillar. Amherst: University of Massachusetts Press.

Dethier, V.G. 1989. Patterns of locomotion of polyphagous caterpillars in relation to foraging. Ecol. Entomol. 14:375–386.

Dethier, V.G., and J.H. Kuch. 1971. Electrophysiological studies of gustation in lepidopterous larvae. I. Comparative sensitivity to sugars, amino acids, and glycosides. Z. Vgl. Physiol. 72:343–363.

Dethier, V.G., and L.M. Schoonhoven. 1968. Evaluation of evaporation by cold and humidity receptors in caterpillars. Insect Physiol. 14:1049–1054.

Downes, W. 1920. The life-history of *Apateticus crocatus* Uhl. (Hemiptera). Proc. Entomol. Soc. B.C. 16:21–27.

Dreisig, H. 1987. The Skallingen peninsula and the insects. Entomol. Medd. 54:9–32.

Duncan, D.P., and A.C. Hodson. 1958. Influence of the forest tent caterpillar upon the aspen forests of Minnesota. For. Sci. 4:71–93.

Dunn, L.H. 1917. The coconut tree caterpillar (*Brassolis isthmia*) of Panama. J. Econ. Entomol. 10:473–488.

Dyar, H.G. 1890. The number of molts in lepidopterous larvae. Psyche 5:420–422.

Edgerly, J.S., and T.D. Fitzgerald. 1982. An investigation of behavioral variability within colonies of the eastern tent caterpillar, *Malacosoma americanum* (Lepidoptera: Lasiocampidae). J. Kans. Entomol. Soc. 55:145–155.

Eggen, D.A. 1987. A comparative study of the parasite complexes of gypsy moth and forest tent caterpillar in New York. Ph.D. dissertation, SUNY College of Environmental Sciences and Forestry, Syracuse, N.Y.

Eickwort, G.C. 1981. Presocial insects. *In* H.R. Hermann, ed., Social insects. Vol. 2, pp. 199–279. New York: Academic Press.

Ennis, T.J. 1976. Sex chromatin and chromosome numbers in Lepidoptera. Can. J. Genet. Cytol. 18:119–130.

Evans, A.C. 1939. The utilization of food by certain Lepidoptera larvae. Part 2. Trans. R. Entomol. Soc. Lond. 89:13–22.

Evans, E.W. 1982. Influence of weather on predator/prey relations: stinkbugs and tent caterpillars. J. N.Y. Entomol. Soc. 90:241–246.

Evans, E.W. 1983. Niche relations of predatory stinkbugs (*Podisus* spp., Pentatomidae) attacking tent caterpillars (*Malacosoma americanum*, Lasiocampidae). Am. Midl. Nat. 109:316–323.

Evans, H.E. 1987. Observations on the prey and nests of *Podalonia occidentalis* Murray (Hymenoptera: Sphecidae). Pan-Pac. Entomol. 63(2):130–134.

Evans, H.F., and P.F. Entwistle. 1987. Viral diseases. *In* J.R. Fuxa and Y. Tanada, eds., Epizootiology of insect diseases, pp. 257–322. New York: John Wiley and Sons.

Fabre, J.H. 1991. The pine processionary. *In* E.W. Teale, ed., The insect world of J. Henri Fabre, pp. 10–24. Boston: Beacon Press.

Fashingbauer, B.A. 1957. The effects of aerial spraying with DDT on wood frogs. Flicker 29:160.

Fashingbauer, B.A., A.C. Hodson, and W.H. Hodson. 1957. The inter-relations of a forest tent caterpillar outbreak, song birds and DDT application. Flicker 29:132–147.

Feeny, P. 1970. Seasonal change in oak leaf tannins and nutrients as a cause of spring feeding by winter moth caterpillars. Ecology 51:565–581.

Felt, E.P. 1899. Notes of the year for New York. U.S. Dep. Agric. Div. Entomol. New Ser. Bull. 20:60–61.

Filip, V., and R. Dirzo. 1985. Life cycle of *Malacosoma incurvum* var. *aztecum* (Lepidoptera: Lasiocampidae) of Xochimilco, Federal District, Mexico. Folia Entomol. Mex. 66:31–45.

Fisher, R.A. 1958. The genetical theory of natural selection. 2nd ed. New York: Dover.

Fiske, W.F. 1903. A study of the parasites of the American tent caterpillar. New Hampshire College Agric. Exp. Stn. Tech. Bull. 6:181–230.

Fitzgerald, T.D. 1976. Trail marking by larvae of the eastern tent caterpillar. Science 194:961–963.

Fitzgerald, T.D. 1980. An analysis of daily foraging patterns of laboratory colonies of the eastern tent caterpillar, *Malacosoma americanum* (Lepidoptera: Lasiocampidae), recorded photoelectronically. Can. Entomol. 112:731–738.

Fitzgerald, T.D. 1982. Exploratory and recruitment trail marking by tent caterpillars. *In* J.R. Matthews and R.W. Matthews, eds., Insect behavior: a source book of laboratory and field exercises, pp. 111–119, 262–263. Boulder, Colo.: Westwood Press.

Fitzgerald, T.D. 1993a. Social caterpillars. *In* N.E. Stamp and T.M. Casey, eds., Caterpillars: ecological and evolutionary constraints on foraging, pp. 372–403. New York: Chapman and Hall.

Fitzgerald, T.D. 1993b. Trail following and recruitment: response of eastern tent caterpillar, *Malacosoma americanum,* to 5β-cholestane-3,24-dione and 5β-cholestane-3-one. J. Chem. Ecol. 19:449–457.

Fitzgerald, T.D. 1993c. Trail and arena marking by caterpillars of *Archips cerasivoranus* (Lepidoptera: Tortricidae). J. Chem. Ecol. 19:1479–1489.

Fitzgerald T.D., and J.T. Costa. 1986. Trail-based communication and foraging behavior of young colonies of the forest tent caterpillar *Malacosoma disstria* Hubn. (Lepidoptera: Lasiocampidae). Ann. Entomol. Soc. Am. 79:999–1007.

Fitzgerald, T.D., and J.S. Edgerly. 1979a. Specificity of trail markers of forest and eastern tent caterpillars. J. Chem. Ecol. 5:564–574.

Fitzgerald, T.D., and J.S. Edgerly. 1979b. Exploration and recruitment in field colonies of eastern tent caterpillars. J. Ga. Entomol. Soc. 14:312–314.

Fitzgerald, T.D., and J.S. Edgerly. 1982. Site of secretion of the trail marker of the eastern tent caterpillar. J. Chem. Ecol. 8:31–39.

Fitzgerald, T.D., and E.M. Gallagher. 1976. A chemical trail factor from the silk of the eastern tent caterpillar, *Malacosoma americanum* (Lepidoptera: Lasiocampidae). J. Chem. Ecol. 2:187–193.

Fitzgerald, T.D., and S.C. Peterson. 1983. Elective recruitment communication by the eastern tent caterpillar (*Malacosoma americanum*). Anim. Behav. 31: 417–423.

Fitzgerald, T.D., and S.C. Peterson. 1988. Cooperative foraging and communication in social caterpillars. BioScience 38:20–25.

Fitzgerald, T.D., and F.X. Webster. 1993. Identification and behavioral assays of the trail pheromone of the forest tent caterpillar *Malacosoma disstria* Hübner (Lepidoptera: Lasiocampidae). Can. J. Zool. 71:1511–1515.

Fitzgerald, T.D., and D.E. Willer. 1983. Tent-building behavior of the eastern tent caterpillar *Malacosoma americanum* (Lepidoptera: Lasiocampidae). J. Kans. Entomol. Soc. 56(1):20–31.

Fitzgerald, T.D., T. Casey, and B. Joos. 1988. Daily foraging schedule of field colonies of the eastern tent caterpillar, *Malacosoma americanum.* Oecologia 76:574–578.

Fitzgerald, T.D., K.L. Clark, R. Vanderpool, and C. Phillips. 1991. Leaf shelter-building caterpillars harness forces generated by axial retraction of stretched and wetted silk. J. Insect Behav. 4:21–32.

Flemion, F., and A. Hartzell. 1936. Effect of low temperature in shortening the hibernation period of insects in the egg stage. Contrib. Boyce Thomson Inst. 8:167–173.

Fowler, S.V., and J.H. Lawton. 1985. Rapidly induced defenses and talking trees: the devil's advocate position. Am. Nat. 126:181–195.

Franclemont, J.G. 1973. Mimallonoidea and Bombycoidea. *In* R.B. Dominick et al., The moths of America north of Mexico. Vol. 20.1. London: E.W. Classey & R.B.D. Publications.

Frankenhuyzen, K. van. 1990. Development and current status of *Bacillus thuringiensis* for control of defoliating forest insects. For. Chron. October:498–507.
Frye, R.D., and D.A. Ramse. 1975. Control of the forest tent caterpillar with microbial agents. N.D. Farm Res. 33:19–22.
Futuyma, D.J., and M.E. Saks. 1981. The effect of variation in host plant on the growth of an oligophagous insect, *Malacosoma americanum*, and its polyphagous relative, *Malacosoma disstria*. Entomol. Exp. Appl. 30:163–168.
Futuyma, D.J., and S.S. Wasserman. 1981. Food plant specialization and feeding efficiency in the tent caterpillars *Malacosoma disstria* and *M. americanum*. Entomol. Exp. Appl. 30:106–110.
Gauld, I.D. 1988. The species of the *Enicospilus americanus* complex (Hymenoptera: Ichneumonidae) in eastern North America. Syst. Entomol. 13:31–53.
Gautreau, E.J. 1964. Unhatched forest tent caterpillar egg bands in northern Alberta associated with late spring frost. Can. Dep. For., For. Entomol. and Pathol. Branch, Bi-monthly Prog. Rep. 20:3.
Ghent, A.W. 1958. Mortality of overstory trembling aspen in relation to outbreaks of the forest tent caterpillar and spruce budworm. Ecology 39:222–232.
Goater, B. 1991. Lasiocampidae. *In* A.M. Emmet and J. Heath, eds., The moths and butterflies of Great Britain and Ireland. Vol. 7, part 2: Lasiocampidae–Thyatiridae with life history chart of the British Lepidoptera, pp. 306–323. Colchester, England: Harley Books.
Golosova, M.A. 1986. Susceptibility of larvae to cytoplasmic polyhedrosis. Zashch. Rast. 2:31–32.
Gorham, R.P. 1923. Insect pests of the year 1922 in New Brunswick. Proc. Arcadian Entomol. Soc. 8:18–22.
Goyer, R.A., G.J. Lehhard, J.D. Smith, and R.A. May. 1987. Estimating the number of eggs per egg mass of the forest tent caterpillar, *Malacosoma disstria*, on three tree species in the southern USA. J. Entomol. Sci. 22:188–191.
Grant, J. 1959. Pine siskins killing forest tent caterpillars. Proc. Entomol. Soc. B.C. 56:20.
Green, G.W., and C.R. Sullivan. 1950. Ants attacking larvae of the forest tent caterpillar, *Malacosoma disstria* Hbn. Can. Entomol. 82:890–899.
Greenblatt, J.A. 1974. Behavioral studies on tent caterpillars. M.S. thesis, University of Michigan, Ann Arbor.
Greenblatt, J.A., and J.A. Witter. 1976. Behavioral studies on *Malacosoma disstria* (Lepidoptera: Lasiocampidae). Can. Entomol. 108:1225–1228.
Gregory, R.A., and P.M. Wargo. 1986. Timing of defoliation and its effect on bud development, starch reserves and sap sugar concentration in sugar maple. Can. J. For. Res. 16:10–17.
Grisdale, D. 1985. *Malacosoma disstria. In* P. Singh and R.F. Moore, eds. Handbook of insect rearing. Vol. 2, pp. 369–379. New York: Elsevier.
Gross, H.L. 1985. The impact of insects and diseases on the forests of Ontario. Can. For. Serv. Info. Rep. O-X-366. Great Lakes Forest Research Centre.
Gross, H.L. 1991. Dieback and growth loss of sugar maple associated with defoliation by the forest tent caterpillar. For. Chron. 67:33–42.
Guilford, T. 1990. The evolution of aposematism. *In* D.L. Evans and J.O.

Schmidt, eds., Insect defenses, pp. 23–62. Albany: State University of New York Press.

Haanstad, J.O., and D.M. Norris. 1992. Altered elm resistance to smaller European elm bark beetle (Coleoptera: Scolytidae) and forest tent caterpillar (Lepidoptera: Lasiocampidae). J. Econ. Entomol. 85:172–181.

Hagen, R.H., and J.F. Chabot. 1986. Leaf anatomy of maples (*Acer*) and host use by Lepidoptera larvae. Oikos 47:335–345.

Hall, P., and J.A. Traniello. 1985. Behavioral bioassays of termite trail pheromones: recruitment and orientation effects of Cembrene-A in *Nasutitermes costalis* (Isoptera: Termitidae) and discussion of factors affecting termite response in experimental contexts. J. Chem. Ecol. 11:1503–1513.

Hall, T.F., S.G. Breeland, and P.K. Anderson. 1969. Use of cherry-laurel foliage for preparation of an effective insect-killing jar. Ann. Entomol. Soc. Am. 62: 242–244.

Halperin, J. 1990. Life history of *Thaumetopoea* spp. (Lep., Thaumetopoeidae) in Israel. J. Appl. Entomol. 110:1–6.

Hamilton, W.D. 1971. Geometry for the selfish herd. J. Theor. Biol. 31:295–311.

Hanec, W. 1966. Cold-hardiness in the forest tent caterpillar, *Malacosoma americanum* Hübner (Lasiocampidae, Lepidoptera). J. Insect Physiol. 12:1443–1449.

Harper, J.D., and L.P. Abrahamson. 1979. Forest tent caterpillar control with aerially applied formulations of *Bacillus thuringiensis* and Dimilin. J. Econ. Entomol. 72:74–77.

Harper, J.D., and J.R. Hyland. 1981. Patterns of forest tent caterpillar defoliation in southwest Alabama—1973–1979. J. Ala. Acad. Sci. 52:25–31.

Harris, T.W. 1841. A report on the insects of Massachusetts injurious to vegetation, pp. 265–272. Cambridge, Mass.

Heinrich, B. 1979. Foraging strategies of caterpillars. Oecologia 42:3325–3337.

Heinrich, B., and T.P. Mommsen. 1985. Flight of winter moths near 0°C. Science 228:177–179.

Herbert, P.D.N. 1983. Egg dispersal patterns and adult feeding behavior in the Lepidoptera. Can. Entomol. 115:1477–1481.

Hildahl, V., and W.A. Reeks. 1960. Outbreaks of the forest tent caterpillar, *Malacosoma disstria* Hübner, and their effects on stands of trembling aspen in Manitoba and Saskatchewan. Can. Entomol. 92:199–209.

Hodson, A.C. 1939a. Biological notes on the egg parasites of *Malacosoma disstria* Hbn. Ann. Entomol. Soc. Am. 32:131–136.

Hodson, A.C. 1939b. *Sarcophaga aldrichi* Parker as a parasite of *Malacosoma disstria* Hbn. J. Econ. Entomol. 32:396–401.

Hodson, A.C. 1941. An ecological study of the forest tent caterpillar, *Malacosoma disstria* Hbn., in northern Minnesota. Univ. Minn. Agric. Exp. Stn. Bull. 148:1–55.

Hodson, A.C. 1977. Some aspects of forest tent caterpillar population dynamics. *In* H.M. Kulman and H.C. Chaing, eds., Insect ecology: papers presented in the H.C. Hodson ecology lectures, pp. 4–16. Univ. Minn. Agric. Exp. Stn. Tech. Bull. 310.

Hodson, A.C., and C.J. Weinman. 1945. Factors affecting recovery from diapause and hatching of eggs of the forest tent caterpillar *Malacosoma disstria* Hbn. Univ. Minn. Agric. Exp. Stn. Bull. 170:1–31.
Hölldobler, B. 1978. Ethological aspects of communication in ants. Adv. Study Behav. 8:75–115.
Hölldobler, B., and E.O. Wilson. 1990. The ants. Cambridge: Harvard University Press.
Holsten, E.H., and A. Eglitis. 1988. Forest insect and disease conditions in the United States. Washington, D.C.: U.S. Department of Agriculture, Forest Service.
Hou, Tao-qian. 1980. The *Malacosoma* of China (Lepidoptera: Lasiocampidae). Acta Entomol. Sin. 23(3):308–313.
Howse, G.M., H.L. Gross, and A.H. Rose. 1981. Ontario Region. *In* Annual report of the Forest Insect and Disease Survey, 1977, pp. 53–69. Ottawa: Canadian Forest Service.
Hübner, J. 1820. Verzeichniss bekannter Schmettlinge [*sic*]. 12:192.
Hughes, G.M. 1965. Locomotion: terrestrial. *In* M. Rockstein, ed., The physiology of Insecta. Vol. 2. New York: Academic Press.
Hundertmark, A. 1937. Das Formunterscheidungsvermögen der Eiraupen der Nonne (*Lymantria monacha*). Z. Vergl. Physiol. 24:563–582.
Iizuka, E. 1966. Mechanisms of fiber formation by the silkworm, *Bombyx mori* L. Biorheololgy 3:141–152.
Inoue, H., S. Sugi, H. Kuroko, S. Moriuti, and A. Kawabe. 1982. Moths of Japan. Vol. 2. Tokyo: Kodansha.
Ishikawa, S. 1969. The spectral sensitivity and the components of the visual system of the stemmata of the silkworm larva, *Bombyx mori* L. (Lepidoptera: Bombycidae). Appl. Entomol. Zool. 4 (2):87–99.
Ives, W.G.H. 1973. Heat units and outbreaks of the forest tent caterpillar, *Malacosoma disstria* (Lepidoptera: Lasiocampidae). Can. Entomol. 105:529–543.
Ives, W.G.H., and J.A. Muldrew. 1978. Preliminary evaluations of the effectiveness of nucleopolyhedrosis virus sprays to control the forest tent caterpillar in Alberta. North. For. Res. Cen., Can. For. Serv., Fish. Environ. Can., Info. Rep. NOR-X-204.
Ives, W.G.H., J.A. Muldrew, and R.M. Smith. 1982. Experimental aerial application of forest tent caterpillar baculovirus. North. For. Res. Cen., Can. For. Serv., Environ. Can., Info. Rep. NOR-X-240.
Jaffe, K. 1980. Theoretical analysis of the communication system for chemical mass recruitment in ants. J. Theor. Biol. 84:589–609.
Janzen, D.H. 1966. Coevolution of mutualism between ants and acacias in Central America. Evolution 20:249–275.
Janzen, D.H. 1984. Natural history of *Hylesia lineata* (Saturniidae: Hemileucinae) in Santa Rosa National Park, Costa Rica. J. Kans. Entomol. Soc. 57:490–514.
Jaques, R.P. 1961. Control of some lepidopterous pests of apple with commercial preparations of *Bacillus thuringiensis* Berliner. J. Insect Pathol. 3:167–182.
Järvi, T.B., B. Sillen-Tullberg, and C. Wiklund. 1981. The cost of being apo-

sematic: an experimental study of predation on larvae of *Papilio machaon* by the great tit, *Parus major*. Oikos 36:267–272.

Johnson, W.T., and O.N. Morris. 1981. Cold fog applications of pesticides for control of *Malacosoma disstria*. J. Arboric. 7:246–251.

Jones, D.A. 1972. Cyanogenic glycosides and their function. *In* J.B. Harborne, ed., Phytochemical ecology, pp. 103–124. London: Academic Press.

Joos, B. 1992. Adaptions for locomotion at low body temperatures in eastern tent caterpillars, *Malacosoma americanum*. Physiol. Zool. 65:1148–1161.

Joos, B., T.M. Casey, T.D. Fitzgerald, and W.A. Buttemer. 1988. Roles of the tent in behavioral thermoregulation of eastern tent caterpillars. Ecology 69: 2004–2011.

Joshi, K.C., and S.B. Agarwal. 1979. Seasonal history of the tent caterpillar *Malacosoma indica* Wlk. in Kumaon hills. Sci. Cult. 45:495–496.

Joshi, K.C., and S.B. Agarwal. 1987. Bionomics of the tent caterpillar *Malacosoma indica* Wlk. Bull. Entomol. 28:1–11.

Karowe, D.N. 1989. Differential effect of tannic acid on two tree-feeding Lepidoptera: implications for theories of plant anti-herbivore chemistry. Oecologia 80:507–512.

Kerr, T.W. 1952. Further investigations of insecticides for control of insects attacking ornamental trees and shrubs. J. Econ. Entomol. 45:209–212.

Kevan, P.G., and R.A. Bye. 1991. The natural history, sociobiology, and ethnobiology of *Eucheria socialis* Westwood (Lepidoptera: Pieridae), a unique and little-known butterfly from Mexico. Entomologist 110:146–165.

Klein, H.D., and A.M. Wenner. 1991. Tiny game hunting: healthy ways to trap and kill the pests in your house and garden. New York: Bantam.

Klopfeinstein, N.B., H.C. McNabb Jr., E.R. Hart, R.B. Hall, R.D. Hanna, S.A. Heuchelin, K.K. Allen, N.Q. Shi, and R.W. Thornberg. 1993. Transformation of *Populus* hybrids to study and improve plant resistance. Silvae Genet. 43:86–90.

Knapp, R., and T.M. Casey. 1986. Thermal ecology, behavior, and growth of gypsy moth and eastern tent caterpillars. Ecology 67:598–608.

Knight, G.A., R.J. Lavigne, and M.G. Pogue. 1991. The parasitoid complex of forest tent caterpillar, *Malacosoma disstria* (Lepidoptera: Lasiocampidae), in eastern Wyoming shelterbelts. Great Lakes Entomol. 24:256–261.

Konyukhov, V.P., and B.G. Kovalev. 1988. Sex pheromone of *Malacosoma neustria* L. females. Bioorg. Khim. 14(2):268–272.

Krieg, A. 1987. Diseases caused by bacteria and other prokaryotes. *In* J.R. Fuxa and Y. Tanada, eds., Epizootiology of insect diseases, pp. 323–356. New York: John Wiley and Sons.

Krieger, R.I., P.P. Feeny, and C.F. Wilkinson. 1971. Detoxification enzymes in the guts of caterpillars: an evolutionary answer to plant defenses? Science 172:579–581.

Krivda, W.V. 1980. House sparrows feeding on *Malacosoma* moths. Blue Jay 38:189–190.

Krywienczyk, J., and T.A. Angus. 1969. Some behavioral and seriological observations on the response of larvae of *Bombyx mori* and *Malacosoma disstria* to *Bacillus thuringiensis*. J. Invert. Pathol. 14:105–107.

Kulman, H.M. 1965. Natural control of the eastern tent caterpillar and notes on its status as a forest pest. J. Econ. Entomol. 58:66–70.

Lajonquière, Yves De. 1972. Espèces et formes asiatiques du genre *Malacosoma* Hübner. 11e contribution a l'étude des Lasiocampidae. Bull. Soc. Entomol. Fr. 77:297–308.

Langston, R.L. 1957. A synopsis of hymenopterous parasites of *Malacosoma* in California. Univ. Calif. Publ. Entomol. 14:1–50.

Laux, W., and J.M. Franz. 1962. über das Auftreten von Individualunterschieden beim Ringelspinner, *Malacosoma neustria* (L.). Z. Angew. Entomol. 50: 105–109.

Leathers, T.D., and S.C. Gupta. 1993. Susceptibility of the eastern tent caterpillar (*Malacosoma americanum*) to the entomogenous fungus *Beauveria bassiana*. J. Invert. Pathol. 61:217–219.

Leius, K. 1967. Influence of wildflowers on parasitism of tent caterpillar and codling moth. Can. Entomol. 99:444–446.

Li, J.L. 1989. Bionomics of *Malacosoma parallela* Staudinger and its control. Kunchong Zhishi 26:344–346.

Lindauer, M. 1967. Communication among social bees. New York: Atheneum.

Lindroth, R.L., and M.S. Bloomer. 1991. Biochemical ecology of the forest tent caterpillar: responses to dietary protein and phenolic glycosides. Oecologia 86:408–413.

List, G.M. 1934. Entomological report for Colorado. Calif. Argic. Monthly Bull. 170:236.

Little, E.L. 1971. Atlas of United States trees. Vol. 1: Conifers and important hardwoods. U.S. Dep. Agric. Misc. Pub. 1146.

Liu, C.L. 1926. On some factors of natural control of the eastern tent caterpillar (*Malacosoma americana* Harris) with notes on the biology of the host. Ph.D. dissertation. Cornell University, Ithaca, N.Y.

Lutz, F.E. 1936. How about the tent caterpillar? Nat. Hist. 37:149–158.

Lyon, R.L., S.J. Brown, and J.L. Robertson. 1972. Contact toxicity of sixteen insecticides applied to forest tent caterpillars reared on artificial diet. J. Econ. Entomol. 65:928–930.

Machowicz-Stefaniak, Z. 1979. Disease symptoms occurring in caterpillars of *Malacosoma neustria* (Lepidoptera) infected by some fungi. Acta Mycol. 15: 145–150.

MacLeod, D.M., and D. Tyrrell. 1979. *Entomophthora crustosa* n. sp. as a pathogen of the forest tent caterpillar, *Malacosoma disstria* (Lepidoptera: Lasiocampidae). Can. Entomol. 111:1137–1144.

Magnoler, A. 1985. Field evaluation of a baculovirus against the larvae of *Malacosoma neustria* L. in Sardinia. Dif. Piante 8:329–337.

Mansingh, A. 1974. Studies on insect dormancy. II. Relationship of cold-hardiness to diapause and quiescence in the eastern tent caterpillar, *Malacosoma americanum* (Fab.) (Lasiocampidae: Lepidoptera). Can. J. Zool. 52: 629–637.

Manter, J.A. 1945. DDT for tent caterpillar. J. Econ. Entomol. 38:615.

Maple, J.D. 1937. The biology of *Ooencyrtus johnsoni* (Howard), and the role of the egg shell in the respiration of certain encyrtid larvae (Hymenoptera). Ann. Entomol. Soc. Am. 30:123–154.

Martignoni, M.E., and P.J. Iwai. 1986. A catalog of viral diseases of insects, mites and ticks. 4th ed. U.S. Dep. Agric. For. Serv. Gen. Tech. Rep. PNW-195.

Martin, J., and J. Serrano. 1984. Taxonomía, citotaxonomía y biología de *Malacosoma aplicola* y *M. castrensis* de la Península Iberíca. Eos Rev. Esp. Entomol. 60:175–189.

Maschwitz, U., and M. Mühlenberg. 1975. Zur Jagdstrategie einiger orientalischer *Leptogenys*-Arten (Formicidae: Ponerinae). Oecologia 20:65–83.

Mason, W.R.M. 1979. A new *Rogas* (Hymenoptera: Braconidae) parasite of tent caterpillars (*Malacosoma* spp., Lepidoptera: Lasiocampidae) in Canada. Can. Entomol. 111:783–786.

McAtee, W.L. 1927. The relation of birds to woodlots in New York State. Roosevelt Wild Life Bull. 4(1):1–152.

McEvoy, P.B. 1984. Increase in respiration rate during feeding in larvae of the cinnabar moth, *Tyria jacobaeae*. Physiol. Ecol. 9:191–195.

McNabb, Jr., H.S., E.R. Hart, R.B. Hall, S.A. Heuchclin, N.B. Klopfenstein, K.K. Allen, and R.W. Thornberg. 1992. Use of the proteinase inhibitor II gene for increased pest resistance in poplars. *In* D.C. Allen and L.P. Abrahamson, eds., Proc. North American Forest Insect Work Conf., p. 89. PNW-GTR-294. Denver: USDA Forest Service.

Miao, J.C., M. Zhang, X.Y. Pan, J.L. Yu, and C.Z. Gao. 1986. Study on the aerial application of spraying 20% colloidal suspension of Dimilin against tent caterpillar and pine caterpillars. J. Northeast For. Univ. China 14:11–16.

Milstead, J.E., M. Kirby, D. Pinnock, N. Farmer, B. Nelson, D. Odom, and T. Rozek. 1980. Evaluation of commercial formulations of *Bacillus thuringiensis* Berliner for the control of tent caterpillar infesting oaks. FAO Plant Prot. Bull. 28:72–74.

Mitchell, R.G. 1990. Seasonal history of the western tent caterpillar (Lepidoptera: Lasiocampidae) on bitterbush and currant in central Oregon. J. Econ. Entomol. 83:1492–1494.

Moore, L.V., J.H. Myers, and R. Eng. 1988. Western tent caterpillars prefer the sunny side of the tree, but why? Oikos 51:321–326.

Morris, O.N. 1969. Susceptibility of several forest insects of British Columbia to commercially produced *Bacillus thuringiensis*. II. Laboratory and field pathogenicity tests. J. Invert. Pathol 13:285–295.

Morris, O.N. 1972. Susceptibility of some forest insects to mixtures of commercial *Bacillus thuringiensis* and chemical insecticides, and sensitivity of the pathogen to the insecticides. Can. Entomol. 104:1419–1425.

Morris, R.C., and L.W. Orr. 1962. Caterpillars-fish-trees. Proc. Soc. Am. For.: 172–173.

Morris, R.F. 1972. Predation by wasps, birds and mammals on *Hyphantria cunea*. Can. Entomol. 104:1581–1591.

Morse, R.F., and C.D. Howard. 1898. Poisonous properties of wild cherry leaves. New Hampshire Agric. Exp. Stn. Bull. 56:112–123.

Muggli, J.M. 1974. Sex identification of *Malacosoma disstria* pupae (Lepidoptera: Lasiocampidae). Ann. Entomol. Soc. Am. 67:521–522.

Muggli, J.M., and W.E. Miller. 1980. Instar head widths, individual biomass and development rate of forest tent caterpillar, *Malacosoma disstria* (Lepidoptera: Lasiocampidae), at two densities in the laboratory. Great Lakes Entomol. 13(4):207–209.

Myers, J.H. 1978. A search for behavioural variation in first and last laid eggs of western tent caterpillar and an attempt to prevent a population decline. Can. J. Zool. 56:2359–2363.
Myers, J.H. 1981. Interactions between western tent caterpillars and wild rose: a test of some general plant herbivore hypotheses. J. Anim. Ecol. 50:11–25.
Myers, J.H. 1988. Can a general hypothesis explain population cycles of forest Lepidoptera? Adv. Ecol. Res. 18:179–242.
Myers, J.H. 1990. Population cycles of western tent caterpillars: experimental introductions and synchrony of fluctuations. Ecology 71:986–995.
Myers, J.H. 1993. Population outbreaks in forest Lepidoptera. Am. Sci. 81:240–261.
Myers, J.H., and J.N.M. Smith. 1978. Head flicking by tent caterpillars: a defensive response to parasite sounds. Can. J. Zool. 56:1628–1631.
Myers, J.H., and K.S. Williams. 1984. Does tent caterpillar attack reduce the food quality of red alder foliage? Oecologia 62:74–79.
Myers J.H., and K.S. Williams. 1987. Lack of short or long term inducible defenses in the red alder–western tent caterpillar system. Oikos 48:73–78.
Nielsen, D.G. 1990. Evaluation of biorational pesticides for use in arboriculture. J. Arboric. 16:82–88.
Nielsen, D.G., and M.J. Dunlap. 1991. Choke cherry, efficacy of selected insecticides against eastern tent caterpillar, Wayne Co., Ohio. Insectic. & Acaricide Tests 16:269–270.
Nordin, G.L. 1975. Transovarial transmission of *Nosema* sp. infecting *Malacosoma americanum*. J. Invert. Pathol. 25:221–228.
Nordin, G.L. 1976. Influence of natural *Nosema* sp. infections on field populations of *Malacosoma americanum* (Lepidoptera: Lasiocampidae). J. Kans. Entomol. Soc. 49:32–40.
Nothnagle, P.J., and J.C. Schultz. 1987. What is a forest pest? *In* P. Barbosa and J.C. Schultz, eds., Insect outbreaks, pp. 59–80. New York: Academic Press.
Ohnishi, E., S. Takahashi, H. Sonobe, and T. Hayashi. 1968. Crystals from cocoons of *Malacosoma neustria testacea*. Science 160:783–784.
Oliver, A.D. 1964. Control studies of the forest tent caterpillar, *Malacosoma disstria*, in Louisiana. J. Econ. Entomol. 57:157–160.
Ovcharov, D. 1984. Effect of phytoncides on bacterial preparations used for biological control in forestry. Gorskostop. Nauka 21:81–86.
Page, M., and R.L. Lyon. 1973. Toxicity of eight insecticides applied to the western tent caterpillar. J. Econ. Entomol. 66:995–997.
Palaniswamy, P., M.D. Chisholm, E.W. Underhill, D.S. Reed, and S.J. Peesker. 1983. Disruption of forest tent caterpillar (Lepidoptera: Lasiocampidae) orientation to baited traps in aspen groves by air permeation with (5*Z*,7*E*)-5,7-dodecadienal. J. Econ. Entomol. 76:1159–1163.
Papaj, D.R., and M.D. Rausher. 1983. Individual variation in host location by phytophagous insects. *In* S. Ahmad, ed., Herbivorous insects: host-seeking behavior and mechanisms, pp. 77–124. New York: Academic Press.
Peigler, R.S. 1993. Wild silks of the world. Am. Entomol. 39:151–161.
Percy, J.E., and J. Weatherston. 1971. Studies of physiologically active arthro-

pod secretions. IX. Morphology and histology of the pheromone-producing glands of some female Lepidoptera. Can. Entomol. 103:1733–1739.

Perry, D.F., and R.A. Fleming. 1989. *Erynia crustosa* zygospore germination. Mycologia 8:154–158.

Peterson, S.C. 1985. Chemical trail communication and foraging ecology of eastern tent caterpillars, *Malacosoma americanum*. Ph.D. dissertation, State University of New York at Stony Brook, Stony Brook.

Peterson, S.C. 1986a. Breakdown products of cyanogenesis: repellancy and toxicity to predatory ants. Naturwissenschaften 73:627–628.

Peterson, S.C. 1986b. Host specificity of trail marking to foliage by eastern tent caterpillars, *Malacosoma americanum*. Entomol. Exp. Appl. 42:91–96.

Peterson, S.C. 1987. Communication of leaf suitability by gregarious eastern tent caterpillars (*Malacosoma americanum*). Ecol. Entomol. 12:283–289.

Peterson, S.C. 1988. Chemical trail marking and following by caterpillars of *Malacosoma neustria*. J. Chem. Ecol. 14:815–823.

Peterson, S.C., and T.D Fitzgerald. 1991. Chemoorientation of eastern tent caterpillars to trail pheromone, 5β-cholestane-3,24-dione. J. Chem. Ecol. 17: 1963–1972.

Peterson, S.C., N.D. Johnson, and J.L. LeGuyader. 1987. Defensive regurgitation of allelochemicals derived from host cyanogenesis by eastern tent caterpillars. Ecology 68:1268–1272.

Pinhey, E.C.G. 1975. Moths of southern Africa. Cape Town: Tafelberg Publishers.

Pinkham, J.D, R.D. Frye, and R.B. Carlson. 1984. Toxicities of *Bacillus thuringiensis* against forest tent caterpillar. (Lepidoptera: Lasiocampidae). J. Kans. Entomol. Soc. 57:672–674.

Prentice, R.M. 1963. Forest Lepidoptera of Canada recorded by the forest insect survey. Dep. For., For. Entomol. Pathol. Branch 3:284–293.

Price, P.W. 1987. The role of natural enemies in insect populations. *In* P. Barbosa and J.C. Schultz, eds., Insect outbreaks, pp. 287–312. New York: Academic Press.

Raske, A.G. 1975. Cold-hardiness of first instar larvae of the forest tent caterpillar, *Malacosoma disstria* (Lepidoptera: Lasiocampidae). Can. Entomol. 107:75–80.

Retnakaran, A., and L. Smith. 1976. Greenhouse evaluation of PH 60–40 activity on the forest tent caterpillar. Can. For. Serv. Bi-monthly Res. Notes 32:2.

Retnakaran, A., L. Smith, B. Tomkins, and J. Granett. 1979. Control of forest tent caterpillar, *Malacosoma disstria* (Lepidoptera: Lasiocampidae), with Dimilin. Can. Entomol. 111:841–846.

Rhoades, D.F. 1983. Responses of alder and willow to attack by tent caterpillars and webworms: evidence for pheromonal sensitivity of willows. Am. Chem. Soc. Symp. Ser. 208:55–68.

Rhoades, D.F. 1985. Offensive-defensive interactions between herbivores and plants: their relevance in herbivore population dynamics and ecological theory. Am. Nat. 125:205–238.

Riley, C.V., and L.O. Howard. 1890–1891. Insect life. Vol. 3, pp. 477–478.

Robertson, J.L., and N.L. Gillette. 1973. Western tent caterpillar: contact toxicity of ten insecticides applied to the larvae. J. Econ. Entomol. 66:629–630.

Robison, D.J. 1993. The feeding ecology of the forest tent caterpillar, *Malacosoma disstria* Hübner, among hybrid poplar clones, *Populus* spp. Ph.D. dissertation, University of Wisconsin, Madison.

Robison, D.J., B.H. McCown, and K.F. Raffa. 1994. Responses of gypsy moth (Lepidoptera: Lymantriidae) and forest tent caterpillar (Lepidoptera: Lasiocampidae) to transgenic poplar, *Populus* spp., containing a *Bacillus thuringensis* d-endotoxin gene. Environ. Entomol. 23:1030–1041.

Roden, D.B., J.R. Miller, and G.A. Simmons. 1992. Visual stimuli influencing orientation by the larval gypsy moth, *Lymantria dispar* (L.). Can. Entomol. 124:287–304.

Roessingh, P. 1989. The trail following behaviour of *Yponomeuta cagnagellus*. Entomol. Exp. Appl. 51:49–57.

Roessingh, P. 1990. Chemical marker from silk of *Yponomeuta cagnagellus*. J. Chem. Ecol. 16:2203–2216.

Roessingh, P., S.C. Peterson, and T.D. Fitzgerald. 1988. The sensory basis of trail following in some lepidopterous larvae: contact chemoreception. Physiol. Entomol. 13:219–224.

Roland, J. 1993. Large-scale forest fragmentation increases the duration of tent caterpillar outbreak. Oecologia 93:25–30.

Romanova, Y.S. 1972. Entomophages of *Malacosoma neustria* and their regulatory importance in the foci of the Moscow district in 1954–70. Zool. Zh. 51:1158–1195.

Romanova, Y.S., and V.A. Lozinsky. 1958. Experiments on the practical use of egg parasites of *Malacosoma neustria* in forest conditions. Zool. Zh. 37: 542–547.

Root, R.R. 1966. The avian response to a population of the tent caterpillar, *Malacosoma constrictum* (Stretch). Pan-Pac. Entomol. 42:48–53.

Rose, A.H. 1958. The effect of defoliation on foliage production and radial growth of quaking aspen. For. Sci. 4:335–342.

Ross, G.N. 1964. Life history studies of Mexican butterflies. I. Notes on the early stages of four papilionids from Catemaco, Veracruz. J. Res. Lepid. 3: 9–18.

Ruggiero, M.A., and H.C. Merchant. 1986. Energy budget for a population of eastern tent caterpillars (Lepidoptera: Lasiocampidae) in Maryland. Environ. Entomol. 15:795–799.

Runcie, C.D. 1987. Behavioral evidence for multicomponent trail pheromone in the termite *Reticulitermes flavipes* (Kollar) (Isoptera: Rhinotermitidae). J. Chem. Ecol. 13:1967–1978.

Sample, B.E. 1992. Temporal separation of flight time of two sympatric *Malacosoma* species. Environ. Entomol. 21:628–631.

Sampson, R.A., and H.N. Nigg. 1992. *Furia crustosa,* fungal pathogen of forest tent caterpillar in Florida. Fla. Entomol. 75:280–284.

Saunders, A.A. 1920. Birds and tent caterpillars. Auk 37:312–313.

Saxena, K.N., and S. Goyal. 1978. Orientation of *Papilio demoleus* larvae to coloured solutions. Experientia 34:35–36.

Schaefer, P.W., R.W. Fuester, R.J. Chianese, L.D. Rhoads, and R.B. Tichenor Jr. 1989. Introduction and North American establishment of *Coccygomimus disparis* (Hymenoptera: Ichneumonidae), a polyphagous pupal parasite of Lepidoptera, including gypsy moth. Environ. Entomol. 18(6):1117–1125.

Schmid, J.M., P.A. Farrar, and I. Ragenovich. 1981. Length of western tent caterpillar egg masses and diameter of their associated stems. Great Basin Nat. 41:465–466.

Schoonhoven, L.M. 1968. Chemosensory bases of host plant selection. Ann. Rev. Entomol. 13:115–136.

Schoonhoven, L.M. 1972. Plant recognition by lepidopterous larvae. Symp. R. Entomol. Soc. Lond. 6:87–99.

Schoonhoven, L.M., and V.G. Dethier. 1966. Sensory aspects of host-plant discrimination by lepidopterous larvae. Arch. Neerl. Zool. 16:497–530.

Schroeder, L.A. 1978. Consumption of black cherry leaves by phytophagous insects. Am. Nat. 100:294–306.

Schroeder, L.A. 1986. Changes in tree leaf quality and growth performance of lepidopteran larvae. Ecology 67:1628–1636.

Schroeder, L.A., and J. Lawson. 1992. Temperature effects on the growth and dry matter budgets of *Malacosoma americanum*. J. Insect Physiol. 38:743–749.

Schroeder, L.A., and M. Malmer. 1980. Dry matter, energy and nitrogen conversion by Lepidoptera and Hymenoptera larvae fed leaves of black cherry. Oecologia 45:63–71.

Schultz, J.C. 1983. Tree tactics. Nat. Hist. 92(5):12–25.

Schultz, J.C., and I.T. Baldwin. 1982. Oak leaf quality declines in response to defoliation by gypsy moth larvae. Science 217:149–151.

Schultz, J.C., P.J. Nothnagle, and I.T. Baldwin. 1982. Seasonal and individual variation in leaf quality of two northern hardwoods tree species. Am. J. Bot. 69:753–759.

Schultz, P.B. 1989. Forest tent caterpillar, its management as an urban pest in Virginia. J. Arboric. 15:92–93.

Schwenke, W. 1978. Die Forstschädlinge Europas: Ein Handbuch in fünf Bänden. Hamburg: Verlag Paul Parey.

Scriber, J.M. 1977. Limiting effects of low leaf-water content on the nitrogen utilization, energy budget, and larval growth of *Hyalophora cercropia* (Lepidoptera: Saturniidae). Oecologia 28:269–287.

Scudder, S.H. 1889. The butterflies of the Eastern United States and Canada with special reference to New England. 1, Cambridge. Published by the author.

Sealy, S.G. 1979. Extralimital nesting of bay-breasted warblers: response to forest tent caterpillars? Auk 96:600–603.

Sealy, S.G. 1980. Reproductive responses of northern orioles to a changing food supply. Can. J. Zool. 58:221–227.

Sedivy, J. 1978. Group effect in the common lackey moth (*Malacosoma neustrium* L.). Sb. UVTIZ (Ustav Vedeckotech. Inf. Zemed.) Ochr. Rostl. 14:137–142.

Segarra-Carmona, A., and P. Barbosa. 1983. Nutrient content of four rosaceous hosts and their effects on development and fecundity of the eastern tent caterpillar, *Malacosoma americanum* (Fab.) (Lepidoptera: Lasiocampidae). Can. J. Zool. 61:2868–2872.

Sehnal, F., and H. Akai. 1990. Insect silk glands: their types, development and function, and effects of environmental factors and morphogenic hormones on them. Int. J. Insect Morphol. Embryol. 19:79–132.

Sharp, D. 1970. Insects. New York: Dover.

Shcherbakova, L.N., D. V. Ovcharov, and A.M. Aukshtikal'nene. 1981. The susceptibility of lackey moths to bacterial preparations in relation to food-plant. *In* A.M.J. Aukshtikal'nene, Noveishie dostizheniya lesnoi entomologii (po materialam USh s''ezda VED, Lil'nyus, 9–13 oktyabrya 1979 g.), pp. 174–178. Institut Zoologii I Parazitologii Academii Nauk Litovskoi SRR.

Shepherd, R.F. 1979. Comparison of the daily cycle of adult behavior of five forest Lepidoptera from western Canada, and their response to pheromone traps. Mitt. Schweiz. Entomol. Ges. 52:157–168.

Shepherd, R.F., and C.E. Brown. 1971. Sequential egg-band sampling and probability methods of predicting defoliation by *Malacosoma disstria* (Lasiocampidae: Lepidoptera). Can. Entomol. 103:1371–1379.

Shiga, M. 1976a. A quantitative study on food consumption and growth of the tent caterpillar *Malacosoma neustria testacea* Motschulsky (Lepidoptera: Lasiocampidae). Bull. Fruit Tree Res. Stn. Ser. A 3:67–86.

Shiga, M. 1976b. Effect of group size on the survival and development of young larvae of *Malacosoma neustria testacea* Motschulsky (Lepidoptera: Lasiocampidae) and its role in the natural population. Kontyu 44:537–553.

Shiga, M. 1977. Population dynamics of *Malacosoma neustria testacea* (Lepidoptera: Lasiocampidae): stabilizing process in a field population. Res. Popul. Ecol. 18:284–301.

Shiga, M. 1979. Population dynamics of *Malacosoma neustria testacea* (Lepidoptera, Lasiocampidae). Bull. Fruit Tree Res. Stn. Ser. A 6:59–168.

Sillen-Tullenberg, B., and E.H. Bryant. 1983. The evolution of aposematic coloration in distasteful prey: an individual selection model. Evolution 37:993–1000.

Simchuk, P.A. 1980. The effect of microsporidians of *Pleistophora carpocapsae* and *P. schubergi* on the growth, development and mortality of caterpillars of *Malacosoma neustria* silkworm. Parazitologiya 14:158–163.

Singleton-Smith, J., and B.J.R. Philogene. 1981. Structure and organization of the stemmata in the larvae of five species of Lepidoptera. Rev. Can. Biol. 40:331–341.

Sippell, W.L. 1957. A study of the forest tent caterpillar, *Malacosoma disstria* Hbn., and its parasite complex in Ontario. Ph.D. dissertation, University of Michigan, Ann Arbor.

Sippell, W.L. 1962. Outbreaks of the forest tent caterpillar, *Malacosoma disstria* Hübner, a periodic defoliator of broad-leaved trees in Ontario. Can. Entomol. 94:408–416.

Sippell, W.L., J.E. Macdonald, and A.H. Rose. 1964. Forest insect and disease conditions in Ontario. *In* Annual report of the Forest Insect and Disease Survey, 1964, pp. 51–54. Ottawa: Canadian Forest Service.

Slingerland, M.V. 1899. Emergency report on tent caterpillars. Cornell Univ. Agric. Exp. Stn. Bull. 170:553–564.

Smeathers, D.M., E. Gray, and J.H. James. 1973. Hydrocyanic acid potential of black cherry leaves as influenced by aging and drying. Agron. J. 65:775–777.

Smith, J.D., and R.A. Goyer. 1985. Rates of parasitism and sex ratios of *Ablerus clisiocampae* and *Ooencyrtus clisiocampae* egg parasites of the forest tent caterpillar. J. Entomol. Sci. 20:189–193.

Smith, J.D., and R.A. Goyer. 1986. Population fluctuations and causes of mortality for the forest tent caterpillar, *Malacosoma disstria* (Lepidoptera: Lasiocampidae), on three different sites in southern Louisiana. Environ. Entomol. 15:1184–1188.

Smith, J.D., R.A. Goyer, and J.P. Woodring. 1986. Instar determination and growth and feeding indices of the forest tent caterpillar, *Malacosoma disstria* (Lepidoptera: Lasiocampidae), reared on tupelo gum, *Nyssa aquatica* L. Ann. Entomol. Soc. Am. 79:304–307.

Smitley, D.R., T.W. Davis, and K.A. Kearns. 1991. Eastern tent caterpillar control, Michigan, 1990. Insectic. & Acaricide Tests 6:268.

Snodgrass, R.E. 1922. The tent caterpillar. Smithson. Rep. 328–362.

Snodgrass, R.E. 1930. Insects, their ways and means of living. Smithson. Misc. Collect. Sci. Ser. 5.

Snodgrass, R.E. [1935] 1993. Principles of insect morphology. Reprint, Ithaca: Cornell University Press.

Snodgrass, R.E. 1961. The caterpillar and the butterfly. Smithsonian Misc. Collect. 143(6).

Srivastava, A.S., and H.F. Wilson. 1947. Benzene hexachloride as a fumigant and a contact insecticide. J. Econ. Entomol. 40:569–571.

Stacey, L., R. Roe, and K. Williams. 1975. Mortality of eggs and pharate larvae of the eastern tent caterpillar *Malacosoma americanum* (F.) (Lepidoptera: Lasiocampidae). J. Kans. Entomol. Soc. 48:521–523.

Stairs, G.R. 1964. Dissemination of nuclear polyhedrosis virus against the forest tent caterpillar *Malacosoma disstria* (Hübner) (Lepidoptera: Lasiocampidae). Can. Entomol. 96:1017–1020.

Stairs, G.R. 1965. Artificial initiation of virus epizootics in forest tent caterpillar populations. Can. Entomol. 97:1059–1062.

Stairs, G.R. 1966. Transmission of virus in tent caterpillar populations. Can. Entomol. 98:1100–1104.

Stairs, G.R. 1972. Pathogenic microorganisms in the regulation of forest insect populations. Ann. Rev. Entomol. 18:355–372.

Stamp, N.E. 1980. Egg deposition patterns in butterflies: why do some species cluster their eggs rather than deposit them singly? Am. Nat. 115:367–380.

Stamp, N.E. 1982. Behavioral interactions of parasitoids and Baltimore checkerspot caterpillars, *Euphydryas phaeton*. Environ. Entomol. 11:100–104.

Stamp, N.E. 1984. Interactions of parasitoids and checkerspot caterpillars, *Euphydryas* spp. (Nymphalidae). J. Res. Lepid. 23:2–18.

Stark, E.J., and J.D. Harper. 1982. Pupal mortality in forest tent caterpillar (Lepidoptera: Lasiocampidae): causes and impact on populations in southwestern Alabama. Environ. Entomol. 11:1071–1077.

Stehr, F.W. 1987. Lasiocamidae (Bombycoidea). *In* F.W. Stehr, ed., Immature insects, pp. 511–513. Dubuque: Kendall/Hunt.

Stehr, F.W., and E.F. Cook. 1968. A revision of the genus *Malacosoma* Hübner in North America (Lepidoptera: Lasiocampidae): systematics, biology, immatures, and parasites. U.S. Natl. Mus. Bull. 276.

Stelzer, M.J. 1965. Susceptibility of the Great Basin tent caterpillar, *Malacosoma fragile* (Stretch), to a nuclear-polyhedrosis virus and *Bacillus thuringiensis* Berliner. J. Invert. Pathol. 7:122–125.

Stelzer, M.J. 1967. Control of a tent caterpillar, *Malacosoma fragile incurva*, with an aerial application of a nuclear-polyhedrosis virus and *Bacillus thuringiensis*. J. Econ. Entomol. 60(1):38–41.

Stelzer, M.J. 1968. The Great Basin tent caterpillar in New Mexico: life history, parasites, disease, and defoliation. U.S. Dep. Agric. For. Serv. Res. Paper Rm-39.

Stoetzel, M.B. 1989. Common names of insects and related organisms. College Park, Md.: Entomological Society of America.

Struble, D.L. 1970. A sex pheromone in the forest tent caterpillar. J. Econ. Entomol. 63:295–296.

Sullivan, C.R., and G.W. Green. 1950. Reactions of the eastern tent caterpillar, *Malacosoma americanum* (F.), and of the spotless fall webworm, *Hyphantria textor* Harr., to pentatomid predators. Can. Entomol. 82:52–53.

Sullivan, C.R., and W.G. Wellington. 1953. The light reactions of larvae of the tent caterpillars *Malacosoma disstria* Hbn., *M. americanum* (Fab.), and *M. pluviale* (Dyar) (Lepidoptera: Lasiocampidae). Can. Entomol. 85:297–310.

Swaine, J.M. 1918. Tent caterpillars. Can. Dep. Agric. Div. Entomol. Circ. 1.

Sweetman, H.L. 1940. The value of hand control for the eastern tent caterpillars *Malacosoma americana* (Fab.) and *Malacosoma disstria* (Hbn.) (Lepidoptera: Lasiocampidae). Can. Entomol. 72:245–264.

Takahashi, S.Y., G. Suzuki, and E. Ohnishi. 1969. Origin of oxalic acid in Ca oxalate crystals in the Malpighian tubes of the tent caterpillar *Malacosoma neustria testacea*. J. Insect Physiol. 15:403–407.

Talhouk, A.S. 1975. Contributions to the knowledge of almond pests in east Mediterranean countries. I. Notes on *Eriogaster amygdali* Wilts. (Lepid., Lasiocampidae) with a description of a new subspecies by E.P. Wiltshire. Z. Angew. Entomol. 78:306–312.

Teillon, H.B., B.S. Burns, and R.S. Kelley. 1982. Forest insect and disease conditions in Vermont. Montpelier: Vt. Dep. For., Parks, and Recreation.

Teillon, H.B., B.S. Burns, and R.S. Kelley. 1984. Forest insect and disease conditions in Vermont. Montpelier: Vt. Dep. For., Parks, and Recreation.

Thompson, C.G. 1958. A polyhedrosis virus for control of the Great Basin tent caterpillar, *Malacosoma fragile*. Trans. I. Int. Conf. Insect Pathol. Biol. Control, Prague, pp. 201–204.

Thomson, H.M. 1959. A microsporidian parasite of the forest tent caterpillar, *Malacosoma disstria* Hbn. Can. J. Zool. 37:217–221.

Tietz, H.M. 1972. Index to the described life histories, early stages and hosts of the macrolepidoptera of the continental United States and Canada. Vol. 1. Sarasota, Fla.: Allyn Museum of Entomology.

Tilman, D. 1978. Cherries, ants and tent caterpillars: timing of nectar production in relation to susceptibility of caterpillars to ant predation. Ecology 59: 686–692.

Topoff, H., J. Mirenda, R. Droual, and S. Herrick. 1980. Behavioural ecology of mass recruitment in the army ant *Neivamyrmex nigrescens.* Anim. Behav. 28:779–789.

Tostowaryk, W. 1971. Relationship between parasitism and predation of diprionid sawflies. Ann. Entomol. Soc. Am. 64:1424–1427.

Tothill, J.D. 1918. The meaning of natural control. Proc. Entomol. Soc. N.S. 4: 10–14.

Tutt, J.W. 1900. A natural history of the British Lepidoptera: a text-book for students and collectors. Vol. 2. London: Swan Sonnenschein & Co.

Tutt, J.W. 1906a. A natural history of the British Lepidoptera: a text-book for students and collectors. Vol. 5. London: Swan Sonnenschein & Co.

Tutt, J.W. 1906b. A natural history of the British Lepidoptera: a text-book for students and collectors. Vol. 7. London: Swan Sonnenschein & Co.

Tutt, J.W. 1907. A natural history of the British Lepidoptera: their world-wide variation and geographical distribution: a text-book for students and collectors. Vol. 9. London: Swan Sonnenschein & Co.

Tyrell, D., and I.S. Ben-Ze'ev. 1990. An emended description of *Furia crustosa* (Entomophthorales: Entomophthaceae). Mycotaxon 37:211–215.

Underhill, E.W., M.D. Chisholm, and W. Steck. 1980. (*E*)-5,(*Z*)-7-dodecadienal, a sex pheromone component of the western tent caterpillar, *Malacosoma californicum* (Lepidoptera: Lasiocampidae). Can. Entomol. 112:629–631.

van der Linde, R.J. 1968. Einfluss des Zustandes der Nahrungspflanze auf die Entwicklung des Ringelspinners (*Malacosoma neustria* L.) Z. Angew. Entomol. 62:386–394.

Vander Meer, R.K., F. Alvarez, and C.S. Lofgren. 1988. Isolation of the trail recruitment pheromone of *Solenopsis invicta.* J. Chem. Ecol.14:825–838.

Vander Meer, R.K., C.S. Lofgren, and F.M. Alvarez. 1990. The orientation inducer pheromone of the fire ant *Solenopsis invicta.* Physiol. Entomol. 15:483–488.

Wagge, J.K., and J.M. Bergelson. 1985. Differential use of pin and black cherry by the eastern tent caterpillar *Malacosoma americanum* Fab. (Lepidoptera: Lasiocampidae). Am. Midl. Nat. 113:45–55.

Wallner, W.E. 1971. Suppression of four hardwood defoliators by helicopter application of concentrate and dilute chemical and biological sprays. J. Econ. Entomol. 64:1487–1490.

Weaver, W.P. 1986. Ovipositional choices of *Malacosoma californicum fragile,* a subspecies of western tent caterpillar. *In* C.A. Hall and D.J. Young, eds., Natural history of the White-Inyo range, eastern California and western Nevada and high altitude physiology, pp. 114–118. University of California, White Mountain Research Station Symposium, Vol. 1.

Weed, C.M. 1898. The winter food of the chickadee. New Hampshire College Agric. Exp. Stn. Bull. 54:85–98.

Weiss, H., E.E. McCoy, and W.M. Boyd. 1944. Group motor response of adult and larval forms of insects to different wave-lengths of light. J. N.Y. Entomol. Soc. 52:27–43.

Wellington, W.G. 1950. Effects of radiation on the temperatures of insects and habitats. Sci. Agric. 30:209–234.

Wellington, W.G. 1957. Individual differences as a factor in population dynamics: the development of a problem. Can. J. Zool. 35:293–323.
Wellington, W.G. 1962. Population quality and the maintenance of nuclear polyhedrosis between outbreaks of *Malacosoma pluviale* (Dyar). J. Insect Pathol. 4:285–305.
Wellington, W.G. 1965. Some maternal influences on progeny quality in the western tent caterpillar, *Malacosoma pluviale* (Dyar). Can. Entomol. 97: 1–14.
Wellington, W.G. 1974. Tents and tactics of caterpillars. Nat. Hist. 83:64–72.
Wellington, W.G., C.R. Sullivan, and G.W. Green. 1951. Polarized light and body temperature level as orientation factors in the light reactions of some hymenopterous and lepidopterous larvae. Can. J. Zool. 29:339–351.
Westwood, J.O. 1836. IX. Descriptions of the nest of a gregarious species of butterfly from Mexico. Trans. Entomol. Soc. Lond. 38–44.
Wetzel, B.W., H.M. Kulman, and J.A. Witter. 1973. Effects of cold temperatures on hatching of the forest tent caterpillar, *Malacosoma disstria* (Lepidoptera: Lasiocampidae). Can. Entomol. 105:1435–1149.
Weyh, R., and U. Maschwitz. 1978. Trail substance in larvae of *Eriogaster lanestris* L. Naturwissenschaften 65:64.
Weyh, R., and U. Maschwitz. 1982. Individual trail marking by larvae of the scarce swallowtail *Iphiclides podalirius* L. (Lepidoptera: Papilionidae). Oecologia 52:415–416.
Wheeler, W.M. 1910. Ants: their structure, development and behavior. New York: Columbia University Press.
Wheeler, W.M. 1923. Social life among the insects. New York: Harcourt, Brace.
Whorton, J. 1974. Before silent spring: pesticides and public health in pre-DDT America. Princeton: Princeton University Press.
Wiklund, C., and T. Järvi. 1982. Survival of distasteful insects after being attacked by naive birds: a reappraisal of the theory of aposematic coloration evolving through individual selection. Evolution 36:998–1002.
Williams, J.L. 1939. The mating and egg laying of *Malacosoma americana* (Lepid.: Lasiocampidae). Entomol. News 50(2):45–50; (3):69–72.
Williams, J.L. 1940. The anatomy of the internal genitalia and the mating behavior of some lasiocampid moths. J. Morphol. 67:411–437.
Wilson, E.O. 1971. The insect societies. Cambridge: Harvard University Press.
Wilson, E.O. 1975. Sociobiology: the new synthesis. Cambridge: Harvard University Press.
Wilson, G.G. 1977. Effects of the microsporidian *Nosema disstriae* and *Pleistophora schubergi* on the survival of the forest tent caterpillar, *Malacosoma disstria* (Lepidoptera: Lasiocampidae). Can. Entomol. 109:1021–1022.
Wilson, G.G. 1979. Effects of *Nosema disstriae* (Microsporidia) on the forest tent caterpillar, *Malacosoma disstria* (Lepidoptera: Lasiocampidae). Proc. Entomol. Soc. Ont. 110:97–99.
Wilson, G.G. 1984. Pathogenicity of *Nosema disstriae, Pleistophora schubergi* and *Vairimorpha necatrix* (Microsporidia) to larvae of the forest tent caterpillar, *Malacosoma disstria*. Parasitenkunde 12:763–767.
Wilson, G.G., and W.J. Kaupp. 1977. Application of *Nosema disstriae* and *Pleis-*

tophora schubergi (Microsporidia) against the forest tent caterpillar in Ontario, 1977. Report FPM-X-4. Sault Ste. Marie, Ontario: Forest Pest Management Institute.

Winch, F.E., and R.R. Morrow. 1962. Production of maple syrup and other maple products. Cornell Univ. Ext. Bull. 974.

Winston, M.L. 1987. The biology of the honey bee. Cambridge: Harvard University Press.

Witter, J.A., and H.M. Kulman. 1969. Estimating the number of eggs per egg mass of the forest tent caterpillar, *Malacosoma disstria* (Lepidoptera: Lasiocampidae). Mich. Entomol. 2:63–71.

Witter, J.A., and H.M. Kulman. 1972. A review of the parasites and predators of tent caterpillars (*Malacosoma* spp.) in North America. Univ. Minn. Agric. Exp. Stn. Tech. Bull. 289.

Witter, J.A., and H.M. Kulman. 1979. The parasite complex of the forest tent caterpillar in northern Minnesota. Environ. Entomol. 8:723–731.

Witter, J.A., H.M. Kulman, and A.C. Hodson. 1972. Life tables for the forest tent caterpillar. Ann. Entomol. Soc. Am. 65:25–31.

Witter, J.A., W.J. Mattson, and H.M. Kulman. 1975. Numerical analysis of a forest tent caterpillar (Lepidoptera: Lasiocampidae) outbreak in northern Minnesota. Can. Entomol. 107:837–854.

Wolfe, K.L. 1988. *Hylesia acuta* (Saturniidae) and its aggregate larval and pupal pouch. J. Lepid. Soc. 42:132–137.

Work, R.W. 1985. Viscoelastic behaviour and wet supercontraction of major ampullate silk fibers of certain orb-web-building spiders (Araneae). J. Exp. Biol. 118:379–404.

Young, A.M. 1983. On the evolution of egg placement and gregariousness of caterpillars in the Lepidoptera. Acta Biotheor. 32:43–60.

Young, A.M. 1986. Natural history notes on *Brassolis isthmia* Bates (Lepidoptera: Brassolinae) in northeastern Costa Rica. J. Res. Lepid. 24:385–392.

Zarin'sh, I., and G. Eglite. 1985. Antioxidants: promising compounds for the protection of entomopathogenic viruses from UV-rays. Tr. Latv. Skh. Akad. 222:15–21.

Index

The color plates follow page 16.